Wealth out of Food Processing Waste

High amounts of waste are generated daily in various food processing industries, including seed, pomace, pit, peel, germ, husk, broken pulses, sludge, skin, bones, blood, feathers, wash water, and spent residue, among others. Several tons of generated waste can be effectively used to manufacture or recover such value-added by-products as fibers, antioxidants, proteins, vitamins and minerals, biofilms, fertilizers, and animal feed. While food processing–generated waste may lead to health and environmental hazards, it is critical to identify proper protocols to recover valuable ingredients from waste, thereby creating wealth in the society.

Wealth out of Food Processing Waste identifies and describes the proper protocols to recover valuable ingredients from waste, thereby creating wealth in society. The effective utilization of waste can generate income for the entrepreneur, lead to more employment for society, enhance fertility of soil, reduce environmental pollution, conserve resources, and help augment national economies to a greater extent.

Key Features

- Provides in-depth knowledge about conversion of waste derived from various food processing industries into various value-added products.
- Highlights the extraction of antioxidants and functional food ingredients from industrial food waste.
- Presents current and emerging trends using biotechnological approaches for conversion of waste into various value-added products.

This book provides food industry personnel, scientists, food engineers, biotechnologists, research scholars, and students with strategies for effective utilization of waste from various food processing industries.

Nutraceuticals: Basic Research and Clinical Applications

Series Editor: Yashwant Pathak, PhD

Bioactive Peptides: Production, Bioavailability, Health Potential and Regulatory Issues
edited by *John Oloche Onuh, M. Selvamuthukumaran, Yashwant Pathak*

Nutraceuticals for Aging and Anti-Aging: Basic Understanding and Clinical Evidence
edited by *Jayant Lokhande, Yashwant Pathak*

Marine-Based Bioactive Compounds: Applications in Nutraceuticals
edited by *Stephen T. Grabacki, Yashwant Pathak, Nilesh H. Joshi*

Applications of Functional Foods and Nutraceuticals for Chronic Diseases: Volume I
edited by *Syam Mohan, Shima Abdollahi, and Yashwant Pathak*

Flavonoids and Anti-Aging: The Role of Transcription Factor Nuclear Erythroid 2-Related Factor2
edited by *Karam F.A. Soliman and Yashwant Pathak*

Molecular Mechanisms of Action of Functional Foods and Nutraceuticals for Chronic Diseases: Volume II
edited by *Shima Abdollahi, Syam Mohan and Yashwant Pathak*

Plant Food Phytochemicals and Bioactive Compounds in Nutrition and Health,
edited by *John O. Onuh, Yashwant V. Pathak*

Anxiety, Gut Microbiome, and Nutraceuticals: Recent Trends and Clinical Evidence
edited by *Yashwant V. Pathak, Sarvadaman Pathak, and Con Stough*

Wealth out of Food Processing Waste: Ingredient Recovery and Valorization
edited by *M. Selvamuthukumaran*

For more information about this series, please visit: www.crcpress.com/Nutraceuticals/book-series/CRCNUTBASRES

Wealth out of Food Processing Waste

Ingredient Recovery and Valorization

Edited by M. Selvamuthukumaran

CRC Press is an imprint of the
Taylor & Francis Group, an **informa** business

Designed cover image: © shutterstock

First edition published 2024
by CRC Press
2385 NW Executive Center Drive, Suite 320, Boca Raton FL 33431

and by CRC Press
4 Park Square, Milton Park, Abingdon, Oxon, OX14 4RN

CRC Press is an imprint of Taylor & Francis Group, LLC

© 2024 selection and editorial matter, M. Selvamuthukumaran; individual chapters, the contributors

Reasonable efforts have been made to publish reliable data and information, but the author and publisher cannot assume responsibility for the validity of all materials or the consequences of their use. The authors and publishers have attempted to trace the copyright holders of all material reproduced in this publication and apologize to copyright holders if permission to publish in this form has not been obtained. If any copyright material has not been acknowledged please write and let us know so we may rectify in any future reprint.

Except as permitted under U.S. Copyright Law, no part of this book may be reprinted, reproduced, transmitted, or utilized in any form by any electronic, mechanical, or other means, now known or hereafter invented, including photocopying, microfilming, and recording, or in any information storage or retrieval system, without written permission from the publishers.

For permission to photocopy or use material electronically from this work, access www.copyright.com or contact the Copyright Clearance Center, Inc. (CCC), 222 Rosewood Drive, Danvers, MA 01923, 978–750–8400. For works that are not available on CCC please contact mpkbookspermissions@tandf.co.uk

Trademark notice: Product or corporate names may be trademarks or registered trademarks and are used only for identification and explanation without intent to infringe.

ISBN: 978-1-032-21064-3 (hbk)
ISBN: 978-1-032-21605-8 (pbk)
ISBN: 978-1-003-26919-9 (ebk)

DOI: 10.1201/9781003269199

Typeset in Garamond
by Apex CoVantage, LLC

I wholeheartedly thank

God

My Family

My Friends

and

Everyone

*Who has encouraged and supported
me to complete this book*

M. Selvamuthukumaran

Contents

Preface xv

About the Editor xvii

List of Contributors xix

1 Waste Management Techniques for Various Food Processing Industries 1

Preethi Ramachandran, Sweta Rai, Sabbu Sangeeta and Akansha Khati

1.1	Introduction	1
1.2	Waste and Sources of Food Waste Generation	4
	1.2.1 Waste	5
1.3	Classification of Waste	5
1.4	Food Waste and Sources of Food Waste	7
	1.4.1 Agricultural Waste	7
	1.4.2 Residential Waste	8
1.5	Food Processing Industries	8
	1.5.1 Commercial Waste	9
	1.5.2 Institutional Waste	9
1.6	Characterization of Food Waste in Food Processing Industries	9
1.7	Composition of Food Waste	12
1.8	Impact of Food Waste	13
	1.8.1 Environmental	13
	1.8.2 Economic Impact of Food Waste	14
	1.8.3 Social Impact of Food Waste	14

1.9 Food Waste Management 15
1.10 Conventional Methods of Food Waste
      Management 17
      1.10.1 Animal Feed.. 17
      1.10.2 Composting.. 17
      1.10.3 Landfills... 18
      1.10.4 Anaerobic Digestion of Food Waste 18
1.11 Waste Management Techniques Adopted in
      Various Food Processing Sectors 19
      1.11.1 Grain Processing....................................... 19
      1.11.2 Pulse Processing....................................... 21
      1.11.3 Fruit and Vegetable Processing................... 22
      1.11.4 Dairy Processing 23
      1.11.5 Meat and Poultry Processing...................... 25
      1.11.6 Seafood Processing 25
      1.11.7 Edible Oil Processing................................. 26
      1.11.8 Spices Processing..................................... 28
      1.11.9 Sugar Processing...................................... 30
      1.11.10 Confectionery Processing........................... 31
1.12 Challenges of Food Waste Management in India 31
1.13 Conclusion and Future Thrust...................................... 32

2 Value Addition of Cereal and Millet Processing Industrial Waste 41

*Sweta Rai, Preethi Ramachandran, and
Sabbu Sangeeta*

2.1 Introduction... 41
2.2 By-Products from Cereal Grains 44
2.3 Extraction of Wheat Bran.. 45
2.4 Functionality of Bran 47
2.5 Stabilization of Bran... 48
2.6 Fortification of Bran.. 48
      2.6.1 Impact on Organoleptic Acceptability
              and Nutritional Fortification of Bran............... 48
2.7 Rice 49
      2.7.1 By-Products from Rice................................ 50
      2.7.2 Stabilization of Rice Bran for
              Value Addition 51
      2.7.3 Extraction of Protein from Rice Bran 51
      2.7.4 Rice Bran Oil (RBO) 52
      2.7.5 Rice Husk... 52
2.8 Corn 53
      2.8.1 Gluten Meal Extracted from Corn................. 53
      2.8.2 Total Dietary Fiber..................................... 54

	2.8.3	Zein Protein	54

2.8.3 Zein Protein 54
2.8.4 Germ Protein 54
2.8.5 Corn Steep Liquor 55
2.8.6 Corn Starch 55
2.8.7 Corn Germ Oil 55
2.8.8 Nutritional Benefits of Germ Oil in
       Value-Added Foods 57
2.8.9 Novel Methods of Oil Extraction from
       Cereal Germ 57
2.9 By-Products from Barley Processing 60
2.9.1 Barley Flakes 60
2.9.2 Barley Malt 61
2.9.3 Use of Barley in Value-Added Products 61
2.10 Millets 62
2.10.1 The Primary Millet Processing
       By-Products Encompass 63
2.10.2 Value Addition from By-Products of Millets 63
2.10.3 Antioxidants and Fiber Extraction from
       By-Products of Cereal and Millets
       Processing 68
2.11 Conclusion 69

3 Value Addition of Oilseed Processing Industrial Waste 73

M. Selvamuthukumaran

3.1 Introduction 73
3.2 Production of Oilseed Cake Meal 73
3.3 Oilseed Cake Meal Fortification 74
3.4 Enzyme Production from Oilseed Meal 74
3.5 Antibiotic and Antimicrobials Production
    from Oilseeds 75
3.6 Mushroom Production from Oilseed Meal 75
3.7 Hulls 75
3.8 Wastewater Sludge 76
3.9 Conclusions 76

4 Value Addition of Pulse Processing Industrial Waste 78

M. Selvamuthukumaran

4.1 Introduction 78
4.2 Bioactive Constituents from By-Products of Pulses 78
4.3 Heath Benefits of Consuming Pulse-Based
    By-Products 79

4.4   Extraction of Proteins from Wastes of Pulse
Processing Industry.. 80
4.5   Fortification..81
4.6   Conclusions 81

5 Value Addition of Fruit Processing Industrial Waste 84

Merve Aydin, Vildan Eyiz, and İsmail Tontul

5.1   Introduction.. 84
5.2   The Wastes of the Fruit Processing Industry................ 85
     5.2.1   Citrus Wastes 85
     5.2.2   Dark-Colored Fruit Wastes.......................... 88
     5.2.3   Other Fruit Waste Sources Rich in
Valuable Functional Compounds 89
5.3   Valorization Approaches for Fruit Wastes 93
5.4   Green Extraction and Recovery Techniques.................94
     5.4.1   Basics of Novel Extraction Techniques 94
     5.4.2   Extraction of Essential Oil from
Citrus Peels 101
     5.4.3   Extraction of the Colorants from
Fruit Varieties...102
     5.4.4   Extraction of Flavoring and Aroma
Ingredients 109
     5.4.5   Extraction of Functional Ingredients (Phenolics,
Antioxidants, Pectin, Dietary Fiber, etc.)
from Fruit Pit, Seed, Peel, and Pomace 109
5.5   Efficient Utilization of Pomace for Effective
Fortification.. 115
5.6   Production of Enzymes from By-Products of
Fruit Processing Industries...120
5.7   Circular Bio-Economy and Techno-Economic
Prospects 121
5.8   Legislation and Regulations 122
5.9   Future Trends..122
5.10 Conclusions 123

6 Value Addition of Vegetable Processing Waste 147

Anil S. Nandane

6.1   Introduction.. 147
6.2   Methods of Waste Conversion................................ 149
     6.2.1   Thermal Conversion Methods........................ 149
     6.2.2   Chemical Conversion Methods 149

6.2.3 Biological Conversion Methods 150
6.2.4 Other Conversion Methods 150
6.3 Bioconversion of Vegetable Wastes into Different
Value-Added Products 150
6.3.1 Bioactive Compounds 151
6.3.2 Phenolic Compounds 151
6.3.3 Enzymes 152
6.3.4 Pigments 153
6.3.5 Dietary Fiber 153
6.3.6 Bioenergy 154
6.3.7 Single-Cell Protein (SCP) 156
6.4 Challenges for the Extraction of Natural Food
Ingredients from Fruit and Vegetable By-Products 156
6.4.1 Availability of Raw Materials at
Different Locations 157
6.4.2 Low Concentrations of Natural Ingredient
Vegetable By-Products 157
6.4.3 The Variable Sources and Characteristics of
Fruit and Vegetable By-Products 158

7 Recent and Novel Technology Used for the Extraction and Recovery of Bioactive Compounds from Fruit and Vegetable Waste 162

Rohini Dhenge, Massimiliano Rinaldi,
Tommaso Ganino, and Karen Lacey

7.1 Introduction 162
7.2 Bioactive Substances Are Extracted from Fruits and
Vegetables 165
7.3 New and Green Extraction Techniques 167
7.3.1 Ultrasound-Assisted Extraction (UAE) 168
7.3.2 Hydrodynamic Cavitation-Assisted
Extraction (HCAE) 169
7.3.3 Microwave-Assisted Extraction (MAE) 170
7.3.4 Supercritical Fluid Extraction (SFE) 171
7.3.5 Extraction with Instant Controlled Pressure
Drop 173
7.3.6 Pressurized Hot Water Extraction 173
7.3.7 Pressurized Liquid Extraction 174
7.3.8 Pulsed Electric Field 176
7.3.9 Pulsed Ohmic Heating
Extraction (POHE) 177
7.3.10 High Hydrostatic Pressure Assisted
Extraction 178

Contents   xi

7.3.11 Dielectric Barrier Discharge Plasma
Extraction (DBDE)...................................... 179
7.3.12 Enzyme-Assisted Extraction........................ 180
7.4 Conclusion and Future Prospects 180

8 Value Addition of Spices Processing Industrial Waste 188

*Sabbu Sangeeta, Sweta Rai, Preethi
Ramachandran, Poonam Yadav, and
Gaurav Chandola*

8.1 Introduction...188
8.2 Waste Utilization of Major Spices...........................190
8.3 Chili (Capsicum annum)..191
8.4 Turmeric 194
8.5 Pepper (Piper nigrum L.).......................................195
8.6 Ginger (Zingiber officinale)................................196
8.7 Cardamom (Elettaria cardamomum) 197
8.8 Waste Utilization of Minor Spices............................199
8.9 Coriander (Coriandrum sativum L.).........................199
8.10 Tamarind...199
8.11 Cumin...201
8.12 Clove (Syzygium aromaticum) 202
8.13 Mustard..204
8.14 Celery (Apium graveolens L.)................................205
8.15 Cinnamon (Cinnamomum zeylanicum)....................206
8.16 Fenugreek..208
8.17 Onion (Allium cepa) and Garlic (Allium sativa) 209

9 Value Addition of Confectionery Processing Industrial Waste 224

M. Selvamuthukumaran

9.1 Introduction...224
9.2 Preparation of Biodiesel from Confectionery
Industrial Wastes 224
9.3 Wastes from Cocoa Processing...............................225
9.4 Production of Poly(3-hydroxybutyrate) from
Cocoa Bean Shell...225
9.5 Ingredients from Cocoa Bean Shell 225
9.6 Bioactive Components from Cocoa Bean Shell 226
9.7 Utilizing Cocoa Bean Shell as a Corrosion Inhibitor ...227
9.8 Conclusions 227

10 Value Addition of Dairy Processing Industrial Waste 230

Arjun Mohanakumar, Rohini Vijay Dhenge, Amar Shankar, and Anandhu T S

10.1 Introduction... 230
    10.1.1 Overview of Dairy Industry by FAO 230
    10.1.2 Importance of Managing and Valorizing
            Dairy Processing Industrial Waste 232
10.2 Valorization Techniques for Dairy Processing
    Industrial Waste 233
10.3 Biological Processes..234
10.4 Conversion of Whey into Essential By-Products:
    Recent Insights...234
    10.4.1 Whey Processing 234
10.5 Biorefinery Products: Cheese Whey Valorization 236
10.6 Case Study on Ghee Residue: A Recent Concern
    in Dairy Industries...237
    10.6.1 M.Tech Dissertation 237
10.7 Conclusion...241

11 Value Addition of Meat and Poultry Processing Industrial Waste 244

M. Selvamuthukumaran

11.1 Introduction..244
11.2 Poultry Waste as Manure...244
11.3 Processing of Feathers 245
11.4 Collagen 246
11.5 Waste from Livestock Meat Processing.......................246
11.6 Conclusions 246

12 Value Addition of Fish Processing Industrial Waste 250

Ritu Agrawal, Akansha Khati and Sabbu Sangeeta

12.1 Introduction...250
12.2 Collagen 251
12.3 Gelatin..252
12.4 Insulin...253
12.5 Chitin and Chitosan 254
12.6 Fish Albumin 256
12.7 Fish Protein Concentrate...256
12.8 Bioactive Peptide 257

12.9   Beche-de-mer 259
12.10 Isinglass...259
12.11 Squalene 260
12.12 Shark Fin Rays 261
12.13 Conclusion 262

Index 263

Preface

Large amounts of waste are generated every day in various food processing industries. Waste is generated by various food processing sectors, including fruits and vegetables, which generate waste like seeds, pomaces, pits, and peels; cereal and millet, which generates waste including bran, germ, and husk; pulse, viz., hull and broken pulses; oilseed, such as meal and cake; dairy, which includes whey, solid waste like sludge and wastewater as a liquid waste; meat, fish and poultry, which includes skin, bones, blood, feathers and wash water; spice industry waste as spent residue and confectionery manufacturing industrial waste, such as syrup, wash water, etc. Several tons of generated waste can be effectively used to manufacture or recover value-added by-products like fibres, antioxidants, proteins, vitamins and minerals, biofilms, fertilizers and animal feed. The generated waste may lead to health as well as environmental hazards. Therefore, it is mandatory to identify the proper protocols in order to recover valuable ingredients from waste, thereby promoting wealth in the society. The effective utilization of waste can generate income for the entrepreneur; it also creates employment opportunities, helps enhance the fertility of soil, reduces environmental pollution, conserves resources and helps augment the national economy to a greater extent.

This book deals with the extraction of dietary fibres and antioxidants from by-products of the cereal processing industry. It describes the recovery and purification of protein isolates and concentrates from oilseed cake and meal obtained from various oilseed processing industries. It deals with the utilization of whey from different paneer, khoa and cheese processing industries for developing protein concentrates and also explains the recovery and extraction of calcium supplements from such industrial waste. The book describes the removal of fat and oil from beef slaughtering plant wastewater and presents the steps involved in the conversion of slaughterhouse waste into intermediate products like meat bone meal (MBM), dicalcium phosphate (DCP) and bicalphos (BCP).

I would like to express my sincere thanks to all the contributors; without whose continuous support this book would not have seen daylight. I would also like to express my gratitude towards Mr. Steve Zollo, Randy Brehm and all other CRC Press staff who have made every continuous cooperative effort to make this book a great standard publication at a global level.

M. Selvamuthukumaran

About the Editor

Dr. M. Selvamuthukumaran is currently a Professor, Department of Food Science & Technology, Hamelmalo Agricultural College, Eritrea. He was a visiting Professor at Haramaya University, School of Food Science & Postharvest Technology, Institute of Technology, Dire Dawa, Ethiopia. He received his PhD in Food Science from the Defence Food Research Laboratory affiliated with the University of Mysore, India. His core area of research is processing of underutilized fruits for development of antioxidant-rich functional food products. He has transferred several technologies to Indian firms as an outcome of his research work. He has also received several awards and citations for his work and has published several international papers and book chapters in the area of antioxidants and functional foods. He has also guided several national and international postgraduate students in the area of food science and technology.

Contributors

Ritu Agrawal
College of Fisheries, GBPUA&T
Pantnagar, India

Anandhu T S
Department of Food & Drug
University of Parma
Parma, Italy

Merve Aydin
Department of Food Engineering
Necmettin Erbakan University
Konya, Türkiye

Gaurav Chandola
Food Science and Technology
College of Agriculture, GBPUA&T
Pantnagar, Uttarakhand, India

Rohini Vijay Dhenge
Dipartimento di Scienze degli
  Alimenti e del Farmaco
Università degli studi di Parma
Parma, Italy

Vildan Eyiz
Necmettin Erbakan University
Faculty of Engineering
Department of Food Engineering

Konya, Türkiye

Tommaso Ganino
Department of Food & Drug
University of Parma
Parma, Italy

Akansha Khati
Department of Food Science and
  Technology
College of Agriculture, GBPUAT
Pantnagar, Uttarakhand, India

Karen Lacey
Department of Food & Drug
University of Parma
Parma, Italy

Arjun Mohanakumar
Department of Food & Drug
University of Parma
Parma, Italy

Anil S. Nandane
Department of Food Processing
  Technology
A.D. Patel Institute of Technology
New Vallabh Vidyanagar, Anand,
  Gujarat, India

Sweta Rai
Department of Food Science and
  Technology
College of Agriculture, GBPUA&T
Pantnagar, Uttarakhand, India

Preethi Ramachandran
Food Science and Technology
College of Agriculture, GBPUA&T
Pantnagar, Uttarakhand, India

Massimiliano Rinaldi
Department of Food & Drug
University of Parma
Parma, Italy

Sabbu Sangeeta
Department of Food Science and
  Technology

College of Agriculture, GBPUA&T
Pantnagar, Uttarakhand, India

Amar Shankar
Department of Food & Drug
University of Parma
Parma, Italy

İsmail Tontul
Necmettin Erbakan University
Faculty of Engineering and
  Architecture
Department of Food Engineering
Konya, Turkiye

Poonam Yadav
Food Science and Technology
College of Agriculture, GBPUA&T
Pantnagar, Uttarakhand, India

Waste Management Techniques for Various Food Processing Industries

Preethi Ramachandran, Sweta Rai, Sabbu Sangeeta and Akansha Khati

1.1 Introduction

The global demand for food is increasing widely due to the ever-increasing population. With the highest mountain range in the world, the Himalayas, to its north, the Thar Desert to its west, the Ganges Delta to its east and the Deccan Plateau in the south, India is home to vast agro-ecological diversity. Four out of the 34 global biodiversity hotspots and 15 WWF global 200 eco-regions fall fully or partly within India. Having only 2.4% of the world's land area, India harbors around 8% of all recorded species, including over 45,000 plant and 91,000 animal species. India is the second-largest producer of food next to China. It is the world's largest producer of milk, pulses and jute, and ranks second in the production of rice, wheat, sugarcane, groundnut, vegetables, fruits and cotton. It is also one of the leading producers of spices, fish, poultry, livestock and plantation crops (Agricultural Statistics at a Glance, 2022; Government of India, 2019). Figure 1.1 gives an outlay of primary producers of various major crops in 2020. On one hand, we feel proud, as we are one of the largest producers of foods in the world, and there has been a huge rise in the production of various food commodities in last few years (Table 1.1). But on the other hand, we must be concerned about the fact that we are also among those countries where food waste is pervasive. Of the total food produced in the country, only 2% is processed, and a large portion is wasted without being consumed.

India has made vast progress in providing food security for its people and has become largely self-reliant in agriculture. The food processing industry (FPI) is one area which has the potential to add value to farm output, create alternate employment opportunities, improve exports and strengthen the domestic supply chain. India is the sixth-largest food and grocery market in the world

Table 1.1 Production Statistics of Various Commodities in India (NABCONS, 2022)

Sr. No.	Name of the crop/commodity	Production (million MT) 2018–19	2019–20	2020–21
1.	Cereals	263.14	274.48	285.28
2.	Pulses	22.08	23.03	25.46
3.	Oilseeds	31.52	33.22	35.95
4.	Fruits	97.97	102.08	102.48
5.	Vegetables	183.17	188.28	200.45
6.	Plantation crops	16.37	16.12	16.63
7.	Sugarcane	405.42	370.50	405.40
8.	Spices	9.43	10.14	11.12
9.	Milk	187.70	198.40	209.96
10.	Meat	8.10	8.60	8.80
11.	Eggs (billions)	103.32	114.40	122.05
12.	Fish	13.57	14.16	14.50

(*Source:* Directorate of Economics and Statistics, Department of Agriculture and Farmers Welfare, MoA&FOOD WASTE, GoI)

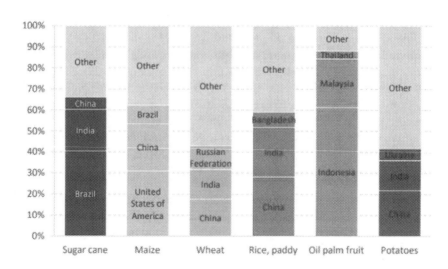

Figure 1.1 Global production of main primary crops in the year 2020 by main producers.

(Source: FAO, 2021)

(Law et al., 2019). Food and food products are the biggest consumption category in India, with spending on food accounting for nearly 21% of India's GDP and with a market size of $181 billion. The industry produces several food products, such as meat, poultry, fisheries, fruits, vegetables, spices, milk and milk products, alcoholic beverages, plantations and grains. It also manufactures cocoa products and chocolates, confectionery, mineral water, soya-based items and high-protein foods. India's food processing sector is a sunrise sector which has gained prominence in recent years. Major processed food products exported from India include processed fruits and juices, pulses, guar gum, groundnuts, milled products, cereal preparations, oil meals and alcoholic beverages. It can be seen that the production and processing of all categories of food and beverage products have increased over the years. But at the same time, high post-harvest losses in the supply chain always remain a cause for concern for planners and policymakers. Due to inadequate infrastructural facilities and poor post-harvest management, the post-production losses in food commodities in India is estimated to be 68.90 MT in terms of production and Rs. 149503.10 crore (NABCONS, 2022) in terms of monetary loss (Table 1.2).

Various sectors of the food processing industry are engaged in the production of different types of processed products. The essential raw material for processed products mainly constitutes biological materials, which are of plant, animal or aquatic origin. The process of converting raw material into processed products generates waste in solid and liquids phases, such as stalks, shells, peels and cores of fruits and vegetables; stalks, husks and bran in rice milling; trimmings, bones, hides, offals and shells in from farm and aquatic animals and spillages, whey, etc. in the dairy industry. This waste is a direct loss for producers and

Table 1.2 Absolute Losses in Physical Quantity and Monetary Value by Category (NABCONS, 2022)

S. No.	Category	Production (million MT)	Quantity lost (million MT) NABCONS (2020–21)	Monetary loss NABCONS (₹ in Crore)	% Contribution of the total loss (in monetary value)
1.	Cereals	281.28	12.49	26000.79	17.02
2.	Pulses	21.55	1.37	9289.21	6.08
3.	Oilseeds	37.27	2.11	10924.97	7.15
4.	Fruits	90.82	7.36	29545.07	19.34
5.	Vegetables	164.74	11.97	27459.08	17.97
6.	Plantation crops (including sugarcane) and spices	426.13	30.59	16412.56	10.74
7.	Livestock produce (milk, meat and fish)	232.86	3.01	29871.41	21.70
Total		**1254.65**	**68.90**	**149503.10**	
8.	Eggs*	122110	7363	3287.32	
	Grand Total			**152790.42**	

*For eggs, production in millions and price per eggs were taken

manufacturers and also plays a major role in creating environmental pollution if not disposed of properly. The generation of waste causes economic losses as well as creates requirements for additional cost input for their management.

The waste generated from food processing industries are either not utilized or predominantly utilized as animal feed, fertilizer and in preparation of by-product to a limited extent. Most of the agricultural by-products produced from fruit and vegetable processing are used as animal feed. Approximately 45% of fruits and vegetables are wasted worldwide, which can be used for the extraction of starch, pectin, natural coloring agents and fat (Lau et al., 2021). Wheat and maize bran along with waste generated during oil extraction from plant sources are utilized for the production of feed for animals and poultry. These feeds are highly rich in protein. Other types of non-edible by-products are also manufactured from waste generated in meat processing industries, such as glue, leather, industrial oil and lubricants (Joshi and Sharma, 2011). Waste generated from spices is used for the extraction of essential oils and oleoresins.

Food waste management in India is still in its early stages. Although there are many ways in which waste generated by various food industries can be utilized for reduction in economic losses and generation of additional income, they are all limited to laboratories and the data are documented only in research papers. No or very little interest has been shown by the industry or government in this area of science.

1.2 Waste and Sources of Food Waste Generation

The food processing industry in India can be segmented into cereals and grains, fruits and vegetables, dairy, beverages, seafood, meat and poultry, spices and condiments, bakery items, pulses and oilseed, confectionery and sugar. The aforementioned domains of the food processing industry manufacture different types of specific products based on the raw materials being used. Therefore, the waste generated by these industries are of specific types depending on the raw materials and products being processed, viz., peels, seeds, pulp etc. from fruits and vegetables; molasses and bagasse from sugar refining; bones, flesh and blood from meat and fish; stillage and other residues from wineries, distilleries and breweries; and dairy wastes such as cheese whey. Many of these contain low levels of suspended solids and low concentrations of dissolved materials (Kosseva, 2013). Hence, the waste management approaches for different domains of food processing differ from each other and involve research and innovation. However, some waste generated in nearly all sectors of the food industry are common, e.g. wash water, equipment cleaning waste and spillage. In such cases, common methods or methods with slight modifications can be adopted for such waste management. The generation and management of waste in different sectors of the food processing industry, along with pollution, other environmental impacts and control strategies thereof are discussed in detail in the following chapters.

1.2.1 Waste

Waste refers to any material, substance or product that is discarded or no longer useful and is subsequently disposed of. Waste is the by-product after the completion of a certain process which physically contains the same substances that are available in the useful product. As per the European Union's Waste Framework Directive (Directive 2008/98/EC) waste means any substance or object which the holder discards or intends or is required to discard. In India, the legal definition of waste is provided by the Environment (Protection) Act, 1986 and the various rules and regulations made under it. The definition of waste is primarily covered under the Hazardous and Other Wastes (Management and Transboundary Movement) Rules, 2016. According to these rules, "waste" means any substance or object which is discarded, rejected, abandoned or is intended to be discarded, rejected or abandoned. There have been huge advancements in the quantity and nature of the waste generated alongside human development and invention in science and technology. Food waste is common throughout the entire food production system, from farms to distribution sites to retailers to the consumer.

Figure 1.2 explains the generation of waste during processing in different unit operations. Waste generation starts right from the procurement of the raw material. Raw material is considered as food waste if it does not meet quality parameters and is unusable or if it spoils during storage. During processing and the production of food products, waste is generated due to process or equipment inefficiency, low-quality raw materials, spillage and careless handling. A lot of processed products are wasted as rejected product during quality control (if the product does not match the company's standard). Cleaning and sanitation processes generate waste in the form of wash water and sanitizing chemicals. Processed products are considered waste during storage, packaging and distribution if they are spoiled by insects or other pests, do not sell on time, expire, spill due to improper handling and packaging etc. Waste at the consumer end is mainly due to improper product preparation or delayed consumption. Measurement of these wastages at different stages of food manufacturing is nearly impossible.

1.3 Classification of Waste

Most commonly waste is classified as biodegradable or non-biodegradable based on its nature of decomposition.

- *Biodegradable waste* is mainly composed of biological entities and is therefore decomposed or broken down by microorganisms and other living organisms. Biodegradables generally include organic wastes. These wastes are degraded by natural factors and abiotic components like microbes, temperature, UV, oxygen etc. Together with other

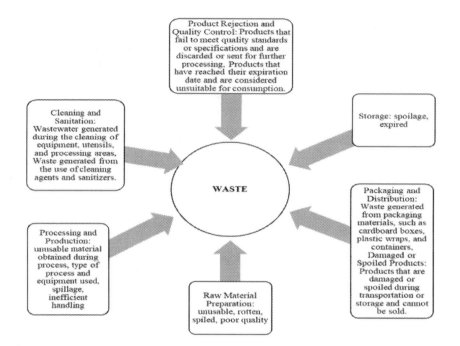

Figure 1.2 Generation of waste during different unit operations in the food processing industry.

abiotic components, microorganisms decompose complicated materials into simple organic matter that eventually suspends and dissipates into the earth. Although it is generally considered that biodegradable waste does not possess any threat to the environment, reality if the disposal of biodegradable waste is not controlled, it leads to various environmental and human impacts, like the release of greenhouse gases and the development of various diseases.
- *Non-biodegradable waste* includes substances that cannot be decomposed or broken down by microorganisms and other living organisms. As a result, the environmental threat they pose is more serious, such as pollution. Non-biodegradables are generally inorganic wastes like plastic bags, cans, bottles and chemicals.

Waste can come from a variety of sources, and it is often classified based on the origin or source of its generation. The main types of waste are:

- *Solid waste*: Examples of solid waste include packaging materials, food waste, paper, plastics, metals, glass, textiles and construction debris.
- *Liquid waste*: Liquid waste refers to wastewater generated from various sources, such as households, industries and agricultural activities.

It includes domestic sewage, industrial effluents and agricultural run-off. Liquid waste may contain pollutants, contaminants or harmful substances that require proper treatment before disposal.

- *Hazardous waste*: Hazardous waste possesses properties that make it potentially harmful to human health or the environment. It includes substances that are toxic, corrosive, flammable or reactive.
- *Biomedical waste*: Biomedical waste refers to waste generated from healthcare facilities, such as hospitals, clinics and laboratories. It includes materials contaminated with blood, bodily fluids or infectious agents, such as used syringes, bandages, pathological waste and laboratory cultures. Due to its potential to spread diseases, biomedical waste requires specialized handling, treatment and disposal procedures to ensure public health and safety.
- *E-waste*: Electronic waste, or e-waste, refers to discarded electronic devices and equipment. It includes items such as computers, mobile phones, televisions, refrigerators and batteries.

According to recyclability, wastes can be classified as:

- *Recyclable*: Recyclable waste refers to materials that can be processed and used to create new products. These materials can go through a recycling process to be turned into raw materials that can be used again, e.g. paper, plastic, glass and metal.
- *Non-recyclable*: Non-recyclable waste refers to materials that cannot be recycled or are not suitable for recycling. These items typically end up in landfills or are incinerated. Examples of non-recyclable waste include soluble solids, broken glass, batteries, paint, chemicals and sanitizers.

1.4 Food Waste and Sources of Food Waste

The generation of food waste is increasing day by day from diverse industrial, agricultural, commercial, domestic and other sources due to changing lifestyles and the fast urbanization of the global population. The Food and Agriculture Organization (FAO) estimated that almost 1.3 billion tons of wasted food are generated annually, which is one-third of the total food production on a global scale (Gustavsson et al., 2011). Food waste leads to the significant loss of other resources, like water, land, manpower and energy. There are several sources, such as food processing industries, agricultural waste and commercial and household kitchens, that generate Food waste.

1.4.1 Agricultural Waste

Agricultural waste includes straw, bagasse, molasses, spent grains, grain husks (rice, maize and wheat), nut shells (walnuts, coconuts and groundnuts), fruit and vegetable skins (potato, jackfruit, pomegranates, bananas and avocados), crop stalks (cotton, plant waste), and animal and bird dung (Dia et. al., 2018). Inappropriate use and handling of refrigeration can also lead to the production of agricultural waste. In India, there are various sources that produce more than 350 million tons of agro-industrial waste annually (Madurwar et al., 2013).

1.4.2 Residential Waste

Urbanization, rapid economic growth and unregulated population growth have intensified food consumption, which has raised the proportion of kitchen garbage production each year (Zhao et al., 2017). Kitchen waste (KW) is a type of anthropogenic organic waste that is typically produced by canteens, restaurants, homes, public catering facilities, factories etc. (Liu et al., 2019). Kitchen wastewater from both commercial and residential kitchens is also produced when food is washed, rinsed, cooked, dishes and cooking utensils are cleaned and when basic housekeeping is performed (Sharma et al., 2020).

1.5 Food Processing Industries

Food processing industries include cereal grains; fruits and vegetables; beverages; dairy; meat, poultry and egg; seafood and edible oil. Cereal grain (wheat, rice, barley, maize, sorghum, millet, oat and rye) production reached 2577.85 million tons globally in 2016 (FAO). According to Anal (2017), the processing of grains and pulses results in significant amounts of by-products, like bran and germ. India is the greatest producer of pulses in the world, and during processing a significant amount of husk is obtained (Parate and Talib, 2015). In the case of fruits and vegetables, waste is produced at several stages of the farm-to-table food supply chain, including those for production, processing, packaging, handling, storage and transportation (Ji et al., 2017). It produces waste only when a consumer removes products from the range of acceptance due to several factors, such as microbial attack (rotting, softening and product surface growth), thermal treatment, biochemical reactions (enzymes, antioxidants, oxygen, phenolic and favonoid compounds), discoloration, wounding or chilling and degree of ripening (Sharma et al., 2020). The largest milk-producing country, India generates 3.739–11.217 million m^3 of effluent waste per year during the processing of milk. A significant amount of whey is also produced as a by-product while making cheese (Parashar et al., 2016). The meat, poultry and egg processing industry produces a substantial quantity of animal by-products (feathers, hairs, skin, horns, hooves, soft meat and bones), slaughterhouse waste (blood residue, protein, detergent residues and high organic matter [carbon, nitrogen and phosphorous]) and wastewater (Adhikari et al., 2018). In seafood and aquatic biotic life processing, approximately 50–70% of raw material is wasted every year (Kumar et al., 2018).

Seafood waste, mainly in the form of crab, shrimp and lobster shells, reaches about 6–8 million tons of waste worldwide. Waste is produced by the edible oil industry during several processing steps, including degumming, neutralization, bleaching, deodorization and oxidative or hydrolytic rancidity (Okino-Delgado et al., 2017). It is reported that the edible oil industry generates 350.9 million tons of de-oiled cake and oil meal waste annually (Chang et al., 2018).

1.5.1 Commercial Waste

Commercial waste is manmade, organic, biodegradable waste in sectors such as supermarkets, supercentres, food wholesale, restaurants and hotels. Supermarkets and supercentres generate 8.7 million tons and food wholesale generates 4 million tons of food waste. Restaurants produce 4–10% of raw food waste (Sharma et al., 2020). The primary source of food waste is reported to be the food left over by restaurant diners, caterers and buffet settings (Pirani and Arafat, 2016) as well as preparing 10–15% of extra food at every event (Afzal et al., 2022).

1.5.2 Institutional Waste

Institutional waste includes hospitals, nursing homes, military mess halls, office buildings, colleges, universities and boarding schools that produce 7.2 million tons of wasted food that is harmful to the environment. Food, however, is a significant part of the daily waste stream that is generated by patients, healthcare professionals and visitors. Food waste generation in educational institutions varies depending on the number of students.

1.6 Characterization of Food Waste in Food Processing Industries

Food waste is complex, consisting of both liquid and solid waste. Food waste generated from different types of food industries has specific characteristics. Therefore, these wastes have different impacts on the environment, and hence the treatment and utilization of these wastes are different. The characterization of food waste creates new opportunities and benefits for reducing food waste through the recovery of valuable components and thereby increasing sustainability. Food waste can be characterized into the following groups:

Fruit and vegetable processing waste: Fruits and vegetables are one of the most consumed commodities; hence, the generation of waste by this industry is also very high. Table 1.3 represents the extent of the waste produced from processing industries of some of the most important fruits and vegetables. Fruits and vegetables used to be used raw or were minimally processed, but with the development of society, changes were necessary and, thus, processing has grown significantly. Canning fruits and vegetables produces the most amount of waste, both solid waste and wastewater. The raw materials undergo peeling, coring, trimming etc. to get a good, finished product, leading to a lot of solid waste. Additionally, water used for cleaning the products adds to wastewater generation. The waste obtained from the fruit and vegetable processing industry is enormously diverse due to the use of a wide variety of fruits and vegetables, the broad range of processes and the multiplicity of the product (William, 2005). Waste can include seeds, leaves, pulp, skin, roots, stones, cores and more. Waste from fruits and vegetables is rich in suspended solids and dissolved organic matter, such as starch and fruit sugar.

Dairy processing waste: Among the different sectors of the food processing industry, dairy is one of the major wastewater-producing industries. Wastewater is produced during the processing of liquid milk, butter, cheese, skim milk powder, ghee etc. In various phases of dairy operation, clean water is used, for example, for milk processing, cleaning, packing and cleaning of the milk duct, can and tanker. Dairy wastewater contains high amounts of organic liquid ingredients, such as whey products, lactose, fat and minerals. The milk waste is largely neutral or slightly alkaline and, because of milk sugar fermentation, is likely to become very acidic very quickly (Hansen and Cheong, 2019). Whey is another major waste obtained during the production of cheese, paneer and butter. Discharging whey as waste causes serious

Table 1.3 Quantities of Processing Waste Generated during Value Addition of Various Fruits and Vegetables (Joshi, 2020)

Sr. no.	Commodity	Percent weight basis
1.	Apple	12–47
2.	Apricot	8–25
3.	Grapefruit	3–58
4.	Orange	3
5.	Peach	11–40
6.	Pear	12–46
7.	Bean, green	5–20
8.	Beet	7–4
9.	Broccoli	20
10.	Cabbage	5–25
11.	Carrot	18–52
12.	Cauliflower	8
13.	Peas	6–79
14.	Potatoes	5
15.	Spinach	10–40
16.	Sweet potato	15
17.	Tomato	5–25

contamination issues owing to its high biological oxygen demand (BOD) (35–40 g/L). This strong BOD is primarily attributed to lactose present in a range of 4.5–5% (Mansoorian et al., 2016).

Meat, fish and poultry waste: A considerable amount of food waste emerges from slaughterhouses and causes a huge amount of pollution. Skin and fur are important by-products of the meat industry and are used in the leather, athletics, cosmetics, edible gelatin and glue sectors. Gelatin is manufactured from bone, while cholesterol obtained from brains, nervous systems, and spinal cords acts as a raw material for the production of vitamin D3. Similarly, the extracts from pig and cattle livers are used as a source for vitamin B12. Insulin is derived from the pancreas and used in treatments for diabetes mellitus (Jayathilakan et al., 2012). Fish waste is a source of protein, minerals and fat. In addition to chemicals such as chitosan, fish protein hydrolysate and fish oil are important items obtained from fish waste (da Rocha et al., 2018). As by-products of the poultry sector, feathers are produced in vast amounts. Unless they are treated, they can become a source of emission threat. Poultry feathers are used in biofuels development after hydrolyzation (Seidavi et al., 2019).

Cereal and grain processing waste: Milling is the major unit operation in any cereal processing industry. Milling of cereals can be categorized into dry and wet milling. Bran and germ are co-products removed by the dry milling process, while wet milling generates by-products such as steep solids, germ and bran (Balandran-Quintana, 2018). The waste generated through cereal processing is non-toxic and can be divided into solid and liquid forms (Kumar et al., 2017). Solid waste contains corn pericarp (CP), corn grits, brewer's spent grain (BSG) and baking industry waste, while liquid waste contains rice milling wastewater, parboiled rice effluent, corn steep liquor (CSL), bakery wastewater and wastewater from the corn tortilla industry etc. Cereal industrial waste consists of various types of organic and inorganic substances. The liquid (effluent) waste colour ranges from yellow to brown with higher values of biological oxygen demand (BOD), chemical oxygen demand (COD), dissolved oxygen (DO), nutrients and minerals (Mukherjee et al., 2016). Cereal industrial waste is not only treated to reduce the risk of environmental pollution, but it also produces some value-added products such as biofuel (biohydrogen, bioethanol, butanol, biogas and biocoal), some industrial valued enzymes, biomass, biofertilizer, proteins, organic acids (citric, succinic and lactic acid), polysaccharides and few others.

Brewery and distillery waste: Brewery wastewater is very special. The brewing process generates alcohol, sugars and proteins, all of which end up in their wastes. Brewery wastewater is rich in nutrients that will significantly conflict with natural environments if released without the appropriate treatment.

Waste from the oil seed industry: The wastes from oilseed processing plants have been significantly increasing from the past few decades. As the oilseeds are majorly processed by either mechanical expression or solvent extraction,

Table 1.4 Characterization of Food Waste Based on Physical and Chemical Characteristics

Characteristics	Description
Solids content	Compared to animal sludge and sewage sludge, plant-based food waste contains more solid matter due to the presence of heavy organic material. The use of solids abundant in organic matter is based on their solubilization and consequent microbiological biodegradation.
C/N ratio and pH of food wastes	Plant-based food waste has a high C/N ratio and pH due to its high lignocellulose content.
Vitamins	B-complex vitamins are present in all animal products, but not in plants especially attributed higher content of riboflavin in dried skimmed milk and whey. Therefore, food waste that contains animal, as well as microorganism, products may have considerable contents of vitamins.
Minerals	Macro- and micronutrients such as phosphorus, potassium, magnesium and calcium are also estimated in food waste.
Amino acids	Animal products (fish, eggs, flush, butter, whey and milk) contain a higher concentration of amino acids compared to the protein supplements of plant-originated food products.

wastes like oilseed meal, seeds, oil sludge, peels etc. are formed during the extraction process. The waste generated from the oil processing industry are rich in protein, polyphenols, colorants, fibres and antioxidants.

Food waste can also be characterized based on its physical and chemical constitution. Several physical (moisture content, bulk density and pH) and chemical (carbon, hydrogen, oxygen, nitrogen, sulphur, particle size, C/N and total carbohydrate) characteristics determine its biodegradability and handling expenses. Detailed description about various characteristics of food waste based on its physical and chemical attributes are presented in Table 1.4.

1.7 Composition of Food Waste

As discussed earlier, food waste is a complex mixture of various organic and inorganic materials that are discarded or lost during the production, processing, distribution and consumption of food. The composition of different food waste (Table 1.5) varies depending on its origin (plant, animal or aquatic), the part of the plant or animal, the method of processing, the type of spoilage etc.

Food production is one of the most resource-intensive industries. Generation of food waste causes degradation of all resources used in production, transport or distribution. Wasting food depletes water, energy, soil and money. The specific consequences of food waste are described in the FAO's *Food Wastage Footprint: Impacts on Natural Resources*. Most often, we categorize them as environmental, economic and social impacts (Figure 1.3).

Table 1.5 Some Common Components Found in Food Waste in Different Sectors of Food Processing Industry

Sr. no.	Food domains	Waste generated	Composition of waste
1.	Fruit and vegetable waste	Peels, skins, cores, seeds and other plant parts that are typically removed before consumption	Fibres, vitamins, minerals, antioxidants, organic acid, sugars and phytochemicals
2.	Dairy products	Expired milk, production residue, spills, wash water sanitizing chemicals	Protein, fat, vitamins (such as vitamin D and B vitamins) and minerals (such as calcium)
3.	Cereal and grains	Stale bread, rice, pasta and other grain-based products	Carbohydrates, some protein and dietary fibre
4.	Meat, poultry and fish	Trimmings, bones, expired or spoiled meat products, blood and blood rinse, hides, hair, eviscerated parts, inedible fats, wash water	Some protein, minerals and fats
5.	Brewery and distillery	Wastewater, spent yeast	Rich in nutrients
6.	Oils and fats	Used cooking oil, grease and fat trimmings	Traces of vitamins (such as vitamin E) and essential fatty acids, press solids and cakes, oil water emulsion, rancid oils, shells of oil seeds

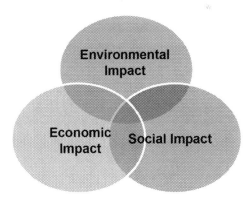

Figure 1.3 Conceptual configuration of the impact of food waste.

1.8 Impact of Food Waste

1.8.1 Environmental

Proper management of waste is fundamental for avoiding environmental pollution, as it may lead to the contamination of air, water and land. An example of water contamination is eutrophication, and for atmospheric pollution there is the ozone layer depletion or global warming potential. Some countries produce nearly 50% of food waste out of the total solid waste generated. The production of food that is not consumed has a significant impact on the environment. Uneaten parts of food in landfills contributes to global warming. Food waste decomposes, producing harmful methane that escapes into the atmosphere. Methane is a poisonous greenhouse gas and is 21 times stronger than carbon dioxide (Seberini, 2020). Processing of food requires more material input and energy. Additionally, the environment is more affected when demand increases for resource-intensive foods. Food loss and waste (FLW) puts water, soil and air at risk because food production and distribution require large amounts of water, land and energy. The reduction in FLW means that it can save resources used for production, processing and transportation, which provides benefits to the environment (Ishangulyyev et al., 2019).

1.8.2 Economic Impact of Food Waste

Both producers and consumers are part of the current economic system. This means that consumer preferences are one of the factors influencing the behaviour of food producers and the generation of waste. The economic impact of food waste is:

- *Financial loss*: Food waste represents a significant loss of money for individuals, households, businesses and governments. Producing, processing, transporting and disposing of food that goes uneaten incurs costs that could have been avoided.
- *Increased food prices*: Food waste puts additional pressure on the food supply chain, leading to increased food prices for consumers. This can exacerbate food insecurity and inequality.
- *Lost economic opportunities*: Efficiently managing food waste can create economic opportunities, such as job creation in waste management, recycling and the development of innovative solutions.

1.8.3 Social Impact of Food Waste

Although enough food is produced in the world, nearly a billion people worldwide suffer from hunger and malnutrition. Many developed countries throw away tons of food that could be consumed in developing countries. These foods are thrown away even though they are still of good quality and safe to consume. Similar problems also occur within individual developed countries. Part of the population has sufficient access to food, up to their surplus, but on the other hand there are socially weaker groups of the population who do not have the opportunity to buy quality food, or sometimes they do not have enough food at all. The social impact of food waste around the world includes:

- *Hunger and food insecurity*: While food is being wasted, millions of people worldwide suffer from hunger and lack access to nutritious

meals. By reducing food waste, more food can be redirected to those in need.
- *Ethical concerns*: Wasting food when others are going hungry raises ethical questions about fairness and social responsibility.
- *Food system inefficiency*: Food waste is indicative of inefficiencies in the food system, from production to consumption. Addressing food waste can help create a more sustainable and equitable food system.

It is important to note that the impacts of food waste may vary across different regions and contexts. According to FAO's study, 54% of the world's food wastage occurs at the upstream process, i.e. during production, post-harvest handling and storage, while 46% of it happens during the downstream process, at the processing, distribution and consumption stages.

1.9 Food Waste Management

Food waste has varyious chemical compositions based on its generation and origin. Thus, food waste contains a mixture of proteins, carbohydrates and lipids. According to the waste hierarchy framework, the most preferable food waste management options respectively are reduction, redistribution, feeding to animals, anaerobic digestion (AD), composting, incineration and, finally, landfilling (Figure 1.4, Figure 1.5). Illegal open dumps and landfills are the primary methods frequently used in food waste management (Adhikari et al., 2006). Based on the current data for food waste treatments in

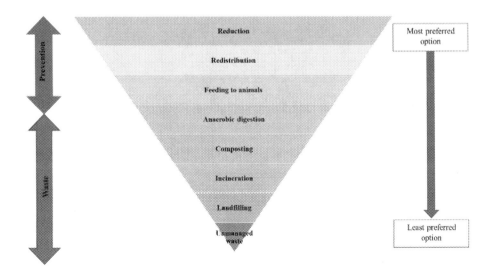

Figure 1.4 *Food waste management hierarchy.*

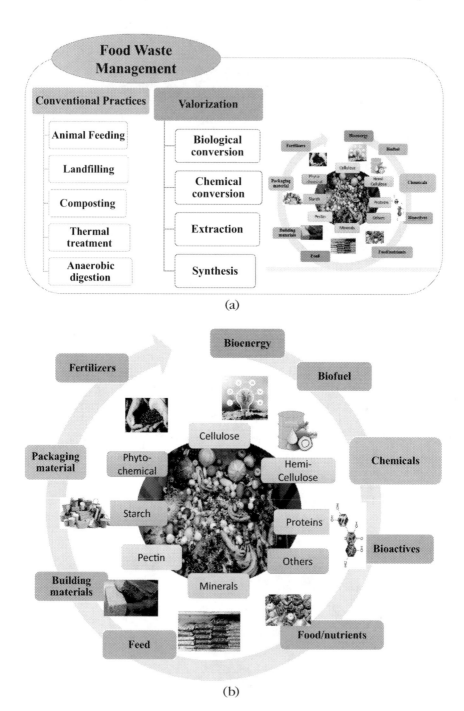

Figure 1.5 Food waste management practices.

developing countries, the most common food waste treatment method is dumping/landfills, which account for over 90% of food waste treatment, and the second most used method is composting, which accounts for 1–6%. Anaerobic digestion is used for the treatment of 0.6% of food waste, whereas other treatments, such as incineration and animal feeding, are rarely used.

Recently, based on the composition of food waste, many food industries have come up with a way to utilize this waste to generate biofuels and value-added products (valorization). Various biofuels such as biomethane, biohydrogen, bioethane etc. can be generated from food waste either biologically or thermochemically based on their composition. In addition, food waste valorization pathways involve both extraction and synthesis bioactive components (Figure 1.5) from food waste and utilization in the production of value-added products. The management food industry waste includes various treatments, such as chemical, physical, mechanical and biological techniques, which have numerous benefits and drawbacks. For example, the application of food processing industry waste as animal feed is the most conventional technique. Proteinaceous and lipid-rich food waste is appropriate for animal feed, while food waste rich in cellulosic composition can be appropriate for cattle feed. On the other hand, the possible occurrence of toxic products that have anti-nutritious influence and unstable compositions of nutrients might threaten the health of both humans and animals.

1.10 Conventional Methods of Food Waste Management

1.10.1 Animal Feed

Animal feed is commonly the best inexpensive valorization route for food supply-chain waste, but in some cases it is restricted by governing problems besides the characteristics of by-products produced in the process. The legislative laws of Japan, South Korea and Taiwan encourage using food waste, which compose 33%, 81% and 72.1% of total food waste generation respectively, to feed animals (Gen et al., 2006; Kim et al., 2011). The separation and collection of food wastes are not properly practised in developing countries, as almost all of the generated food waste is mixed with municipal solid waste (MSW), which cannot be purified and utilized for animal feeding.

1.10.2 Composting

Composting is an effective approach in treating food waste in emerging nations. Composting is referred to as a general land spreading or injection process and is a promising and long-term approach. It is an eco-friendly process, and it diverts food waste from landfills and decreases planters' requirements (e.g. for fertilizers). In India, presently, over 70 composting treatment amenities are handling mixed MSW, and reprocess nearly 5.9% of the total food waste to produce around 0.0043 billion tons of compost annually. Nearly every composting facility accepts mixed trash, while two units in Vijayawada and Suryapet, India, are known to accept source-separated organic waste (Ranjith et al., 2012). Presently, as per the Pollution Control Department and Ministry of Natural Resources and Environment (2010), around 0.00059 billion tons of food waste has been composted to biofertilizers and biomethane. In most developing countries, there are issues with composting due to impure wastes which are the by-products of source-segregated food waste.

1.10.3 Landfills

A typical waste disposal approach is the landfill, as it is a cost-effective treatment route. A landfill is described as the dumping, solidity and spreading of food waste at suitable locations and comprises four general phases, namely, the hydrolysis or aerobic digestion, hydrolysis and fermentation, acidogenic and methanogenic phases. During this process, the organic matter in food waste is oxidized and decomposed, ultimately causing methane production and the pollution of groundwater. But landfill strategies for the management of food waste has raised considerable ecological issues. Discarding of food waste to landfills creates influences both economically and environmentally through greenhouse gas emissions (methane and carbon dioxide) directly and indirectly. For instance, 4.2 tons of carbon dioxide are released from 1 ton of food waste. This includes emissions to air, water and soil (World Economic Forum, 2011). The global competitiveness report 2011–2012. Various policies have focused on food waste treatment approaches to divert food waste from landfills. Generally, the best strategies aim to shift food waste management from landfills to avoidance, reutilization and recovery. Thus, a stable strategy comprises joint actions, for example, (1) food waste landfilling is banned, (2) taxes on landfilling boost diversion and enhance the use of alternate managements, (3) expansion of composting or AD options, (4) expansion of the requisite setup and (5) establishing an inclusive management system.

1.10.4 Anaerobic Digestion of Food Waste

Since 2006, several affluent nations in Asia and the European Union have used anaerobic digestion (AD) extensively for food waste treatment (Abbasi et al., 2012). Contrarily, it is noted that AD is still not widely used as a significant therapeutic strategy for food waste control in developing nations. A number of institutions and NGOs (non-governmental organizations) in China and India have set up various anaerobic digesters on a domestic and commercial scale to improve AD technology. India, for instance, opened biogas facilities that are used by several institutions and adopted AD on a trial basis.

1.11 Waste Management Techniques Adopted in Various Food Processing Sectors

Effective waste treatment strategies enhance the environment, health, and economy. However, more than two-thirds of the waste generated end up in landfills and dump yards. This is due to the costs associated with treating the waste, which ranges from $35 to $100 per ton. Though waste treatments have a positive effect on the environment, during the treatment certain harmful chemicals could be released. For example, incineration releases dioxins, which pollute the atmosphere (Zhou et al., 2022). Looking at the waste treatment beyond conventional approaches, it is the need of the hour, where the essence is to reduce or approach a zero-emission strategy. Several countries are looking towards net-zero by 2050, which can only be achieved when waste management also achieves zero-emissions. Most of the waste treatments are not economically feasible for a variety of reasons, including waste-disposal costs being too high; in fact, many countries do not have waste collection and processing because of high costs and carbon tax.

Food industries are growing rapidly due to globalization and population increase and are providing a wider range of food products to satisfy the needs of the consumers. However, these industries generate huge amounts of by-products and wastes, which consist of high amounts of organic matter, leading to problems regarding disposal, environmental pollution and sustainability (Russ and Pittroff, 2004). In addition, there is the loss of biomass and valuable nutrients that can be used for developing value-added products. Food industries are currently focusing on solving the problems of waste management and recycling by valorization, i.e. utilization of the by-products and discarded materials and developing new value-added products from them for commercial applications. Waste valorization is an interesting new concept that offers a range of alternatives for the management of waste other than disposal or landfilling. Valorization allows the exploration of reusing nutrients in the production of main products, and thus highlights the potential gains that can be achieved. The potential for turning food waste into valuable resources minimizes environmental impact and promotes sustainability. Appropriate technologies that focus on their reuse for the creation of valuable products, whose costs exceed the costs of reprocessing, should be considered. Innovations in technology, processing methods and market demand play a pivotal role in determining the feasibility and viability of these value-added products. Waste management in different food sectors include the following:

1.11.1 Grain Processing

Cereals are cultivated for edible grains and composed of the endosperm, germ and bran. They constitute a major portion of the staple diet in most of the countries of the world. India is a leading producer of cereals, and its processing generates huge amounts of waste and by-products each year. The utilization of cereal by-products could reduce the environment and economic burden and significantly increase the sustainability of our diets and nutrition. This addresses a step forward to waste reduction in the food chain, but also the production of different value-added food products. Cereals have a large volume of solid wastes resultant from processing. Presently, they are used as animal fodder or are returned to the harvested field for land application. Several value-added foods are produced by the incorporation of cereals by-products such as bread, biscuits, cookies, breakfast cereals and pasta (Verma and Mogra, 2013). By-products of cereals processing are steep, germ, bran, husk, spent grains, middling and the endospermic tissue aleurone layer. These grain segments are a rich source of bio-functional particles, fibre, vitamins, lignans, minerals, phytoestrogens and phenolic composite. The grain by-products are rich in numerous compounds like ferulic acid, β-glucans, arabinoxylans, sterols, polyphenols, oryzanol and vitamins, which can hamper the presence diseases like diabetes, cardiac diseases, cancer etc. (Salazar-López et al., 2019). Therefore, these by-products have different possible uses in nutrition and non-food industries (Papageorgiou and Skendi, 2018).

Rice bran contains 11–18% fat, 11–17% protein, 10–14% dietary fibre, 9% ash and 45–60% nitrogen-free extract. They are a rich source of micronutrients like magnesium, potassium, iron, manganese and B vitamins, and they are a good source of choline and inositol (Hoffpauer et al., 2005). Brans can be used as an ingredient in the formulation of products, resulting in various desirable improvements like texture modification and enhanced food stability during production and storage. Rice bran is recommended in preparing bakery products like breads, muffins, pancakes, cookies, cakes, pies, extruded snacks and breakfast cereals (Sharma et al., 2004). Full-fat rice bran contains 16.2–18.5% of oil and is used for various purposes after extracted. Bran wax is a secondary by-product of rice bran and obtained during the dewaxing process of oil refining from crude rice bran oil. It is being widely used as an edible coating material to extend them-shelf life of fruits and vegetables (Shi et al., 2017; Zhang et al., 2017). Milling wheat leads to a production of white flour and a side product called wheat bran. The amount of wheat bran is dependent on the rate of extraction during milling. Wheat bran is largely used to enhance the fibre content of processed foods, as it is a well-known source of dietary fibre. The incorporation of different levels of wheat bran (0–20%) into residual flour was used to develop mineral-enriched brown flour (Butt et al., 2004). Barley bran fractions are more suitable in biscuits, as they have excellent sensory properties and good baking characteristics. Biscuits made with 100% barley bran contain an average of 2.2% β-glucan and are labelled as high-fibre products (Nagel-Held et al., 1997). Oat bran was added in wheat bread and patties and also used as a fat replacer in processed meats. The chief fraction of alimentary fibre present in oats that has great importance for human health is β-glucan.

Corn husks are the major solid waste obtained from processing sweet corn. Traditionally they are used for animal feed and/or land application. But according to one study the corn husks could be used as the latent substrate for the manufacture of citric acid from *Aspergillus niger* (Hang and Woodams, 2000) and also could be used as the probable resource of pigment anthocyanins (Li et al., 2008). Rice-husk flour has also been used for thermoplastic polymer composite, with polypropylene as the matrix, as the reinforcing filler. Cereal germs composed of fat, protein, crude fibre, dietary fibre, vitamins and minerals are well known for their contributions for value addition to foods and feeds. Germ contains 2–3% of the total weight of grain kernel. Wheat germ contains high amounts of vitamin E and lipids, so it is used for oil production, while defatted wheat germ is a superior source of protein and amino acids, albumins, globulins, glycine, aspartic acid, glutamic acid, arginine, leucine, alanine, proline and lysine, with lower concentrations of isoleucine, valine, methionine and arginine (Brandolini and Hidalgo, 2012). The by-products of malting are malt sprouts, which consist of sprouts, roots and malt hulls and are known as a source of protein. Radosavljević et al. (2019) concluded that malting by-products could be used for the production of lactic acid and also for animal feed. Spent grains can be used as substrate for the cultivation of microorganisms and further for the production of enzymes like alpha-amylase by *Bacillus licheniformis*, *Bacillus subtilis*, xylanase by *Streptomyces avermitilis* and *Aspergillus awamori*, feruloyl esterase by *Aspergillus oryzae* and *Streptomyces avermitilis* and cellulase by *Trichoderma reesei* (Mussatto, 2009). Other uses of malt by-products are adsorbent for removing volatile organic compound emissions, as a source of biogas or the removal of organic material from waste matter.

1.11.2 Pulse Processing

The major by-products generated by the milling of pulses are 6–13% brokens, 7–12% powder and germ mixture and 4–14% husk. These by-products act as a potential resource of various biologically active compounds that can be utilized in food industries, as they have favourable technological aspects and/or nutritional content. The utilization of pulses by-products in bakery and other value-added food materials can improve its nutritional value (Tiwari et al., 2011). With an increase in pulse processing, its by-product disposal also poses an emerging problem, as the materials obtained from plants are prone to spoilage caused by microorganisms, thereby restricting its utilization. Moreover, the cost of drying, storing and shipping by-products are also some of its limiting factors from the economical viewpoint. Hence, these agro-industrial by-products are mostly used as a feed for animals or utilized in the manufacturing of fertilizers.

The by-products of pulses have abundant dietary fibre, polyphenols and minerals with high antioxidative potential, which makes them suitable for manufacturing food products (Luzardo-Ocampo et al., 2019). Tiwari et al. (2011) used the dehulled flour of *Cajanus cajan* L., commonly called pigeon pea

(PPDF) and pigeon pea by-product flour (PPBF), in place of wheat flour to make biscuits. The fruitful supplementation of pulse materials in baked products was found to increase the nutritional value of the final product, along with providing an appropriate use of by-products. Another study showed that the incorporation of pea hull (3%) to wheat flour could slightly help in increasing water retention along with the resistance of dough, and also decrease dough elasticity (Wang et al., 2002). Saini et al. (2017) carried out research to assess the effect of incorporating chickpea husks on physical, sensory, and nutritional characteristics, along with the antioxidant nature of the formulated biscuits. The seed coat of pulse contains a variety of polyphenolic compounds, such as flavonols, flavanols, flavone glycosides and proanthocyanidins (polymeric and oligomeric), which contribute to its antioxidant potential. The husk of black gram is used to extract the enzyme peroxidase, which is highly thermostable and has abundant biomedical, analytical and industrial applications (Ajila and Prasada Rao, 2009).

1.11.3 Fruit and Vegetable Processing

Fruits and vegetables are some of the most consumed commodities globally, accounting for more than 42% of total food wastage. The waste composition generated during fruit and vegetable processing may be in the form of peels, seeds, crop, leaves, straw, stems, roots or tubers, depending on the raw material chosen. Depending on plant species and tissues, the waste obtained from fruit and vegetable processing possesses a wide variety of properties. For instance, peels and seeds are high in phytochemical compounds, making them a possibility in food flavouring agents and preservation compounds (Sridhar et al., 2021). Similarly, vegetal tissues rich in carotenoids, vitamins and fibres possess antioxidant and anti-diabetic properties, thereby helping to prevent diseases and disorders (Gowe, 2015; Yusuf, 2017).

Pre-treatment of fruit and vegetable wastes allow them to be used as substrates in fermentative bioprocesses to obtain bioethanol, biobutanol, biomethane, biohydrogen or organic acids due to their high (20–30% dry weight) carbohydrate content (Albuquerque, 2003). Another way to valorize these residues is by employing them as a substrate to extract bioactive compounds by means of diverse separation techniques. Among the main fruit wastes commonly used for bioethanol production are bananas, citrus fruits, apples and pineapples. Banana waste is considered a suitable substrate to produce bioethanol since it has low lignin content (Pourbafrani et al., 2010). Grapefruit residues (husk, seeds and membrane residues) contain glucose, sucrose and fructose, which can be directly fermented by Saccharomyces cerevisiae to produce ethanol. Potato, carrot, cassava and onion waste are also vegetable residues commonly investigated as substrates for the production of bioethanol.

Another biofuel that is important due to its high energy content (122 kJ g−1) is biohydrogen. This is a very promising energy resource, since producing hydrogen by biological methods requires lower energy consumption than thermo-chemical and electrochemical methods. In addition, during the combustion of biohydrogen, water vapour and energy are released instead of greenhouse gases. Carbohydrate-rich raw materials with low amounts of nitrogen and requiring minimal pre-treatment are suitable for biohydrogen production. Mixtures of fruits and vegetables such as peppers, onions, potatoes, eggplants, carrots, cabbage, cucumbers, citrus, pears, apples and grapes have been used as substrates to produce biohydrogen by dry fermentation under thermophilic and mesophilic conditions.

Bioactive compounds are substances that can play a key role in reducing the risk of certain diseases. As a result of these health-enhancing activities, research and studies on bioactive compounds have increased considerably in recent years. Fruit and vegetable waste contains bioactive fractions that usually include carbohydrates, proteins, lipids and secondary metabolites. Fruit and vegetable seeds are rich in proteins whose hydrolysis favours the release of bioactive peptides with pharmacological activity. In addition, carotenoids, sterols or fatty acids are some examples of the bioactive lipids present in FVW, such as citrus, mango, or tomato seeds. For these reasons, the use of theses wastes as a source of bioactive polysaccharides has generated increasing interest in recent years. Within the family of bioactive compounds, antimicrobial compounds, antioxidant compounds and enzymes are important groups.

1.11.4 Dairy Processing

The dairy industry has an important position among the food industries with developed technology and widespread products, such as cheese, yogurt, butter, cream and ice cream. Milk and other processed dairy products can form the main components of food waste in significant amounts due to incidents on the farm (e.g. spillage), handling food for some analyses, transportation loss (e.g. damaged packaging) and obtaining poor-quality products. In addition to these wastes, dairy industry by-products (cheese whey, spilled milk and curd chunks) are also potential pollutants due to the presence of milk residues. Among them, the excessive amount of whey (approximately 9 L from 1 kg of cheese production) released during cheese production to food waste creates an important environmental problem. Other wastes and by-products containing a high level of fat also make it difficult to treat these wastes. Furthermore, the wastes have high COD and BOD levels due to the milk proteins and lactose they contain. Dairy industry wastes and by-products are potential raw materials to produce a variety of bio-based products while simultaneously reducing their carbon footprint.

Dairy waste can be broadly classified into two types: (1) wastewater, i.e. effluent, and (2) solid waste. Dairy processing for various products such as cheese, yogurt, butter etc. generate in waste streams with complex compositions. The compounds produced from the dairy waste are very useful in the food, agriculture, petroleum, cosmetics and pharmaceutical industries and for sustainable development. In one study, sludge generated from the dairy industry was assessed for the cultivation of various strains of *Rhizobia* (Singh et al., 2013). Dairy industry waste, both wastewater and sludge, is a promising substrate for the production of hydrogen (Patel et al., 2018) and methane (CH4).

Dairy industry by-products are potential sources containing lactose, protein and fat. Dairy waste and related by-products can be purified and used in other industries through reverse osmosis (RO), drying, hydrolysis, ion exchange, nanofiltration (NF), ultrafiltration (UF) and electrodialysis (Ryan and Walsh, 2016). Bioconversion of milk, whey and other dairy products into value-added products is an essential solution in the modern era. Despite their polluting nature, dairy industry by-products are an excellent source of lactose and some other nutrients that can support microbial growth for the production of bio-products (Figure 1.6).

Dairy industry by-products with high levels of chemical oxygen demand (COD) can be used as a possible feedstock for biogas production via anaerobic digestion (AD) (Kozłowski et al., 2019). Produced biogas can be consumed as a thermal energy for steam production in the cheese manufacturing process. Moreover, solid and liquid fractions of the anaerobic digesters contain valuable nutrients that can be used as soil fertilizers (Kavacik and Topaloglu, 2010). Dairy wastewater, whey, whey powder and milk powder have been used to produce hydrogen (Chandra et al., 2018). In the production of ethanol from lactose-containing dairy waste, microorganisms that can naturally consume lactose can be used, or a chemical hydrolysis can be performed using the enzyme β-galactosidase (Guimarães et al., 2010). Dairy industry wastes such as skim milk and whey can be considered suitable media for lab cultivation. Many researchers have shown that L(+)isomer of lactic acid were produced instead of D(−) isomer of lactic acid when bacterial strains (*Lactobacillus*

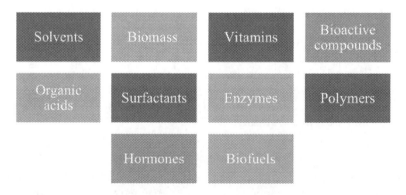

Figure 1.6 *Latent bio-products produced by using waste from the dairy industry.*

casei, L. paracasei subsp. *tolerans, L. rhamnosus, L. lactis,* and *Streptococcus thermophiles*) were cultivated in whey media (Ghasemi et al., 2017).

1.11.5 Meat and Poultry Processing

The majority of the waste in the meat industry is produced during slaughtering. Slaughterhouse waste consists of the portion of a slaughtered animal that cannot be sold as meat or used in meat products. Such waste includes bones, tendons, skin, the contents of the gastro-intestinal tract, blood and internal organs. These vary with each type of animal (Sielaff, 1996; Grosse, 1984). Efficient utilization of meat by-products is important for the profitability of the meat industry. It has been estimated that 11.4% of the gross income from beef and 7.5% of the income from pork come from the by-products. In the past, by-products were a favourite food in Asia, but health concerns have led to an increased focus on non-food uses, such as pet foods, pharmaceuticals, cosmetics and animal feed (Rivera et al., 2000).

The literature indicates that by-products (including organs, fat or lard, skin, feet, abdominal and intestinal contents, bone and blood) of cattle, pigs and lambs represents 66%, 52% and 68% of the live weight respectively. Animal blood has a high level of protein and heme iron, and it is an important edible by-product (Wan et al., 2002). Blood is used in food as an emulsifier, a stabilizer, a clarifier, a colour additive and as a nutritional component (Silva and Silvestre, 2003). Animal hides have been used for shelters, clothing and as containers by human beings since prehistoric times. Examples of finished products from the hides of cattle and pigs, and from sheep pelts, are leather shoes and bags, rawhide, athletic equipment, reformed sausage casing and cosmetic products, sausage skins, edible gelatine and glue (Benjakul et al., 2009). Gelatin is produced by the controlled hydrolysis of a water insoluble collagen derived from protein. It is made from fresh raw materials (hides or bone) that are in an edible condition. Both hides and bones contain large quantities of collagen. Gelatin extracted from animal skins and hides can be used for food (Choa et al., 2005). The raw material can also be rendered into lard. Collagen from hides and skins also has a role as an emulsifier in meat products because it can bind large quantities of fat. Gelatin is added to a wide range of foods, as well as forming a major ingredient in jellies and aspic (Jamilah and Harvinder, 2002). Gelatin is also widely used as a stabilizer for ice cream and other frozen desserts. High-bloom gelatin is added as a protective colloid to ice cream, yogurt and cream pies. Approximately 6.5% of the total production of gelatin is used in the pharmaceutical industry. Most of it is used to make the outer covering of capsules.

1.11.6 Seafood Processing

Fish waste is a great source of minerals, proteins and fat. Potential utilization of fish scraps from five marine species (white croaker, horse mackerel, flying fish, chub mackerel and sardine) to produce fish protein hydrolysate by enzymic treatment was investigated by Khan et al. (2003) and indicated that fish protein hydrolysate could be used as a cryoprotectant to suppress the denaturation of proteins of lizard fish surimi during frozen storage. Ohba et al. (2003) reported that collagen or keratin contained in livestock and fish waste may be converted to useful products by enzymic hydrolysis, providing new physiologically functional food materials. Enzymes and bioactive peptides obtained from fish waste or by-catch can be used for fish silage, fish feed or fish sauce production (Gildberg, 2004). There are several alternative uses of fish processing waste, like fish mince, fish gelatin, fish as a source of nutraceutical ingredients, fishmeal and the possible use of fish and protein concentrate as a food source.

The production of organic acids and amino acids from fish meat by subcritical water hydrolysis would be an efficient process for recovering useful substances from organic waste discovered from fish markets (Yoshida et al., 1999). Fish waste has considerable potential to be used as poultry and animal feed. The potential of fish waste effluent as a fermentation medium for the production of antibacterial compounds by lactic acid bacteria was evaluated by Tahajod and Rand (1996). Fish collagens are of interest to the food processing industry, as they are used to produce gelatin, which is extracted from the collagen. Chitosan, produced from shrimp and crab shells, has shown a wide range of applications from the cosmetic to pharmaceutical industries (Arvanitoyannis and Kassaveti, 2008). One of the important applications of chitosan is the removal of proteinaceous matter in the food industry. Chitosan can potentially be used as a food preservative in food packaging materials since it has film-forming ability with antimicrobial properties. Chitosan has wide-spectrum antimicrobial activity against bacteria, yeast and fungi (Rabea et al., 2003). It may be used in various food preservation applications, such as direct addition into food, direct application of chitosan film or coatings onto food surfaces, addition of chitosan sachets into packages and use of chitosan-incorporated packaging materials.

1.11.7 Edible Oil Processing

Oilseeds are a major agricultural commodity. Oil plays an important role in maintaining human health and growth. For the production of good-quality oil, the oilseeds have to be cleaned before the refining processes, and there should be effective refining steps, including cleaning of seeds, mechanical expression, solvent extraction, degumming, neutralization, bleaching and deodorization. These processes result in good-quality oil, along with a huge quantity of by-products. Disposal of these by-products leads to environmental pollution and causes many other problems. Instead, there are many possible ways to utilize the by-products, like husks, hulls, cake, pomace or meal, gum, soap stock, spent bleach earth/clay, and distillate. Oilseed cake/meal is a by-product obtained in any oil extraction industry when recovering crude oil from the oilseeds through mechanical expression or solvent extraction. Usually, the meal obtained through solvent extraction is not considered human friendly, due to the chemical treatments it undergoes. The nutritional quality of defatted oil cake is very rich since only the oil from the whole seed is removed through the extraction process. Therefore, all other nutritional qualities from the oilseed are mostly retained in the cake. Oilseed cake is rich in proteins, fibre content, antioxidants, colourants and other health beneficial components. The majority of the industries utilize oilseed cake as animal feed. However, there is tremendous scope for it to be utilized in recovering more beneficial and value-added products.

Oilseed cake can be utilized to enhance the nutritional profile of gluten-free foods (Radočaj et al., 2014). The application of proteins extracted from oilseed cakes is increasing lately in the food, pharmaceutical and cosmetic industries (Moure et al., 2006; Rodrigues et al., 2012). Soybean seeds are an excellent source of plant-based proteins, containing all the essential amino acids and having the distinction of generating the highest amount of oil cake with respect to other plants. Soybean oil cake is highly utilized in the preparation of value-added products like isolates, concentrates, defatted flour etc. and are used to develop various products. In addition to food applications, the protein extracts are also being used in the development of biodegradable films for food packaging. The protein from soy seed cake is found to be very suitable for producing good-quality smooth films with greater flexibility and transparency (Gontard and Guilbert, 1994). The protein isolates also have the potential to be used as adhesives (Mullen et al., 2015).

Hull is another important waste by-product and more often is utilized as a fuel for generating heat. Before using them, the hulls are modified lightly by adding some binders and glue-like substances to make small solid blocks (briquettes) suitable for boilers. The hull of a purple-seeded sunflower can be used to extract a red dye, which is approved as a food colourant by the US FDA (Seiler and Gulya, 2015). The hull of these sunflower seeds consists of anthocyanin content of 6–16 g/kg of the hull, which is almost equivalent to that of grape skin and beet pulp. Pectin can also be extracted from the oilseed head. The amount of pectin present in a sunflower seed head after the removal of the seed is 150–250 g/kg (Seiler and Gulya, 2015). Unlike the high methoxyl pectin extracted from apples and citrus fruits, sunflower pectin is low methoxyl. This is favourable in making the low-sugar jellies that are gaining attention among diet food markets nowadays.

The wastewater obtained from the oil processing plant is very toxic and polluted because of high levels of BOD, COD, dissolved solids, oil and fat residues, organic nitrogen and ash residues. At the same time, the wastewater is also highly nutritious. Hence, important co-products like sugars, sugar alcohol, polymer building blocks, polymerins, phenols, sterols, biogas, isoflavones, biodiesel etc. can be recovered from the wastewaters of an oil processing plant. The solid wastes such as stems, pods, leaves, damaged/spoiled kernels, dirt, small stones, extraneous seeds, stalks, coir etc. are also obtained from oilseed processing industries and can be either utilized as boiler feed to generate steam or converted into compost. It is also sometimes, albeit rarely, landfilled in some small-scale oilseed processing industries. Even charcoal can be produced by subjecting this plant material to pyrolysis (Lua and Guo, 1999).

Refining of oil after extraction also generates a considerable amount of waste that can act as a raw material for producing value-added products. Gum is a by-product which is obtained after the degumming process in the refining of crude oil. The gum is crude lecithin, which further has to be purified to remove carbohydrates, proteins and other contaminants. The purified gum/lecithin has various applications, including food uses (bakery and confectionary, emulsifiers, release agents, instant powders, encapsulation etc.), animal feed, food supplements and pharmaceuticals and other applications in cosmetics, paint industries as a lubricant, releasing agents, emulsifiers, adhesives and absorbents (Clarke, 2007). The next step in the refining process is neutralization or deacidification, in which degummed oil is continuously mixed with diluted sodium hydroxide to remove free fatty acids (FFA), phosphatides and other unsaponifiable materials. FFA are the most valued components, and their composition varies depending on certain factors such as the source of the oil, the condition of the oilseeds while processing, oil removal methods, the type of solvent used, the duration of extraction and the condition during the refining process. Lipids are a major application in animal feed and are now being used to grow microorganisms. Palm oil and coconut oil are important fatty acid sources for soap production. Soap stock has been used in the production of lipase by *Aspergillus niger* by a solid-state fermentation method. Soap stock is used as a carbon source for the production of carotenoid-producing yeast (*Rhodotorula rubra*). The application of soap stock has increased carotenoid yield up to 5.36-fold compared to the production using glucose (Alipour et al., 2017). Soap stock having high unsaturated FFAs had been used for enzymatic production of natural epoxides in microchannel bioreactors. The yield of epoxidation enhanced 2.8-fold (Mashhadi et al., 2018). Deodorizer distillate is a very cheap source of several health beneficial components, like tocopherols, sterols, FFAs etc. These valuable components are being used in different foods, pharmaceuticals and cosmetics (Sherazi et al., 2016).

1.11.8 Spices Processing

Major products that are obtained from spices are spice oils and oleoresins. About 80–90% of the waste is generated from spice oil and oleoresins is bulk spice residue, which does not exhibit any commercial value (Sowbhagya, 2019). It is desirable to look for a possible solution from which hindrance to utilize industrial waste for food applications and in turn preventing pollution can be addressed. Due to a number of health attributes associated with dietary fibres (44–62%), there has been tremendous interest in utilizing spice spent residues obtained through chilli, cumin, coriander and pepper with their prime focused application among baked stuffs (Mathew, 2004). Spent spices are full of functional constituents, viz. dietary fibre, protein, vitamins, polyphenols and other divalent cations of human interest to improve body metabolism (Murphy et al., 1978; Maikki, 2004). Despite numerous food applications of spice spent, it also finds its way as a composite in bioactive film preparation through its characteristics to enhance techno-functional features, such as tensile strength and thermal ability among bioactive films (Chethana et al., 2014; Meyer et al., 2011). This way, the spice waste cannot only be better consumed but can also contribute to minimizing environmental hazards.

Cumin spent residue, after oil and oleoresin extraction, has been reported to be rich in dietary fibre and exhibit significant water retention (Sowbhagya et al., 2007). The underutilized spent cumin, with good nutrient composition, can be a good substitute for the conventional foodstuffs in health food formulations and can find application in therapeutic foods. Thus, spent cumin has great potential for producing healthy and functional products that might also strengthen its industrial use as protein supplement (Milan et al., 2008). Isolation and characterization of starch from ginger spent has been carried out (Madenenei et al., 2011). Starch isolated from ginger spent had high gelatinization properties and low in-vitro starch digestibility (45%). This property of starch isolated from ginger spent could be made use of in the preparation of foods at high temperatures for the stabilization of texture. Spent ginger is rich in carbohydrates, and the fermentation of carbohydrates by Saccharomyces cerevisiae resulted in the production of bioethanol (Konar et al., 2013). An antioxidant compound from ginger spent was isolated. Methanolic extract of ginger spent when partitioned with butanol, 3,5-diacetoxy-7-(3,4-dihydroxy phenyl) heptanes was obtained, possessing significant antioxidant properties (Sajjad Khan et al., 2006) which could be made use of in food formulations. The turmeric spent rich in starch fraction can be used as a functional ingredient in food industries and in the preparation of biofilms for packaging (Maniglia et al., 2015).

After the extraction of mustard oil, the mustard meal can be used as a high-quality organic fertilizer (Balesh et al., 2005). Clove contains a significant amount of potassium (0.6 g/100 g), calcium (0.27 g/100 g) and magnesium (0.09 g/100 g). After the extraction of clove oil and oleoresin, the spent cloves can be used as a novel source of fibre, protein and minerals (Al-Jasass and Al-Jasser, 2012). Enzyme application to spice spent to remove the unwanted portion of the spent, in turn increasing the concentration of value-added components, could be a new area of study to utilize spice spent in a more effective way in food applications. One of the major applications of spice spent as a potential source of functional ingredients could be in the bakery industry wherein the protein, fibre and mineral content of the bakery products could be enhanced, resulting in many physiological health benefits.

1.11.9 Sugar Processing

From the cane and beet harvesting stage, all through to the milling stage and down to the sugar refining stage (which includes numerous separation processes), the sugar industry generates vast amounts of waste materials and by-products which are rich in chemicals and have great potential for use in numerous other processes. By-products of the sugar industry include cane trash, sugarcane bagasse, press mud, molasses and wastewater. Cane trash consists of the leaves and other extraneous matter which are found together with the sugar cane stalk during harvesting. Besides being availed as cattle feed (Mahala et al., 2013) and for cane field blanketing purposes, studies focusing on the utilization of cane trash along with bagasse for heat and electricity required to make sugar mill energy self-sufficient along with surplus electricity generation have been reported (Khatiwada et al., 2016). Production of 2G biofuels, such as lignocellulosic ethanol from cane trash in combination with bagasse, has also been explored (Chandel et al., 2012; Krishnan et al., 2010). Sugar cane molasses (SCM) is the viscous liquid produced after sucrose crystals have been removed from the concentrated juice by centrifugation. SCM is widely used in many industries ranging from foods to plastics to agro industries and many more for the production of many value-added products.

Molasses is the by-product of the sugar manufacturing process primarily used in the production of ethanol because of its low cost, availability and ability to produce fermentable sugars without any prior treatment. One litre of ethanol can be produced from approximately 4 kg of molasses, though this can vary depending on the extraction method and sugar content of the molasses (Rasmey et al., 2018). Rum is and alcoholic beverage produced from SCM. It is a fairly tasteless and neutral spirit produced from the fermentation and distillation of molasses and sugar juice or syrup, traditionally produced by the West Indies and Central Americas. SCM is known to contain a large number of polyphenols, making it a very good antioxidant agent and a potential agent in the prevention of several chronic diseases that involve oxidative stress (Iwuozor, 2019). In the food industry, SCM is commonly used in foods like cookies or pies, barbecue sauces and gingerbread, among others, because of its special characteristic flavour and sweet aroma (Mulye, 2019). SCM extract can reduce hazardous compound formation during meat processing (Cheng et al., 2021). It also has potential as a natural functional ingredient capable of modifying carbohydrate metabolism and contributing to glycaemic index reduction of processed foods and beverages (Wright et al., 2014). Moreover, SCM can also be used in the production of succinic acid (Chan et al., 2012), citric acid, butanol and sorbitol, among others.

The dark brownish amorphous residue obtained during the clarification of cane juice is called press mud, and it is generated in about 3% of sugar cane processed. The extraction of residual sugar, wax and protein from sugar cane press mud has been studied (Partha and Sivasubramanian, 2006). Press mud has also been used as a substrate in solid-state fermentation for the production of citric and lactic acids. The organic components in press mud make it a possible source for biogas production by anaerobic digestion. Sugar cane bagasse (SCB) is the fibrous lignocellulosic residue of sugar cane after it has been crushed for the extraction of its juice used for sugar and ethanol production (Ajala et al., 2021). Aside from the generation of heat in sugar production plants, many value-added products have been produced from SCB, with many of them directly linked to its high content of crystalline cellulose and fibre.

1.11.10 Confectionery Processing

Confectionery industries contribute immensely towards global food supply. These industries are major enterprises across the world and are still expanding rapidly. Liquid and solid wastes generated from confectionery industries contain substantial amount of lipids and carbohydrates. These are non-edible in nature and need to be managed from the circular economy perspective. A typical analysis of confectionery waste solids (%w/w) present that it is a source of 55% sucrose, 16% glucose, 22% starch, 3.5% gelatine (protein), 2% caramel, 1% organic acids and 0.5% coconut. This translates to 76% sugar, 22% starch and 3.5% protein, which makes it an ideal feedstock for the manufacture of bio-based products of higher value than bioenergy, such as alcohols, enzymes, organic acids, SCP, pigments and platform chemicals.

1.12 Challenges of Food Waste Management in India

Among the main challenges that need to be tackled are the quantification of wasted food along the value chain, systematic reporting of wasted food and the evaluation of the causes of food waste. The current status of solid waste management in India is poor because the best and most appropriate methods from waste collection to disposal are not being used. There is a lack of training, and the availability of qualified waste management professionals is limited. There is also a lack of accountability in current waste management systems throughout India. Food waste management in India faces several challenges due to a combination of socio-economic, cultural and infrastructural factors. These challenges contribute to significant food wastage at various stages of the supply chain, from production to consumption. Some key challenges include:

1. *Lack of infrastructure*: Insufficient infrastructure for food storage, transportation and distribution leads to post-harvest losses. Inadequate cold storage facilities and transportation systems result in the spoilage of perishable goods before they reach consumers.
2. *Inefficient supply chain*: India's supply chain for food products is often fragmented and unorganized, resulting in poor coordination between producers, distributors and retailers. This leads to delays, improper handling and increased chances of spoilage.

3. *Lack of education and awareness*: Many consumers and businesses lack awareness of the consequences of food waste and how to effectively manage it. This contributes to careless attitudes towards food consumption and disposal.
4. *Cultural factors*: In India, there is a cultural inclination to prepare fresh meals, leading to over-preparation and the subsequent disposal of excess food. Additionally, religious and cultural practices, such as sharing and offering food, can lead to food waste.
5. *Inadequate harvesting techniques*: Poor harvesting techniques and lack of access to modern technology can result in losses during the agricultural production phase. Improper handling, storage and transportation also contribute to waste.
6. *Lack of proper sorting and processing*: Waste segregation practices are often lacking, making it difficult to recover edible food from the waste stream. This prevents surplus food from reaching those in need.
7. *Lack of incentives*: Many businesses and individuals may not have sufficient incentives to reduce food waste. The economic and environmental costs of waste disposal are often not fully internalized by waste generators.
8. *Regulatory challenges*: The absence of comprehensive regulations and standards for food waste management hinders the development of effective strategies and initiatives.
9. *Inadequate food redistribution systems*: Although there are efforts to redistribute surplus food to charities and those in need, there are challenges related to logistics, safety and the proper handling of donated food.
10. *Urbanization and changing lifestyles*: Rapid urbanization and changing lifestyles have led to a greater reliance on convenience foods and increased eating out, which can contribute to larger portions and more food waste.
11. *Lack of data and monitoring*: Limited data on food waste makes it difficult to assess the scale of the problem accurately and design targeted interventions.

Addressing these challenges requires a multifaceted approach involving government policies, awareness campaigns, investment in infrastructure, technology adoption, collaboration between stakeholders and innovative solutions for food redistribution and waste reduction. Limited environmental awareness combined with low motivation has inhibited innovation and the adoption of new technologies that could transform waste management in India. Public attitudes to waste are also a major barrier to improving waste management in India.

1.13 Conclusion and Future Thrust

The issue of food waste has emerged as a critical challenge on a global scale, with its far-reaching implications encompassing environmental sustainability, economic efficiency, and social equity. As we reflect on the strides made in food waste management, it becomes evident that concerted efforts have yielded tangible results. However, the journey is far from over, and the convergence of innovation, collaboration and awareness is crucial to propel food waste management into a more sustainable and effective future.

It is undeniable that progress has been achieved in tackling food waste at various stages of the supply chain. From farm to fork, initiatives to prevent food waste at its source have gained traction. Producers are employing better harvesting practices, and retailers are enhancing inventory management to reduce overstocking. Consumers are becoming more mindful of their consumption habits, driven by heightened awareness of the environmental toll of food waste. The establishment of food recovery and redistribution networks has bridged the gap between surplus food and those in need, addressing not only waste but also food insecurity. The innovation in technology, too, has been a game-changer. From smart inventory systems that optimize stock levels to apps that facilitate surplus food donation, technology has streamlined food waste reduction efforts. Composting and anaerobic digestion processes are diverting organic waste from landfills, contributing to a more circular approach to waste management. These combined efforts underscore the progress made, but the challenges that persist require an unwavering commitment to future strategies.

Looking forward, the future thrust in food waste management presents a multifaceted approach that addresses both the immediate and long-term aspects of the challenge. Technological innovations are set to take centre stage, with the development of sensor-based systems and artificial intelligence aiding real-time monitoring of food quality and shelf life. Such advancements can revolutionize inventory management, allowing for accurate demand forecasting and minimizing excess production. Furthermore, the integration of blockchain technology can enhance transparency in supply chains, curbing inefficiencies and reducing the likelihood of food waste. The circular economy models are anticipated to gain prominence, where the concept of waste as a resource is embraced. Businesses are likely to adopt closed-loop systems, wherein food waste is repurposed for energy generation, animal feed or nutrient-rich compost, reducing the strain on conventional waste disposal methods.

Policy and regulation are also pivotal components of the future thrust. Governments are expected to enforce more stringent regulations on food waste, mandating reporting and incentivizing waste reduction practices. Tax incentives for food donations to charities and shelters can bolster food recovery efforts, creating a more symbiotic relationship between surplus food generators and communities in need. Collaborative partnerships among stakeholders are essential for a holistic approach. Governments, businesses, non-governmental organizations and research institutions need to synergize their efforts, sharing expertise and resources to drive meaningful change.

Promoting behavioural change among consumers remains a cornerstone of the future thrust in food waste management. Education campaigns, social media influence and incentivized programmes can inspire individuals to adopt conscious consumption habits and embrace portion control, proper storage techniques and creative ways to utilize leftovers. A global perspective is also imperative, recognizing that food supply chains transcend borders. International collaboration can lead to the exchange of best practices, innovative solutions and a collective understanding of the complexities inherent in food waste management.

In conclusion, the strides made in food waste management are commendable, reflecting the collective commitment to a more sustainable future. However, the road ahead is marked by challenges that demand dynamic and innovative solutions. The fusion of cutting-edge technology, circular economy models, robust policies, collaborative partnerships, consumer education and global cooperation will be the driving forces in shaping the future of food waste management. As we march towards a world with minimized food waste, optimized resources and a just distribution of nourishment, the lessons learned and the actions taken today will serve as the foundation upon which this vision is realized.

References

Abbasi, T., Tauseef, S. M. and Abbasi, S. A. 2012. Anaerobic digestion for global warming control and energy generation—An overview. Renewable Sustainable Energy Review 16 (5): 3228–3242. https://doi.org/10.1016/j.rser.2012.02.046.

Adhikari, B. B., Chae, M. and Bressler, D. C. 2018. Utilization of slaughterhouse waste in value-added applications: Recent advances in the development of wood adhesives. Polymers 10 (2): 176.

Adhikari, B. K., Barrington, S. and Martinez, J. 2006. Predicted growth of world urban food waste and methane production. Waste Management Research 24 (5): 421–433.

Afzal, N., Basit, A., Daniel, A., Ilyas, N., Imran, A., Awan, Z. A., Papargyropoulou, E., Stringer, L. C., Hashem, M., Alamri, S. and Bashir, M. A. 2022. Quantifying food waste in the hospitality sector and exploring its underlying reasons—A case study of Lahore. Pakistan Sustainable 14 (11): 6914. https://doi.org/10.3390/su14116914.

Agricultural Statistics at a Glance. 2022. Government of India Ministry of Agriculture & Farmers Welfare Department of Agriculture & Farmers Welfare Economics & Statistics Division. https://agricoop.nic.in/Documents/CWWGDATA/Agricultural_Statistics_at_a_Glance_2022_0.pdf (Accessed on: 01.06.2023).

Ajala, E., Ighalo, J., Ajala, M., Adeniyi, A. and Ayanshola, A. 2021. Sugarcane bagasse: A biomass sufficiently applied for improving global energy, environment and economic sustainability. Bioresources and Bioprocessing 8 (1): 1–25.

Ajila, C. M. and Prasada Rao, U. J. S. 2009. Purification and characterization of black gram (Vigna mungo) husk peroxidase. Journal of Catalysis B Enzymology 60: 36–44.

Albuquerque, P. 2003. MSc Thesis, Universidade Federal de Santa Catarina Brazil.

Alipour, S. et al. 2017. β-carotene production from soap stock by loofa-immobilized Rhodotorula rubra in an airlift photobioreactor. Process Biochemisty 54: 9–19. https://doi.org/10.1016/j.procbio.2016.12.013.

Al-Jasass, F. M. and Al-Jasser, M. S. 2012. Chemical composition and fatty acid content of some spices and herbs under Saudi Arabia conditions. The Scientific World Journal: 859892.

Anal, A. K. 2017. Food processing by-products and their utilization: Introduction. Food Process Prod Their Utilization 29: 1–10.

Arvanitoyannis, IS., and Kassaveti, A. 2008. Fish industry waste: treatments, environmental impacts, current and potential uses. *Int J Food Sci Technol.* 43(4):726–745. doi: 10.1111/j.1365-2621.2006.01513.x.

Balandran-Quintana, R. R. 2018. Recovery of proteins from cereal processing byproducts. In: Sustainable Recovery and Reutilization of Cereal Processing by-Products. Elsevier, Kidlington UK, pp. 125–157.

Balesh, T. F., Zapata, B. I., Aune, L. and Sitaula, B. 2005. Evaluation of mustard meal as organic fertilizer on tef (Eragrostis tef (Zucc) Trotter) under field and greenhouse conditions. Nutrient Cycling in Agro Ecosystems 73: 49–57.

Benjakul, S., Oungbho, K., Visessanguan, W., Thiansilakul, Y. and Roytrakul, S. 2009. Characteristics of gelatin from the skins of bigeye snapper, Priacanthus tayenus and Priacanthus macracanthus. Food Chemistry 116 (5): 445–451. doi: 10.1016/j.foodchem.2009.02.063.

Brandolini, A. and Hidalgo, A. 2012. Wheat germ: Not only a by-product. International Journal of Food Science and Nutrition 63 (S1): 71–74.

Butt, M. S., Qamar, M. I., Anjum, F. M., Abdul, A. and Randhawa, M. A. 2004. Development of mineral enriched brown flour by utilizing wheat milling by-products. Nutrition Food Science 34: 161–165.

Chan, S., Kanchanatawee, S. and Jantama, K. 2012. Production of succinic acid from sucrose and sugarcane molasses by metabolically engineered Escherichia coli. Bioresource Technology 103 (1): 329–336.

Chandel, A. K., da Silva, S. S., Carvalho, W. and Singh, O. V. 2012. Sugarcane bagasse and leaves: Foreseeable biomass of biofuel and bio-products. Journal of Chemical Technology & Biotechnology 87 (1): 11–20.

Chandra, R., Castillo-Zacarias, C., Delgado, P. and Parra-Saldívar, R. 2018. A biorefinery approach for dairy wastewater treatment and product recovery towards establishing a biorefinery complexity index. Journal of Clean Production 183: 1184–1196. https://doi.org/10.1016/j.jclepro.2018.02.124.

Chang, FC, Tsai, MJ, Ko, CH. (2018). Agricultural waste derived fuel from oil meal and waste cooking oil. *Environmental Science and Pollution Research* 25(6): 5223–30. DOI 10.1007/s11356-017-9119-x.

Cheng, Y., Yu, Y., Wang, C., Zhu, Z. and Huang, M. 2021. Inhibitory effect of sugarcane (Saccharum officinarum L.) molasses extract on the formation of heterocyclic amines in deep-fried chicken wings. Food Control 119: 107490.

Chethana, M., Madhukar, B. S., Somashekar, R. and Hanta, S. 2014. The influence of ginger spent loading on mechanical, thermal and micro structural behaviours of polyurethane green composites. Journal of Composite Materials 48: 2251–2264.

Choa, S. M., Gub, Y. S. and Kima, S. B. 2005. Extracting optimization and physical properties of yellow fin tuna (Thunnus albacares) skin gelatin compared to mammalian gelatins. Food Hydrocolloids 19 (2): 221–229. doi: 10.1016/j.foodhyd.2004.05.005.

Clarke, Z. 2007. Lecithin. xPharm: The Comprehensive Pharmacology Reference (1): 1–3. https://doi.org/10.1016/B978-008055232-3.62016-1.

da Rocha, M., Alemán, A., Romani, V. P., López-Caballero, M. E., Gómez-Guillén, M. C., Montero, P. and Prentice, C. 2018. Effects of agar films incorporated with fish protein hydrolysate or clove essential oil on flounder (Paralichthys orbignyanus) fillets shelf-life. Food Hydrocolloids 81: 351–363. https://doi.org/10.1016/j.foodhyd.2018.03.017.

Dai, Y, Sun, Q., Wang, W., Lu, L., Liu, M., Li, J., Yang, S., Sun, Y., Zhang, K., Xu, J., Zheng, W. (2018). Utilizations of agricultural waste as adsorbent for the removal of contaminants: A review. *Chemosphere*. 211: 235–253. https://doi.org/10.1016/j.chemosphere.2018.06.179

FAO. 2021. Production: Crops and livestock products. In: FAO. Rome. Cited March 2022. www.fao.org/faostat/en/#data/QCL (Accessed on: 01.06.2023).

Gen, M., Altiparmak, F. and Lin, L. 2006. A genetic algorithm for two-stage transportation problem using priority-based encoding. OR Spectrum 28: 337–354.

Ghasemi, M., Ahmad, A., Jafary, T., Azad, A. K., Kakooei, S., Wan Daud, W. R. and Sedighi, M. 2017. Assessment of immobilized cell reactor and microbial fuel cell for simultaneous cheese whey treatment and lactic acid/electricity production. International Journal of Hydrogen 42 (14): 9107–9115. https://doi.org/10.1016/j.ijhydene.2016.04.136.

Gildberg, A. 2004. Enzymes and bioactive peptides from fish waste related to fish silage, fish feed and fish sauce production. Journal of Aquatic Food Product Technology 13 (2): 3–11. doi: 10.1300/J030v13n02_02.

Gontard, N. and Guilbert, S. 1994. Bio-packaging: Technology and properties of edible and/ or biodegradable material of agricultural origin. In: Food Packaging and Preservation. Springer, Boston, MA, pp. 159–181.

Government of India. 2019. Annual Report-2018–19. New Delhi: Ministry of Food Processing Industries.www.mofpi.gov.in/sites/default/files/eng_mofpi_annual_report_201819.pdf (Accessed on: 01.06.2023).

Gowe, C. 2015. Review on potential use of fruit and vegetables by-products as a valuable source of natural food additives. Food Science and Quality Management 45: 47–61.

Grosse, C. 1984. Absatz und Vermarktungsmoglichkeiten fur Schlachtneben-produkte und Schlachtabfalle in der Bundesrepublik Deutschland. Dissertation, Universitat Bonn, Institut fur Agrarpolitik, Marktforschung und Wirstschaftssoziologie, Marz.

Guimarães, P. M. R., Teixeira, J. A. and Domingues, L. 2010. Fermentation of lactose to bio-ethanol by yeasts as part of integrated solutions for the valorisation of cheese whey. Biotechnology Advance 28 (3): 375–384. https://doi.org/10.1016/j.biotechadv.2010.02.002.

Gustavsson, J., Cederberg, C., Sonesson, U., Van Otterdijk, R. and Meybeck, A. 2011. Global Food Losses and Food Waste. Food and Agriculture Organization of the United Nations, Rome.

Hang, Y. D. and Woodams, E. E. 2000. Corn husks: A potential substrate for production of citric acid by Aspergillus niger. LWT-Food Science and Technology 33 (7): 520–521.

Hansen, C. L. and Cheong, D. Y. 2019. Agricultural waste management in food processing. In: Kutz, M. (ed.) Handbook of Farm Dairy and Food Machinery Engineering, vol 1, 3rd edn. Academic Press, pp. 673–716. https://doi.org/10.1016/B978-0-12-814803-7.00026-9.

Hoffpauer, D. W., Light Heart, L. C. C. and Crowley, L. A. 2005. New application for whole rice bran. Cereal Foods World 50: 173–174.

Ishangulyyev, R., Kim, S. and Lee, S. H. 2019. Understanding food loss and waste-why are we losing and wasting food? Foods. Jul 29 8 (8): 297. doi: 10.3390/foods8080297. PMID: 31362396; PMCID: PMC6723314.

Iwuozor, K. O. 2019. Qualitative and quantitative determination of anti-nutritional factors of five wine samples. Advanced Journal of Chemistry-Section A 2 (2): 136–146.

Jamilah, B. and Harvinder, K. G. 2002. Properties of gelatins from skins of fish-black tilapia (Oreochromis mossambicus) and red tilapia (Oreochromis nilotica). Food Chemistry 77 (3): 81–84. doi: 10.1016/S0308-8146(01)00328-4.

Jayathilakan, K., Sultana, K., Radhakrishna, K. and Bawa, A. S. 2012. Utilization of byproducts and waste materials from meat, poultry and fish processing industries: A review. Journal of Food Science and Technology 49: 278–293. https://doi.org/10.1007/s13197-011-0290-7.

Ji, C., Kong, C. X., Mei, Z. L. and Li, J. 2017. A review of the anaerobic digestion of fruit and vegetable waste. Applied Biochemistry and Biotechnology 183 (3): 906–922.

Joshi, V. K. 2020. Fruit and vegetable processing waste management—An overview. International Journal of Food Fermentation Technology 10 (2): 67–94.

Joshi, V. K. and Sharma, S. K. 2011. Food processing industrial waste-present scenario. In: Joshi, V. K. and Sharma, S. K. (eds.) Food Processing Waste Managements: Treatment and Utilization Techniques. NIPA Publication, New Delhi, India. p. 130.

Kavacik, B. and Topaloglu, B. 2010. Biogas production from co-digestion of a mixture of cheese whey and dairy manure. Biomass Bioenergy 34 (9): 1321–1329. https://doi.org/10.1016/j.biombioe.2010.04.006.

Khan, M. A. A., Hossain, M. A., Hara, K., Ostomi, K., Ishihara, T. and Nozaki, Y. 2003. Effect of enzymatic fish-scrap protein hydrolysate on gel-forming ability and denaturation of lizard fish Saurida wanieso surimi during frozen storage. Fisheries Science 69 (6): 1271–1280. doi: 10.1111/j.0919-9268.2003.00755.x.

Khatiwada, D., Leduc, S., Silveira, S. and McCallum, I. 2016. Optimizing ethanol and bioelectricity production in sugarcane biorefineries in Brazil. Renewable Energy 85: 371–386.

Kim, J. H., Lee, J. C. and Pak, D. 2011. Feasibility of producing ethanol from food waste. Waste Management 31 (9–10): 2121–2125.

Konar, E. M., Harde, S. M., Kagliwal, L. D. and Singhal, R. S. 2013. Value-added bio-ethanol from spent ginger obtained after oleoresin extraction. Industrial Crops and Products 42: 299–307.

Kosseva, M. R. 2013. Sources, characterization, and composition of food industry wastes. In: Kosseva, M. R. and Webb, C. (eds.) Food Industry Wastes: Assessment and Recuperation of Commodities. Elsevier Science, London, UK. pp. 37–60. http://dx.doi.org/10.1016/B978-0-12-391921-2.00003-2.

Kozłowski, K., Pietrzykowski, M., Czekała, W., Dach, J., Kowalczyk-Juśko, A., Jóźwiakowski, K. and Brzoski, M. 2019. Energetic and economic analysis of biogas plant with using the dairy industry waste. Energy 183: 1023–1031. https://doi.org/10.1016/j.energy.2019.06.179.

Krishnan, C., Sousa, L. D. C., Jin, M., Chang, L., Dale, B. E. and Balan, V. 2010. Alkali-based AFEX pretreatment for the conversion of sugarcane bagasse and cane leaf residues to ethanol. Biotechnology and Bioengineering 107 (3): 441–450.

Kumar, A., Kumar, D., George, N., Sharma, P. and Gupta, N. 2018. A process for complete biodegradation of shrimp waste by a novel marine isolate *Paenibacillus* sp AD with simultaneous production of chitinase and chitin oligosaccharides. International Journal of Biology Macromolecules 109: 263–272.

Kumar, A., Roy, A., Priyadarshinee, R., Sengupta, B., Malaviya, A., Dasguptamandal, D. and Mandal, T. 2017. Economic and sustainable management of wastes from rice industry: Combating the potential threats. Environmental Science and Pollution Research 24: 26279–26296. https://doi.org/10.1007/s11356-017-0293-7.

Lau, K. Q., Sabran, M. R. and Shafie, S. R. 2021. Utilization of vegetable and fruit by-products as functional ingredient and food. Frontiers in Nutrition 8: 661–693. doi: 10.3389/fnut.2021.661693.

Law, C., Green, R., Kadiyala, S., Shankar, B., Knai, C., Brown, K. A. and Cornelsen, L. 2019. Purchase trends of processed foods and beverages in Urban India. Global Food Security 23: 191–204.

Li, C. Y., Kim, H. W., Won, S. R., Min, H. K., Park, K. J., Park, J. Y. and Rhee, H. I. 2008. Corn husk as a potential source of anthocyanins. Journal of Agricultural and Food Chemistry 56 (23): 11413–11416.

Liu, Y., Wachemo, A. C., Yuan, H. and Li, X. 2019. Anaerobic digestion performance and microbial community structure of corn stover in three-stage continuously stirred tank reactors. Bioresource Technology 287: 121339.

Lua, A. C. and Guo, J. 1999. Chars pyrolyzed from oil palm wastes for activated carbon preparation. Journal of Environmental Engineering 125 (1): 72–76.

Luzardo-Ocampo, I., Cuellar-Nuñez, M. L., Oomah, B. D. and Loarca-Piña, G. 2019. Pulse by-products. In: Food Wastes and by-Products, pp. 59–92. https://doi.org/10.1002/9781119534167.ch3.

Madenenei, M. N., Faiza, S., Ramaswamy, R., Guha, M., Pullabhatla, S. 2011. Physico-chemical and functional properties of starch isolated from ginger spent. *Starch/Starke, 63*: 570–578.

Madurwar, M. V., Ralegaonkar, R. V. and Mandavgane, S. A. 2013. Application of agro-waste for sustainable construction materials: A review. Construction Building Materials 38: 872–878. https://doi.org/10.1016/j.conbuildmat.2012.09.011.

Mahala, A., Mokhtar, A., Amasiab, E. and AttaElmnan, B. 2013. Sugarcane tops as animal feed. International Research Journal of Agricultural Science and Soil Science 3 (4): 147–151.

Maikki, Y. 2004. Trends in dietary fiber research and development. Acta Alimentaria 33: 39–62.

Maniglia, B. C., de Paula, R. L., Domingos, J. R., Tapia-Blacido, D. R. 2015. Turmeric dye extraction residue for use in bioactive film production: optimization of turmeric film plasticized with glycerol. *LWT – Food Science and Technology, 64*: 1187–1195.

Mansoorian, H. J., Mahvi, A. H., Jafari, A. J. and Khanjani, N. 2016. Evaluation of dairy industry wastewater treatment and simultaneous bioelectricity generation in a catalyst-less and mediator-less membrane microbial fuel cell. Journal of Saudi Chemical Society 20: 88–100. https://doi.org/10.1016/j.jscs.2014.08.002.

Mashhadi, F., Habibi, A. and Varmira, K. 2018. Enzymatic production of green epoxides from fatty acids present in soap stock in a microchannel bioreactor. Indian Crop Production 113: 324–334. https://doi.org/10.1016/j.indcrop.2018.01.052.

Mathew, A. G. 2004. Future of spices and floral extract. Indian Perfumer 48: 35–40.

Meyer, S. L., Zasada, I. A., Orisajo, S. B. and Morra, M. J. 2011. Mustard seed meal mixtures: Management of meloidogyne incognita on pepper and potential phytotoxicity. Journal of Nematology 43: 7–15.

Milan, K. S. M., Dholakia, H., Kaul Tiku, P. and Prakash, V. 2008. Enhancement of digestive enzymatic activity by cumin (Cuminum cyminum L.) and role of spent cumin as a bio-nutrient. Food Chemistry 110: 678–683.

Moure, A., Sineiro, J., Domínguez, H. and Parajó, J. C. 2006. Functionality of oilseed protein products: A review. Food Research International 39 (9): 945–963.

Mukherjee, C., Chowdhury, R., Sutradhar, T., Begam, M., Ghosh, S. M., Basak, S. K. et al. 2016. Parboiled rice effluent: A wastewater niche for microalgae and cyanobacteria with growth coupled to comprehensive remediation and phosphorus biofertilization. Algal Research 19: 225–236.

Mullen, A. M., Álvarez, C., Pojić, M., Hadnadev, T. D. and Papageorgiou, M. 2015. Classification and target compounds. In: Food Waste Recovery. Academic Press, 125 London Wall, London, UK, pp. 25–57.

Mulye, S. 2019. The effect of distillation conditions and molasses concentration on the volatile compounds of unaged rum. New Zealand: Auckland University of Technology.

Murphy, E. W., Marsh, A. C. and Willis, B. W. 1978. Nutrient content of spices and herbs. Journal of the American Dietetic Association 72: 174–176.

Mussatto, S. I. 2009. Biotechnological potential of brewing industry by-products. In: Biotechnology for Agro-Industrial Residues Utilization. Springer, Dordrecht, pp. 313–326.

NABCONS. 2022. Study to Determine Post-Harvest Losses of Agri Produces in India. www.mofpi.gov.in/sites/default/files/study_report_of_post_harvest_losses.pdf (Accessed on: 01.06.2023).

Nagel-Held, B., Welling, B., Seibel, W., Brack, G. and Schildbach, R. 1997. Production of barley fractions with high nutritive value and their utilization in baked products. 111. Production of semi-hard biscuits with milled barley products. Getreide-Mehl-und-Brot 51: 180–182.

Ohba, R., Deguchi, T., Kishikawa, M., Arayad, F., Morimura, S. and Kida, K. 2003. Physiological functions of enzymatic hydrolysates of collagen or keratin contained in livestock and fish waste. Food Science Technology Research 9(1): 91–93. doi: 10.3136/fstr.9.91.

Okino-Delgado, C. H., Prado, D. Z. D., Facanali, R., Marques, M. M. O., Nascimento, A. S., Fernandes, C. J. D. C., Zambuzzi, W. F. and Fleuri, L. F. 2017. Bioremediation of cooking oil waste using lipases from wastes. PLoS One 12 (10): 0186246.

Papageorgiou, M. and Skendi, A. 2018. Introduction to cereal processing and by-products. In: Sustainable Recovery and Reutilization of Cereal Processing by-Products. Cambridge, US: Woodhead Publishing, pp. 1–25.

Parashar, A., Jin, Y., Mason, B., Chae, M. and Bressler. D. C. 2016. Incorporation of whey permeate a dairy effluent, in ethanol fermentation to provide a zero-waste solution for the dairy industry. Journal of Dairy Science 99 (3): 1859–1867.

Parate, V. R. and Talib, M. I. 2015. Utilization of pulse processing waste (Cajanus cajan husk) for developing metal adsorbent: A value-added exploitation of food industry waste. American Journal of Food Science Technology 3 (1): 1–9.

Partha, N. and Sivasubramanian, V. 2006. Recovery of chemicals from pressmud-a sugar industry waste. Indian Chemical Engineer 48 (3): 160–163.

Patel, S. K. S., Lee, J. K. and Kalia, V. C. 2018. Nanoparticles in biological hydrogen production: An overview. Indian Journal of Microbiology 58: 8–18. https://doi.org/10.1007/s12088-017-0678-9.

Pirani, S. I. and Arafat, H. A. 2016. Reduction of food waste generation in the hospitality industry. Journal of Clean Production 132: 129–145.

Pourbafrani, M., Forgács, G., Horváth, IS., Niklasson, C., Taherzadeh, MJ. 2010. Production of biofuels, limonene and pectin from citrus wastes. *Bioresour Technol. 101*(11):4246–50. doi: 10.1016/j.biortech.2010.01.077.

Rabea, W. I., Badawy, M. E. T., Stevens, C. V., Smagghe, G. and Steurbaut, W. 2003. Chitosan as antimicrobial agent: Applications and mode of action. Biomacromolecules 4 (6): 1457–1465. doi: 10.1021/bm034130m.

Radočaj, O., Dimić, E. and Tsao, R. 2014. Effects of hemp (Cannabis sativa L.) seed oil press-cake and decaffeinated green tea leaves (Camellia sinensis) on functional characteristics of gluten-free crackers. Journal of Food Science 79 (3): C318–C325.

Radosavljević, M., Pejin, J., Pribić, M., Kocić-Tanackov, S., Romanić, R., Mladenović, D. and Mojović, L. 2019. Utilization of brewing and malting by-products as carrier and raw materials in l-(+)-lactic acid production and feed application. Applied Microbiology and Biotechnology 103 (7): 3001–3013.

Ranjith, P. G., Viete, D. R., Chen, B. J. and Perera, M. S. A. 2012. Transformation plasticity and the effect of temperature on the mechanical behaviour of Hawkesbury sandstone at atmospheric pressure. Engineering Geology 151: 120–127.

Rasmey, A.-H. M., Hassan, H. H., Abdulwahid, O. A. and Aboseidah, A. A. 2018. Enhancing bioethanol production from sugarcane molasses by Saccharomyces cerevisiae Y17. Egyptian Journal of Botany 58 (3): 547–561.

Rivera, J. A., Sebranek, J. G., Rust, R. E. and Tabatabai, L. B. 2000. Composition and protein fractions of different meat by-products used for petfood compared with Mechanically Separated Chicken (MSC). Meat Science 55 (5): 53–59. doi: 10.1016/S0309-1740(99)00125-4.

Rodrigues, I. M., Coelho, J. F. and Carvalho, M. G. V. 2012. Isolation and valorisation of vegetable proteins from oilseed plants: Methods, limitations and potential. Journal of Food Engineering 109 (3): 337–346.

Russ, W., and Pittroff, RM. (2004). Utilizing waste products from the food production and processing industries. *Crit Rev Food Sci Nutr. 44*(2):57–62. doi: 10.1080/10408690490263783.

Ryan, M. P. and Walsh, G. 2016. The biotechnological potential of whey. Review Environmental Science Biotechnology 15 (3): 479–498. doi: 10.1007/s11157-016-9402-1.

Saini, P., Kaur, D., Yadav, N., Kumar, R. and Sanghmitra. 2017. Effect of incorporation of chick pea husk on quality characteristics of biscuits: Corporation of chick pea husk on quality characteristics of biscuits. International Journal of Food and Nutrition Science 6 (2): 2320–7876.

Sajjad Khan, M., Salma, K., Deepak, M. and Shivananda, B. G. 2006. Antioxidant activity of a new diarylheptanoid from Zingiber officinale. Pharmacognosy Magazine 2: 254–257.

Salazar-López, N. J., Ovando-Martínez, M. and Domínguez-Avila, J. A. 2019. Cereal/grain by-products. In: Food Wastes and by-Products, pp. 1–34. https://doi.org/10.1002/9781119534167.ch1.

Seberini, A. 2020. Economic, social and environmental world impacts of food waste on society and zero waste as a global approach to their elimination. Globalization and Its Socio-Economic Consequences 2019, SHS Web of Conferences 74: 03010.

Seidavi, A., Zaker-Esteghamati, H. and Scanes, C. G. 2019. Poultry by products. In: By Products from Agriculture and Fisheries. Wiley, pp. 123–146. https://doi.org/10.1002/9781119383956.ch6.

Seiler, G. J. and Gulya, T. J. 2015. Sunflower: Overview. In: Wrigley, C., Corke, H., Seetharaman, K. and Faubion, J. (eds.) Encyclopedia of Food and Grains, vol 1, 2nd edn. Elsevier, Waltham, pp. 247–253.

Sharma, H. R., Chauhan, G. S. and Agarwal, K. 2004. Physico-chemical characteristics of rice bran processed by dry heating and extrusion cooking. International Journal of Food Properties 7: 603–614.

Sharma, P., Gaur, V. K., Kim, S. H. and Pandey, A. 2020. Microbial strategies for bio-transforming food waste into resources. Bioresource Technology 299: 122580.

Sherazi, S. T. H., Mahesar, S. A. and Sirajuddin. 2016. Vegetable oil deodorizer distillate: A rich source of the natural bioactive components. Journal of Oleoresin Science 65 (12): 957–966. https://doi.org/10.5650/jos.ess16125.

Shi, C., Zhang, Y., Lu, Z. and Wang, Y. 2017. Solid-state fermentation of corn-soybean meal mixed feed with Bacillus subtilis and Enterococcus faecium for degrading antinutritional factors and enhancing nutritional value. Journal of Animal Science and Biotechnology 8: 50.

Sielaff, H. 1996. Fleischtechnologie. Behr's Verlag, Hamburg.

Silva, V. D. M. and Silvestre, M. P. C. 2003. Functional properties of bovine blood plasma intended for use as a functional ingredient in human food. LWT- Food Science and Technology 36 (5): 709–718. doi: 10.1016/S0023-6438(03)00092-6.

Singh, A. K., Singh, G., Gautam, D. and Bedi, M. K. 2013. Optimization of dairy sludge for growth of Rhizobium cells. Biomedical Research International 2013: 1–6. https://doi.org/10.1155/2013/845264.

Sowbhagya, H. B. 2019. Value-added processing of by-products from spice industry. Food Quality and Safety May 3 (2): 73–80. https://doi.org/10.1093/fqsafe/fyy029.

Sowbhagya, H. B., Suma, P. F., Mahadevamma, S. and Tharanathan, R. N. 2007. Spent residue from cumin—A potential source of dietary fiber. Food Chemistry 104: 1220–1225.

Sridhar, A., Ponnuchamy, M., Kumar, P. S., Kapoor, A., Vo, D. V. N. and Prabhakar, S. 2021. Techniques and Modeling of Polyphenol Extraction from Food: A review, Environmental Chemistry Letters. Springer International Publishing. https://doi.org/10.1007/s10311-021-01217-8.

Tahajod, A. S. and Rand, A. G. 1996. Seafood waste potential to support antimicrobial compound production by lactic acid bacteria. 1996 IFT Annual Meeting: Book of Abstracts, pp. 31–32, ISSN 1082–1236.

Tiwari, B. K., Tiwari, U., Brennan, C. S., Jagan, M. R., Surabi, A. and Alagusundaram, K. 2011. Utilisation of pigeon pea (Cajanus cajan L) byproducts in preparation of biscuits. LWT-Food Sci Technol. https://doi.org/10.1016/j.lwt.2011.01.018.

Verma, A. and Mogra, R. 2013. Psyllium (Plantago ovata) husk: A wonder food for good health. International Journal of Science and Research 4 (90): 1581–1585.

Wan, Y., Ghost, R. and Cui, Z. 2002. High resolution plasma protein fractionation using ultrafiltration. Desalination 144: 301–306. doi: 10.1016/S0011-9164(02)00332-6.

Wang, J., Rosell, C. M. and Benedito de Barber, C. 2002. Effect of the addition of different fibers on wheat dough performance and bread quality. Food Chemistry 79: 221–226.

William, P. T. 2005. Water Treatment and Disposal. Wally, J. (ed.). Greet Britain, p. 9.

Wright, A.G., Ellis, T.P. & Ilag, L.L. 2014. Filtered Molasses Concentrate from Sugar Cane: Natural Functional Ingredient Effective in Lowering the Glycaemic Index and Insulin Response of High Carbohydrate Foods. *Plant Foods Hum Nutr 69*, 310–316.

Yoshida, H., Terashima, M. and Takahashi, Y. 1999. Production of organic acids and amino acids from fish meat by sub-critical water hydrolysis. Biotechnology Progress 15 (60): 1090–1094. doi: 10.1021/bp9900920.

Yusuf, M. 2017. Agro-Industrial waste materials and their recycled value-added applications: Review. In: Handb. Ecomater. pp. 1–12. https://doi.org/10.1007/978-3-319-48281-1.

Zhang, L., Chen, F., Zhang, P., Lai, S. and Yan, H. 2017. Influence of rice bran wax coating on the physicochemical properties and pectin nanostructure of cherry tomatoes. Food Bioprocess Technology 10 (2): 349–357.

Zhao, J., Liu, Y., Wang, D., Chen, F., Li, X., Zeng, G., Yang, Q. (2017a). Potential impact of salinity on methane production from food waste anaerobic digestion. *Waste Management* 67: 308–314.

Zhou, Y., Kumar, V., Harirchi, S., Vigneswaran, V. S., Rajendran, K., Sharma, P., Tong, Y. W., Binod, P., Sindhu, R., Sarsaiya, S., Balakrishnan, D., Mofijur, M., Zhang, Z., Taherzadeh, M. J. and Awasthi, M. K. 2022. Recovery of value-added products from biowaste: A review. Bioresource Technology 360: 127565–127280. https://doi.org/10.1016/j.biortech.2022.127565.

2

Value Addition of Cereal and Millet Processing Industrial Waste

Sweta Rai, Preethi Ramachandran, and Sabbu Sangeeta

2.1 Introduction

India holds the distinction of being the world's second-largest producer and consumer of cereals, including rice, wheat, and other vital grains. This substantial global demand for cereals creates a favorable environment for the export of Indian cereal products. Among the key cereal crops in India are wheat, rice, maize, and a variety of millets, both major and minor, such as sorghum (jowar), pearl millet (bajra), finger millet (ragi), and foxtail millet.

The Ministry of Agriculture & Farmers' Welfare, under the Government of India, reports that the country ranks among the leading rice producers, estimating a production of around 120 million metric tons in the agricultural year 2020–2021. Wheat is another pivotal cereal crop, with a production of approximately 109 million metric tons in the same period. Maize is mainly cultivated as animal feed, though it is also consumed as a food grain, contributing to a production of around 30 million metric tons in the year 2020–2021. Millets have a significant role in Indian agriculture, particularly in semi-arid and rainfed regions. In the agricultural year 2020–2021, production estimates were around 7 million metric tons of sorghum, 10 million metric tons of pearl millet, 3 million metric tons of finger millet, and 1.3 million metric tons of foxtail millet.

India's distinction as the largest producer and exporter of cereal products worldwide is noteworthy. The value of cereals exported from India during the year 2022–2023 reached Rs. 111,062.37 crore (approximately US$13.86 billion). Rice, including both basmati and non-basmati varieties, constituted the majority share, accounting for 80% of India's total cereal exports during the same period. Meanwhile, other cereals, including wheat, represented a 20% share of the total cereals exported.

DOI: 10.1201/9781003269199-2

Millets, among the earliest crops to be domesticated for food, hold significant cultural and ecological value in India. Apart from their numerous health benefits, millets require relatively low water and inputs, making them environmentally friendly. Recognizing the potential of millets to enhance livelihoods, increase farmers' income, and ensure global food and nutritional security, the Government of India (GoI) has prioritized these "Nutri Cereals." The rebranding of millets as "Nutri Cereals" in April 2018, followed by the declaration of 2018 as the National Year of Millets, aimed to promote their consumption and demand. The United Nations, based on India's proposal and supported by 72 countries, designated the year 2023 as the International Year of Millets on March 5, 2021. This recognition honors the historical significance of millets as the first plants to be cultivated for sustenance. The Food and Agriculture Organization (FAO) of the United Nations hosted an inaugural ceremony for the International Year of Millets (IYM) 2023 on December 6, 2022, in Rome, Italy. The Department of Agriculture & Farmers' Welfare's proactive approach to multi-stakeholder engagement further underscores India's commitment to promoting these important crops.

The amount of waste produced during cereals and millets processing can vary based on several factors, including the specific crop, processing methods, and efficiency of processing technologies (Slavin et al., 2000). However, it is important to note that waste generated during processing can include components like bran, husks, broken grains, chaff, and other by-products. This waste, if not properly managed, can contribute to environmental pollution and resource wastage. Cereal and millet processing industrial waste can have several potential value additions, for example:

- **Bran and hulls**: Bran, which is removed during milling, is a valuable by-product rich in dietary fiber and nutrients. It can be used to produce various value-added products, such as bran-enriched flours, bread, muffins, and breakfast cereals. Similarly, hulls and husks can be used in animal feed, and in some cases they can be processed to extract bioactive compounds for nutraceutical applications.
- **Broken grains and chaff**: Broken grains that result from processing can still be nutritionally valuable. They can be incorporated into products like porridge mixes, snacks, and composite flours. Chaff, the outer protective layer of grains, can be utilized as mulch, fodder, or even as a source of cellulose for certain industrial applications.
- **By-product extraction**: From cereal and millet processing waste, various bioactive compounds can be extracted for use in nutraceuticals. For instance, antioxidants, dietary fibers, and phytochemicals present in the waste can be processed to create functional ingredients for health-promoting supplements.
- **Biogas and energy generation**: Some processing waste can be utilized for biogas generation through anaerobic digestion. The biogas produced can be used as a renewable energy source for cooking, heating, or electricity generation.

- **Livestock feed**: Waste components that are not suitable for direct human consumption can often be utilized as feed for livestock. Husks, hulls, and other non-edible parts can provide a source of nutrition for animals.
- **Composting**: Organic waste from processing can be composted to produce nutrient-rich soil amendments, benefiting agriculture and horticulture.
- **Extraction of functional ingredients**: By employing advanced extraction technologies, waste components can be processed to isolate and concentrate specific bioactive compounds. These compounds can be utilized in nutraceuticals, functional foods, or even pharmaceuticals.
- **Packaging materials**: the fibrous components present in cereal and millet waste can be utilized to produce biodegradable packaging materials. By processing the waste fibers and with suitable binding agents, holistic and sustainable packaging options like trays, plates, and containers can be created, reducing the persistence of non-biodegradable packaging materials.
- **Mushroom cultivation**: Cereal and millet waste, such as straw or husks, can be used as a substrate for growing mushrooms. Oyster and button mushrooms have a high demand and can be cultivated using waste materials. This approach provides an additional revenue stream and promotes circular economy practices.
- **Fertilizers and soil amendment**: Cereal and millet processing waste can be composted and used as an organic soil amendment or fertilizer. The nutrient-rich waste can improve soil quality, enhance water retention, and promote sustainable agricultural practices.
- **Biosorbents**: Biosorption refers to the process of capturing metal ions from aqueous pollutants through the utilization of biomass derived from microbial sources, as well as cellulosic and lignocellulosic materials. Cereal remnants like rice bran, wheat bran, and wheat straw exhibit commendable metal biosorption capabilities. The utilization of cereal waste for biosorption has demonstrated notable metal removal capacities, showcasing diverse affinities for distinct heavy metals.
- **Nanoparticles**: Cereal processing wastes, such as rice bran and wheat husk, have been processed for the development of nanomaterials. The polymeric compounds in these residues, such as cellulose, starch, chitosans, or xylans, have been used to produce nanoparticles.

These are just a few examples of the potential value additions of cereal and millet processing industrial waste (Figure 2.1). The specific opportunities and feasibility may vary depending on the local context, available technology, and market demand. It is important to conduct further research and feasibility studies to determine the most suitable value-addition options for a particular situation.

Efforts to reduce waste: Efforts to reduce waste during cereals and millet processing include optimizing processing technologies to minimize losses,

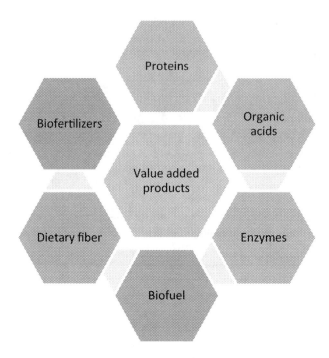

Figure 2.1 *Potential value addition in cereals and millet processing industrial waste.*

developing integrated processing systems that maximize utilization of all components, and implementing sustainable agricultural practices that minimize waste from the start. Utilizing processing waste (Table 2.1) to create value-added products and nutraceuticals not only helps in reducing waste but also contributes to enhancing the overall sustainability of the food and agricultural industries. It is important to implement efficient waste management practices that balance economic, environmental, and social considerations.

2.2 By-Products from Cereal Grains

During the milling process of wheat (*Triticum aestivum*), the bran is obtained, which consists of various layers such as the outer and inner pericarps, testa, hyaline layer, aleurone layer, and a portion of starchy endosperm residue. This bran accounts for approximately 15% of the total grain weight. The extraction of wheat bran from grains is efficiently achieved using a roller mill, which operates at a rapid pace.

Initially, the wheat grains are subjected to tempering, a process where water is sprayed onto the grains to achieve a moisture content of around 15%. The duration of tempering varies based on the hardness of the wheat. During this phase, the pericarp and germ layer of the kernel absorb water,

Table 2.1 Cereal Grain Processing By-Products

Cereals	By-products	Processing methods
Wheat	Bran (25%) and germ (2–3%)	Dry milling
	Starch, fiber, and gluten protein	Wet milling
	Bran, starch, and protein	Malting
	Bran (7–15%)	Pearling
Rice	Paddy husk (20%), bran (8–12%), germ (2%), oil from bran (18–20%), bran layers, fine broken	Dry milling
	Hull bran, oil from rice bran	Wet milling
	Fiber	Malting
	Husk (20%)	Pearling
Maize	Bran (6–7%)	Dry milling
	Gluten feed, gluten meal, fiber (8–11%), germ oil, steeping solids, germ meals, liquefied corn products	Wet milling
Barley	Bran and germ	Dry milling
	Fiber (8%), hull and protein	Wet milling
	Brewers spent grain (85% of total by-products)	Malting
	Hull (30–49%)	Pearling
Oats	Bran	Dry milling
	Bran	Wet milling
	Bran	Pearling

softening the endosperm for subsequent extraction. Tempering prevents the bran from breaking during the separation of the endosperm and bran. The separation process involves segregating the endosperm and germ from the pericarp. The conditioned grains are cracked open using counter-rotating corrugated metal rolls within the roller mill. Subsequently, the broken grains are divided into fractions, including the germ, endosperm, and bran. Wheat bran is a significant by-product of the flour milling process (Slavin et al., 2000). Approximately 90% of it is utilized as animal feed, leaving a mere 10% for utilization in the food industry. In this sector, it serves as a source of dietary fiber in products like baked goods, fried foods, and breakfast cereals.

2.3 Extraction of Wheat Bran

A roller mill can extract wheat bran from grains at an extremely fast rate. Figure 2.2 shows the process flow of extraction of bran. The wheat grains are initially tempered by being sprayed with water until they have a moisture level of about 15% before being put into a tempering bin. The nutritional importance of wheat bran lies in its composition (Table 2.2).

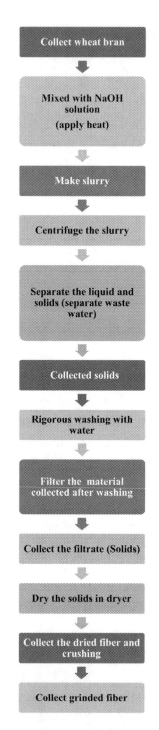

Figure 2.2 *Flow chart for fiber extraction of bran.*

Table 2.2 Nutritional Composition of Wheat Bran

Constituents	Amount (% dm and mg per 100 g)
Moisture	8–13
Ash	4–8
Dietary fiber	33–63
Protein	9–19
Total carbohydrates	60–75
Starch	9–39
Magnesium	530–1030
Phosphorus	900–1500
Iron	2–34
Zinc	8–14
Manganese	1–10
Vitamin E	0.13–9.5
Thiamine (B1)	0.51–1.6
Riboflavin (B2)	0.20–0.80
Pyridoxine (B6)	0.30–1.30
Folate (B9)	0.88–0.80

Source: (Slavin et al., 2000)

Roughly 53% of wheat bran's weight comprises dietary fiber, encompassing elements like xylans, lignin, cellulose, galactan, and fructans. Additionally, wheat bran contains vitamins, minerals, and bioactive compounds such as ferulic acid, flavonoids, carotenoids, lignans, and sterols. These constituents, whether in whole grains or isolated bran, have been shown to influence glycemic and lipid profiles, inflammatory status, as well as hunger and satiety sensations in individuals (Luithui et al., 2019). Furthermore, they might exhibit prebiotic effects, promoting a healthy gut microbiome.

2.4 Functionality of Bran

- **Dietary fiber source**: Bran is rich in dietary fiber, including both insoluble and soluble fibers. Insoluble fiber helps promote regular bowel movements and prevents constipation, while soluble fiber can help lower cholesterol levels and regulate blood sugar levels.
- **Nutrient content**: Bran contains vitamins such as B vitamins (niacin, thiamin, riboflavin, and folate); minerals like iron, magnesium, and zinc; and phytochemicals with antioxidant properties that help protect cells from damage.

- **Gut health**: The dietary fiber in bran acts as a prebiotic, providing nourishment to beneficial gut bacteria and promoting a healthy gut microbiome.
- **Weight management**: Bran's high fiber content can promote a feeling of fullness, potentially aiding in weight management by reducing overeating.

2.5 Stabilization of Bran

Bran, especially from grains like wheat and rice, can contain lipids that are prone to oxidation and rancidity. Stabilization techniques are employed to extend the shelf life and prevent the deterioration of bran.

- **Heat treatment**: Bran can be heat-treated to inactivate enzymes that promote lipid oxidation and reduce the risk of rancidity.
- **Extrusion**: Extrusion cooking involves high-temperature and high-pressure treatment that can modify the structure of bran and improve its stability.
- **Packaging**: Proper packaging, including vacuum sealing or using oxygen barrier materials, can help protect bran from exposure to air and light, which can accelerate lipid oxidation.

2.6 Fortification of Bran

Bran can be fortified with additional nutrients to enhance its nutritional profile and address specific dietary deficiencies. Fortification can include adding vitamins, minerals, and other functional ingredients.

- **Micronutrient fortification**: Bran can be enriched with vitamins and minerals that are commonly deficient in diets, such as iron, folate, and certain B vitamins.
- **Functional ingredients**: Some bran products are fortified with ingredients like omega-3 fatty acids, probiotics, and plant extracts for added health benefits.
- **Customized fortification**: Fortification can be tailored to address specific nutritional needs in different populations, such as targeting specific micronutrient deficiencies.

2.6.1 Impact on Organoleptic Acceptability and Nutritional Fortification of Bran

Incorporating bran into diets can help improve the nutritional quality of foods and promote overall health. However, it is important to note that while bran

Table 2.3 Effect of Bran Fortification on Quality Parameters of Products

Products	Level of incorporation of wheat bran (%)	Effect on product quality
Biscuit	5–30	Low gluten network Increased fiber and protein levels Reduced dough pasting viscosity Increased bran addition decreased the digestible starch and crumbliness of biscuits
Pasta	20–40	Increased ash content Increased dietary fiber and protein Increased chewiness and adhesiveness Harder structure beyond 20% Lower sensory acceptability
Bread	8–9	Increased dietary fiber Decreased crusts score Reduced specific loaf volume
	0–20	Reduced crumb and specific volume Increased crumb moisture content Good sensory attributes
	7–25	Reduced loaf volume Increased springiness Increased hardness

is nutritious, it can also contain anti-nutritional factors that interfere with nutrient absorption (Luithui et al., 2019). Therefore, it is important to process and prepare bran appropriately to maximize its benefits while minimizing potential drawbacks. Table 2.3 elaborates the effect of bran fortification on the quality characteristics of bakery products.

2.7 Rice

Rice (*Oryza sativa*) stands as the world's most cultivated grain, boasting remarkable nutritional, functional, and bioactive attributes. Despite these qualities, rice bran (RB), a by-product of the milling process, remains vastly underutilized. In 2021, global rice production reached a staggering 787 million metric tons. Considering rice's worldwide cultivation, approximately 80 million tons of RB are generated as a secondary product. These RB by-products predominantly find use in the feed industry for various food-related purposes. However, the refined form of RB presents an exciting avenue in the food sector with potential to enhance value addition. During the transformation of brown rice into white rice through milling, RB emerges as a residual output. Industrial-scale milling employs ingenious methods, often resembling a multistage approach, to achieve this conversion. The potential of fresh RB as a consumable is immense due to the multistage milling process, which minimizes mechanical strain and heat accumulation within the grain. With its mild flavor, delicate texture, and rich nutrient content, fresh RB holds great promise. Nevertheless, unprocessed RB cannot serve as a food source due to its rapid rancidity following grinding. The primary culprits behind RB's susceptibility to rancidity are lipophilic enzymes, particularly lipases.

2.7.1 By-Products from Rice

2.7.1.1 Bran

Paddy rice is typically composed of three main components: the starchy endosperm, accounting for 69% of the total (also referred to as total milled rice); the rice hull or husk, constituting 20%; and the bran layer, comprising 11%. Unfortunately, in the rice milling business, by-products like rice husk, rice germ, and bran, which are secondary to the main product, are often treated as waste rather than valuable by-products (Figure 2.3). The commercial utilization of these by-products plays a pivotal role in maintaining the economic stability of the rice milling industry.

2.7.1.2 Rice Bran Utilization

The presence of various anti-nutritional elements like lipases, trypsin inhibitors, hemagglutinin-lectin, and phytates contributes to the underutilization of

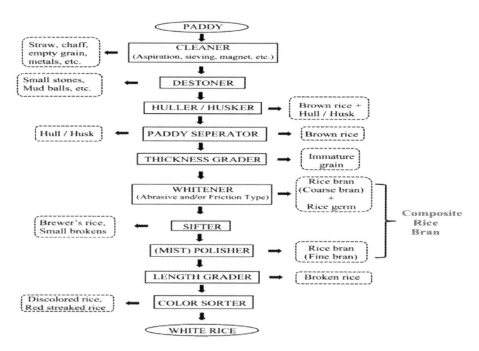

Figure 2.3 *Flow chart for the extraction of rice bran.*

Table 2.4 Nutritional Composition of Rice Bran

Nutrients	Percent amount
Dietary fiber	9–15
Ash	8–10
Crude fat and oil	18–21
Crude protein	14–16
Carbohydrates	33–36
Calcium	0.08–1.4
Phosphorus	1.3–2.9
Oryzanol	0.93–13.8 mg/g
Tocopherols	27–770 µ g/g
Anthocyanins	55 mg/g

rice bran as a value-added food resource. Table 2.4 shows the nutritional composition of rice bran. In the absence of proper stabilization, the lipase enzymes present in fresh rice bran rapidly initiate the hydrolysis of oil, yielding free fatty acids (FFA). Beyond a 5% FFA threshold, rice bran becomes unsuitable for human consumption, and if FFA concentration reaches 12%, it becomes unfit even for livestock feed.

2.7.2 Stabilization of Rice Bran for Value Addition

Several methods have been developed to stabilize rice bran. These techniques encompass processes such as extrusion cooking, fluidized bed drying, parboiling, dry or moist heat treatment, and microwave heating. While most of these technologies suffer from drawbacks like extreme processing conditions potentially damaging valuable bran components, partial enzyme inactivation, and high operational costs, microwave heating emerges as the most efficient and cost-effective approach. It requires less time for preparation and has minimal impact on nutritional content. Another promising method for stabilizing rice bran is ohmic heating.

2.7.3 Extraction of Protein from Rice Bran

Rice bran contains 10–15% protein. Rice bran protein (RBP) possesses distinct nutritional and nutraceutical attributes, consisting of water-soluble, salt-soluble, alcohol-soluble, and alkali-soluble storage proteins. Among cereals, rice protein offers superior nutritional value due to its provision of essential amino acids, such as lysine and threonine. Notably, RBP is recognized for its low allergenic potential, making it suitable for infant formula, protein supplements, and pure protein products. However, due to a lack of commercially viable extraction techniques, RBP concentrates and isolate are not produced on a significant scale. The alkaline extraction method, yielding the highest protein output, remains the most prevalent technique, albeit it generates the potentially harmful by-product lysineo-alanine. Despite the existence of various extraction methods, none have achieved commercial feasibility for RBP extraction. Extraction continues to be challenging due to the bran's limited solubility, phytate content, and fiber content.

RBP hydrolysates find application as dietary supplements, functional additives, and flavor enhancers in an array of products, such as food, coffee whiteners, cosmetics, personal care items, confectionery, and fortified soft beverages and juices. These hydrolysates are also incorporated into savory creations like meat products, soups, sauces, gravies, and more.

2.7.4 Rice Bran Oil (RBO)

Derived from rice bran, rice bran oil (RBO) boasts a delightful, nutty flavor. Its high smoke point of 254°C makes it exceptionally suitable for high-temperature cooking methods, such as stir-frying and deep-frying. Consumption of RBO contributes to the reduction of total plasma cholesterol and low-density lipoproteins, thereby promoting normalized blood cholesterol levels. Furthermore, it exhibits a significant decrease in cholesterol absorption. In comparison to coconut, canola, corn, and peanut oils, RBO demonstrates superior potential in lowering cholesterol. The major therapeutic effects of RBO can be attributed to Oryzanol, a compound that possesses the ability to decrease plasma cholesterol, lower cholesterol absorption, and mitigate early atherosclerosis. RBO boasts an extended shelf life, thermal stability, and impressive oxidative resistance. This oil stands out as the preferred choice among cooking oils due to its optimal fatty acid composition, elevated antioxidant and nutraceutical content, high smoke point, and minimal absorption when used in cooking.

2.7.5 Rice Husk

Constituting 20% of a paddy's weight, rice husk possesses a distinct composition. The major components of rice husk encompass silica (18–22%), cellulose (28–38%), hemicellulose (28%), and lignin (9–20%). In rice mills specializing in parboiled rice production, rice husk predominantly serves as boiler fuel to generate steam. Moreover, it plays a pivotal role as a source of energy for brick kilns, the production of producer gas for internal combustion engines, and electricity generation. The applications extend to the creation of sodium silicate, activated carbon, and furfural, a compound with diverse uses in resin production, oil refineries, and the pharmaceutical sector. The extraction of pure amorphous silica from rice husk is also achievable. This silica variety finds application in cosmetics, toothpaste, and the rubber industry as a reinforcement agent.

2.8 Corn

Corn (*Zea mays*) holds notable importance as a substantial food crop, exerting a considerable impact on human livelihoods. The processing of corn gives rise to an array of cost-effective by-products rich in protein, oil, carbohydrates, and various nutrients. The major by-products of corn are corn gluten meal, corn husk, and corn steep liquor. The primary constituents of corn encompass starch, protein, and fat, which also come brimming with trace elements, such as vitamin A, vitamin E, and selenium. Throughout the processing journey, corn is broken down to yield essential starch products alongside by-products like corn starch syrup.

Extraction of by-products from corn: During the processing of corn, various waste is generated, such as corn husk, germ, etc., as represented in Figure 2.4.

2.8.1 Gluten Meal Extracted from Corn

Corn gluten meal (CGM) boasts an elevated protein content reaching up to 60%, predominantly composed of exceptional plant proteins like zein, globulins, and glutelins. Through enzymatic hydrolysis of corn gluten powder by proteases, minute peptide fragments named corn peptides (CPs) are frequently generated. These CPs exhibit an array of unmatched physiological functions,

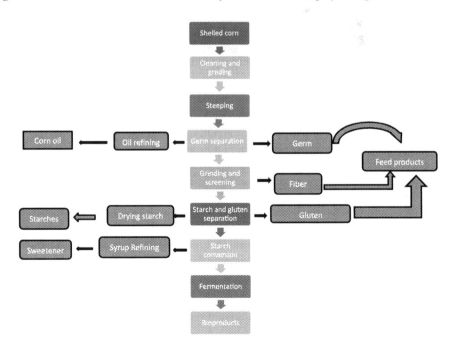

Figure 2.4 Flow chart for the extraction of different by-products from corn processing.

including antioxidant activity, anti-hypertensive properties, and the potential to mitigate alcohol-related effects, among others. Notably rich in hydrophobic amino acids, CPs have the capacity to stimulate glucagon secretion and prove valuable in the formulation of sports drinks and similar high-protein beverage products. The substantial presence of alanine and leucine in CPs results in elevated levels of these amino acids within the bloodstream, which, in turn, enhance the liver's alcohol dehydrogenase and acetaldehyde activity. The applications of CPs are extensive in the realm of health food, as their dehydrogenase activity aids in the metabolism and breakdown of ethanol within the body, contributing to the sobering effects of alcohol. Consequently, CPs are extensively employed in the creation of foods and beverages with attributes such as anti-alcohol effects and liver protection.

2.8.2 Total Dietary Fiber

Corn husks are enriched with 60–70% nutritional fiber sourced from corn, serving diverse roles in both culinary and medicinal domains. This form of dietary fiber, originating from plant tissue, exhibits a remarkable water absorption capability, with maize dietary fiber boasting a water absorption capacity of 1.5 grams of water per gram of powder. Consequently, it can augment fecal volume, accelerate the movement of fecal matter through the human digestive system, effectively cleanse the intestines, and alleviate pressure within the rectum and urinary system. When soluble dietary fiber is introduced, it effectively halts the rapid disintegration of bread, thereby enhancing its longevity. The addition of dietary fiber also contributes to the improved stability and dispersion of beverages.

2.8.3 Zein Protein

Comprising the predominant share of protein within maize husks, zein forms a significant component of CGM. Zein protein possesses a range of advantageous attributes: it is non-toxic, renewable, natural, cost-effective, biocompatible, and biodegradable. This protein comprises four segments, each characterized by distinct peptide chains, molecular dimensions, and solubilities. Commercial zein predominantly consists of the protein alpha-zein. Zein exhibits solubility in alkaline aqueous solutions, acetone, and ethanol concentrations ranging from 60% to 95%. Its applications extend to food encapsulating agents, food emulsifiers, gluten-free products, and materials for food packaging. Zein has garnered extensive utilization within the food industry.

2.8.4 Germ Protein

Predominantly composed of albumin and globulin, two proteins renowned for their exceptional functional attributes, the protein composition within maize germ stands out. The primary method employed for the extraction of corn germ protein involves gradually introducing an alkaline solution to a solution containing maize germ. As the solution's pH approaches approximately 9.5, the protein precipitates, and subsequent standing time permits the protein to be collected through centrifugation. By incorporating corn germ protein, the deficit of lysine and threonine found in wheat protein can be rectified within baked goods. This corn germ protein proves valuable as a nutritious supplement, effectively enhancing the nutritional profile of food.

2.8.5 Corn Steep Liquor

The corn steep processing commences by creating a foam from the grains, leading to the generation of a significant amount of soaking water. The protein network within the corn kernel is disrupted to release the encapsulated starch by dissolving the -S-S link present in the protein structure. The resulting maize steep liquor is a viscous liquid attained through the concentration of the sodium bisulfite aqueous solution utilized for the initial immersion of corn kernels. The principal by-product of the wet method employed for maize starch production is corn steep liquor (CSL). It takes the form of a dense, aromatic slurry characterized by an acidic nature and a yellowish-brown hue. Due to its elevated content of protein, soluble sugars, and sulfide compounds, the potential of maize steep liquor (CSL) remains largely untapped.

2.8.6 Corn Starch

When compared to starch, gluten exhibits a lower density. Employing centrifugation, gluten can be efficiently separated from mill starch and repurposed for animal feeds. In the process of corn starch separation, starch undergoes a series of hydrocloning steps. It is initially diluted and then subjected to eight to fourteen rounds of washing. Following this, another dilution step is executed, followed by a final wash to eliminate any lingering protein traces. This meticulous procedure yields high-quality starch, often attaining a purity level exceeding 99.5%. At this stage, only a minimal 1–2% protein content remains within the starch. The majority of the starch undergoes conversion into glucose and various corn syrups. A small portion is dried and marketed as unmodified corn starch, while an even smaller fraction is subjected to modifications, resulting in the creation of specialty starches.

2.8.7 Corn Germ Oil

Corn germ oil stands as a nourishing and healthful culinary oil. With its distinct golden yellow hue, it boasts clarity and a sweet aroma. In the Western world, it is often referred to as "liquid gold." Comprising 80–85% of the fat content in corn germ oil are unsaturated fatty acids such as oleic, linoleic, and alpha-linolenic acids. Another notable advantage of maize germ oil is its rich content of natural vitamin E, which plays a role in averting atherosclerosis and mitigating the risk of coronary heart disease. The conventional method employed to extract maize germ oil is pressing. This widely used pressing technique, commonly adopted by oil producers, entails mechanical pressure to extract the oil. Figure 2.5 depicted the systemic extraction of oil from corn germ. Expeller-pressed, cold-pressed, and chemical or solvent extraction methods are the traditional methods that have been used for the extraction of oil from corn germs after the obtained oil has been purified through chemical or physical refining (Figure 2.6).

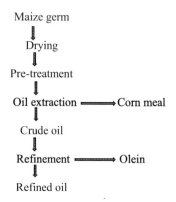

Figure 2.5 *Flow chart for corn oil extraction.*

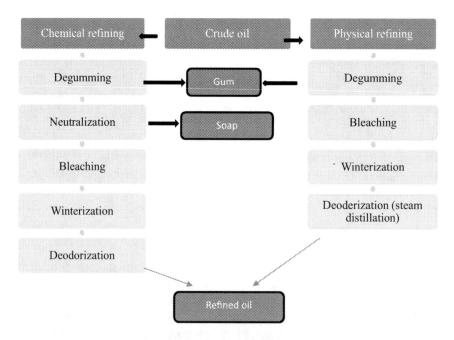

Figure 2.6 *Process flow chart for the corn oil refining process.*

2.8.8 Nutritional Benefits of Germ Oil in Value-Added Foods

Cereals are one of the most significant crops and food sources in the human diet. The cereals processing chain generates huge amounts of agricultural waste; about 12.9% of all food wastes are produced during cereal processing and manufacturing. Generally, cereal waste by-products are straw and cob (produced during the cleaning grain process), bran (produced during the extraction of oil and polishing of rice) (Farcas et al., 2022). Wheat bran, wheat germ, rice bran, rice germ, corn germ, corn bran, barley bran, and brewery spent grain are the major waste of cereal industry that may be exploited to recover bioactive compounds and the production of oil. Cereal germ is required for cereal germination and growth under appropriate environment conditions. Cereal germ is the common waste that is generated during the milling of cereals, and it can be used in the food industry for various purposes. Cereal germ is considered a good source of unsaturated fatty acids (43–64 g/100 g), essential amino acids (24–36 g/100 g), dietary fiber (1–15 g/100 g on dry basis), vitamins (56–2700 mg/kg on dry basis), and many functional phytochemicals, such as flavonoids (0.35 rutin equivalent/100 g dry basis) and sterols (1–6 g/100 g lipid) (Karimi et al., 2020; Rondanelli et al., 2021). Rice germ contains about 25% and corn germ contains 9–50% lipid content; that is higher than the lipid content of soybean (4–24%), which is considered as oil crop (Saldivar et al., 2011). Cereal germ lipid generally contains more than 50% polyunsaturated fatty acids as compared to other cooking oils, such as sesame and peanut oil, which contain higher amounts of saturated and monounsaturated fatty acids (Ghafoor et al., 2017; Rondanelli et al., 2021).

Lipids perform many essential functions in the human bodym including maintaining body temperature and muscle elasticity and facilitating the assimilation of fat-soluble vitamins. The presence of polyunsaturated fatty acids in the daily diet is helpful in reducing cardiovascular disease, diabetes, and other diseases (Dehghan et al., 2017). Cereal germ oil is a valuable component that increases the quality of many foods in terms of nutrition because it contains many functional bioactive substances, such as tocopherols, carotenoids, octacosanol, vitamin B complex, phytosterols, and minerals. For example, wheat germs are rich in carotenoids (~60 mg/kg) and tocopherols (1300–2700 mg/kg), while corn germ is rich in phytosterol (3000 mg/kg) and rice germ is rich in iron (~60 mg/kg) and magnesium (~3000 mg/kg) and has low sodium content (Giordano et al., 2016; Ghafoor et al., 2017; Rondanelli et al., 2021). The nutritional composition of various cereal germs is presented in Table 2.5.

2.8.9 Novel Methods of Oil Extraction from Cereal Germ

There are two categories of cereal germ based on lipid content, i.e., high-fat cereal germs, which contain more than 20% lipid (maize germ, rice germ, and sorghum germ), and low-fat cereal germs, which contain less than 10% lipid (wheat germ and rye germ). Commonly, high-fat germs are utilized for the

Table 2.5 Chemical Composition of Various Cereal Germs

Cereal	Nutritional composition g/100g							References
	Carbohydrate	Starch	Fiber	Protein	Fat	Moisture	Ash	
Common wheat germ	51.30	-	-	28.10	9.60	6.70	-	(Boukid et al., 2018)
Red winter wheat germ	41.74	13.18	-	29.77	11.56	11.33	5.60	(Ling et al., 2019)
Rice germ	41.16	25.50	7.00	18.20	17.46	10.53	5.65	(Rondanelli et al., 2019)
Wet milled corn germ	-	-	-	14–16	40–50	11.57	8.13	(Giordano et al., 2016)
Dry milled corn germ	-	-	-	14–16	20–25	11.57	8.13	(Giordano et al., 2016)
Sorghum germ	-	-	-	18.9	28.1	-	10.4	

extraction of oil by various technologies such as pressing, prepress leaching, supercritical carbon dioxide extraction (Sc-CO2), and water-enzymatic extraction (Abdullah et al., 2020; Nde and Foncha, 2020). The traditional methods of oil extraction produced low yields of oil as well as low quality. Recently, various new methods have been involved in the extraction of germ oil, such as super critical fluid extraction (SFE), ultrasound-assisted extraction (UAE), and microwave-assisted extraction (MAE).

Supercritical fluid is a compound at a temperature and pressure above the critical values (above the critical point) that exhibit properties intermediate between those of gases and liquids. Above the critical temperature of a compound, the pure gaseous component cannot be liquefied regardless of the pressure applied. The critical pressure is the vapor pressure of the gas at the critical temperature. Generally, carbon dioxide (CO_2) is used as supercritical fluid due to its low critical temperature (31.1°C) and critical pressure (73.8 bar). Apart from these, CO_2 is non-flammable, non-corrosive, nontoxic, easily available, inexpensive, safe, and has relatively low viscosity and high molecular diffusivity (Figure 2.7). Good quality and quantity of corn germ oil has been extracted by using SFE (Rebolleda et al., 2012). Similarly, many other researchers may also use SFE for the extraction of germ oil from various cereals (Marinho et al., 2019; Da-wei et al., 2009)

The MAE uses microwave radiation as the source of instantaneous heating (due to dipole rotation and ionic conductance) to the solvent sample mixture, which leads to the fast extraction of components such as oil, bioactive, etc. In most cases the extraction solvent is chosen to absorb microwaves, but for thermo-labile compounds, the microwaves may be absorbed only by the matrix, resulting in heating of the sample and release of the solutes into the cold solvent. Microwave extraction has been used for isolating various

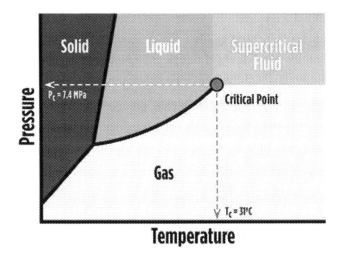

Figure 2.7 *Phase diagram of CO_2 as a super critical fluid.*

components like essential oils (Bayramoglu et al., 2008; Bousbia et al., 2009; Deng et al., 2006).

Ultrasound-assisted extraction is an extraction method based on ultrasonic waves. The ultrasound disintegration of cell structures (lysis) is used for the extraction of intra-cellular compounds. Sonicating liquids at high intensities possess sound waves that propagate into the liquid media, resulting in creating small vacuum bubbles or voids in the liquid due to alteration in pressure. These bubbles collapse due to pressure gradient and cause cavitation, which increase the shear force of material and break the cell mechanically. UAE has been widely applied for the extraction of nutritional material, such as bioactive compounds, e.g., flavonoids (Zhang et al., 2009), flavoring (Da Porto et al., 2009), lipids (Metherel et al., 2009), proteins (Zhu et al., 2009), polysaccharides (Yan et al., 2011), Saponins (Wu et al., 2001) and dibenzylbutyrolactone lignans (Wang et al., 2011).

2.9 By-Products from Barley Processing

Barley (*Hordeum vulgare* L.), one of the earliest cultivated crops, holds a prominent position as a vital grain today. Its cultivation spans various regions across the globe, thriving in locales with temperate climates during the summer and in select areas with temperate to subtropical conditions during the winter. Regarding the morphology of barley grains, the endosperm have the grain composition, constituting 77.2% of the total grain weight. It serves as a pivotal reservoir of protein and starch. The husk, accounting for an average of 13%, follows, succeeded by the pericarp and Testa at 3.3%, and aleurone at 5.5%. These proportions may vary based on barley type and growing circumstances. The germ, composing a third of the barley grain's weight, encompasses proteins, lipids, and tocopherol. In contemporary times, barley frequently finds its place in soups as a staple part of the diet. However, in order to elevate barley consumption to a more substantial role in daily dietary habits, common staples like bread and pasta must be considered. Presently, these staple meals predominantly rely on wheat rather than barley. When incorporating barley flour or milled barley fractions in place of wheat flour for leavened bread, soluble fiber content is boosted, while gluten is reduced. This modification results in a softer bread structure. Enhancing the utilization of barley in a variety of meals could contribute to its broader dietary integration.

2.9.1 Barley Flakes

Barley flakes, also known as rolled barley, are crafted from pre-cleaned and partially de-branned barley grains. These grains undergo a process similar to the production of oat flakes, involving flattening and slicing. Following this, they are steamed and flaked. Despite their swift cooking time, barley flakes are relatively less nutritious compared to hulled barley. Rolled barley flakes come in two distinct variations: pearled and de-hulled. While pearled rolled flakes can be derived from either naked or de-hulled barley, de-hulled rolled flakes are obtained from whole grain barley that has undergone de-hulling. Pearled rolling flakes possess a charming round shape and are slightly smaller in size, lacking visible bran.

2.9.2 Barley Malt

Barley malt is the outcome of the malting process, during which cereal grains are soaked in water, allowed controlled germination, and subsequently dried at elevated temperatures to halt further sprouting. This process yields sprouted grains known as barley malt. Steeping, germination, and kilning are the three pivotal stages in malting. Steeping involves soaking the barley, germination entails opening the grain for fermentation, wherein starch is converted into sugar and eventually alcohol, and kilning serves to dry the barley grain, imparting its final color and flavor. Primarily used in the production of beer, whiskey, and various alcoholic beverages, barley malt is the most favored cereal for this purpose. The variety and shade of malt are influenced by the drying temperature, thus resulting in distinctions like pale, crystal, coffee, chocolate, black, and more. Malting by-products could be used for the production of lactic acid bacteria and for animal feeds. Brewers' spent grains can be used as a substrate for microorganism cultivation and further for the production of enzymes like alpha-amylase by *Bacillus licheniformis, Bacillus subtilis*, etc.

2.9.3 Use of Barley in Value-Added Products

2.9.3.1 Flatbread

The major application for barley flour is in the creation of unleavened bread, commonly known as flatbread. In products relying on fermented cereals, the requirement for high-quality gluten protein is met only by wheat flour. Consequently, barley flour finds its primary use in unleavened bread. In recipes for fermented baked goods, substantial proportions of barley flour are generally avoided. When used, supplementary baking ingredients, such as glycerol monostearate combined with NaCl in chapatti bread, should be incorporated. Barley flour can be incorporated into flatbread recipes with favorable outcomes, as the nature of flatbreads is less dependent on fermentation, and their volume is not a crucial quality parameter. However, the choice of flatbread type, such as chapattis, can impact the acceptable percentage of barley flour. In specific cases, an inclusion of up to 30% barley flour is acceptable. Nevertheless, sensory dissatisfaction may arise as the percentage of barley flour increases.

2.9.3.2 Bakery Products

Barley flour has the potential to be a partial or complete substitute for wheat flour, diluting the gluten protein present in wheat. In contrast to leavened bread, the presence of gluten does not significantly affect the quality of cookies. However, the cookies might acquire a darker shade and softer texture. The incorporation of barley flour not only introduces a distinct quality but also transforms the cookies into a beneficial product, attributed to the inclusion of phenolic compounds with antioxidant properties, beta-glucan, and dietary fiber.

2.9.3.3 Pasta and Noodles

In the production of noodles and pasta, it is essential for the products to exhibit a light-yellow color devoid of dark spots, possess a smooth, non-sticky surface, and display high strength and elasticity. Barley's potential negative effects in pasta production, resulting in attributes like dark color, reduced firmness and elasticity, and increased cooking losses due to limited gluten content, are counteracted by incorporating a small amount of barley flour in wheat flour–based pasta. The optimal ratio was determined to be 20% barley flour. Although the resultant pasta may possess a slightly darker color compared to wheat-based pasta, it retains satisfactory organoleptic qualities and offers a mineral-rich, dietary fiber–enriched, and beta-glucan-loaded alternative.

2.10 Millets

Millets are coarse-grain nutri-cereals that have evolved from traditional to highly nutritious staples. These grains are often hailed as "future crops" due to their exceptional resistance against pests and diseases, as well as their ability to flourish in the challenging conditions of arid and semi-arid regions in Asia and Africa. Millets belong to the category of small-seeded grains. Sorghum (*Sorghum bicolor* L.), pearl millet (*Pennisetum glaucum*), finger millet (Eleusine carocana), proso millet (*Panicum miliaceum*), kodo millet (*Paspalum scrobiculatum*), foxtail millet (*Setaria italica*), and small millet (*Panicum sumatrense*) are the most prevalent and noteworthy millets used for sustenance (Table 2.6). Millets are a remarkably nutrient-rich crop, boasting significant quantities of vitamins and minerals. They stand as an excellent source of dietary fiber, slowly digesting starch, resistant starch, and energy, which contributes to a gradual release of glucose and sustained satiety. In comparison to conventional cereals, millets exhibit a superior fatty acid profile, hold a wealth of protein, and encompass amino acids like methionine and cysteine, which are sulfur-containing.

To enhance nutritional value and sensory attributes, extend shelf life, and remove inedible components, millets commonly undergo preprocessing prior to consumption. Basic processing techniques, such as de-hulling, soaking, germination, roasting, drying, polishing, and milling (size reduction), are employed to render millets suitable for human consumption. Furthermore, millet-based processed foods with augmented value are crafted using modern or secondary

Table 2.6 List of Some Common Millets Available in India

Common name	Seed color
Jowar (*Sorghum vulgare*)	Deep red and brown
Bajra (*Pennisetum typhoids*)	Yellow and white
Foxtail millets (*Setaria italica*)	Yellow, white, brown, red, black
Finger millets (*Eleusine coracana*)	Red to purple
Kodo millets (*Paspalum scorbiculatum*)	Pinkish color
Proso millets (*Panicum miliaceum*)	Cream, white, yellow, red, brown
Little millets (*Panicum miliare*)	Off white and creamish
Barnyard millets (*Echinochloa esculenta*)	Cream and white

processing methods like fermentation, parboiling, frying, puffing, popping, malting, baking, flaking, and extrusion, among others. While these techniques aim to enhance digestibility and nutritional availability, it is noteworthy that a considerable amount of nutrients may be lost during subsequent processing stages.

2.10.1 The Primary Millet Processing By-Products Encompass

2.10.1.1 Bran

Millet bran, extracted during the milling process, constitutes the outer layer of the grain. Abundant in nutritional fiber, as well as B vitamins, iron, zinc, and various other vitamins and minerals, millet bran possesses substantial nutritional value. Its potential applications extend to the food and pharmaceutical industries, where it can serve as a source of bioactive compounds. Additionally, it finds utility as an ingredient in animal feed or as a supplement with nutritional benefits.

2.10.1.2 Flour

The culmination of milling dehulled millet yields millet flour. This versatile product forms the base for an array of food items, including bread, porridge, cakes, and snacks. A significant advantage of millet flour is its gluten-free nature, rendering it an excellent alternative for individuals with celiac disease or gluten intolerance.

2.10.2 Value Addition from By-Products of Millets

2.10.2.1 Sorghum-Based Energy Bars

From the process of flaking sorghum, two valuable by-products emerge: sorghum bran (seed coat) powder and broken flakes. These by-products, characterized by their high fiber, iron, and vitamin content, can be harnessed to create a range of value-added products. An exemplary value-added product resulting from sorghum flaking is the energy bar, crafted using both bran and broken flakes. The bran, after being coarsely pulverized and dried to eliminate moisture, is incorporated into the formulation. Honey syrup serves as a binder and sweetener in addition to the bran and flakes. This ready-to-consume energy bar stands as a nutritious source of energy suitable for regular consumption or emergency situations.

2.10.2.2 Sorghum Bran Peda

A delectable Indian dessert, bran peda, boasts a sweet and enticing flavor profile. The key components include millet bran powder, sugar, ghee, milk powder, and cardamom. The millet bran undergoes roasting and grinding to achieve a fine powder texture. Following this, powdered sugar, milk, and cardamom powder are added, thoroughly combined with the bran powder. Gradual infusion of ghee precedes the formation of small, well-shaped balls. These balls are adorned with almond or cashew nut embellishments, rendering a delightful treat.

2.10.2.3 Zinc-Rich Jowar Vermicelli

An innovation in Indian cuisine, zinc-rich Jowar vermicelli diversifies the traditional landscape of vermicelli-based dishes. Utilizing an extrusion process, processed gingelly seed powder, jowar semolina, and wheat semolina are combined to yield vermicelli high in zinc content. Gingelly seed preparation involves soaking, dehulling, drying, and powdering. Substituting this processed powder for wheat/refined wheat flour vermicelli enhances the nutritional profile of traditional sweets and savory dishes, offering the benefits of zinc fortification (Figure 2.8).

2.10.2.4 Iron-Rich Bajra Pasta

The ready to cook (RTC) pasta, a popular product derived from cold-extruded wheat semolina, exemplifies the merging of cutting-edge technology with culinary innovation. This advancement extends to the realm of sorghum, enabling the creation of commercially viable sorghum-based products. Through the utilization of advanced processing techniques, the amalgamation of processed garden cress seed powder, bajra semolina, and wheat semolina via the extrusion process yields iron-rich bajra pasta. The preparation of garden cress seed involves a meticulous procedure: boiling in three times its volume of water at 90°C, extraction of the gelatinous seed after cooling, subsequent drying in a tray dryer, and ultimately, pulverization.

The integration of this processed garden cress seed powder into pasta production, replacing conventional wheat or refined wheat flour vermicelli, enhances

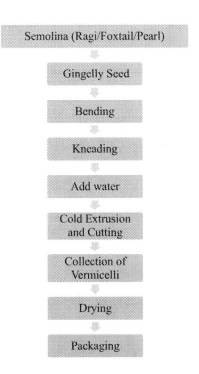

Figure 2.8 Production flow chart of vermicelli.

the nutritional profile of the product. This innovative approach adds an iron-rich dimension to pasta, aligning with the quest for more wholesome dietary choices.

2.10.2.5 Zinc-Rich Ragi Cookies

Cookies, a ubiquitous ready-to-eat food, find their way into the households of all age groups. Gingelly seed undergoes a process that entails soaking overnight, dehulling, subsequent drying in a tray dryer, and finally, reduction to a powdered form. The production of iron-rich ragi cookies involves a sequence of actions: preparing the dough, rolling it out, utilizing an automatic cookie cutter to create various shapes, and baking. Additional stages include creaming (combining fat and sugar powder), mixing with processed gingelly seed powder, and integrating ragi flour into the mixture (Figure 2.9).

2.10.2.6 Modified Starch Preparation from Broken Grains

Cereals are the main raw material used for the isolation and extraction of starch. The purification of starch is more difficult from cereals and its by-products, mainly broken cereals, than from other raw materials due to the lower moisture content of cereal grains. In addition, most of the starch found in cereals is

Figure 2.9 *Production flow chart of zinc rich ragi cookies.*

strongly associated with the protein matrix. Starch processing is an ever-evolving industry, and there are many opportunities to create novel starches that possess innovative functional and value-added terms. Broken rice is the by-product and a key indicator of rice quality. Nearly 14% of rice is broken while milling and during then threshing process. Native starch granules are usually modified to enhance the beneficial properties and eliminate shortcomings, such as low pH, reduced viscosity, easy retrogradation, gelling, and syneresis during cooking, thereby enhancing their applications in the food and non-food sectors. Different modifications/processing treatments also lead to alterations in the functional properties of starch. According to the Food and Nutrition Program (FNP) of the FAO-FNP (1990), a modified starch is a food starch that has one or more of its original physicochemical characteristics altered by treatment in accordance with good manufacturing practice by one of the reaction procedures such as hydrolysis, esterification, etherification, oxidation, and cross-linking. For starches subjected to heating in the presence of acid or with alkali, the alteration (mainly hydrolysis) is considered a minor fragmentation. Different starch modification methods include physical, chemical, and a combination of methods/dual modification. The starch oxidation, acetylation, succinylation, acid/alkali treatment, and cross-linking correspond to chemical modifications, while high hydrostatic pressure (HHP) processing, microwave processing, pulsed electric field (PEF), and cold plasma are novel and environmentally friendly operations.

The method of isolating rice starch is evolved mainly with the separation of starch from different components of rice viz. lipid, protein, and fiber. In rice protein composition, the prolamin, albumin, globulin, and glutelin are found in the following order: 3%, 5%, 12%, and 80%, respectively. These components of rice proteins are easily dissolved in their respective solvents. For instance, albumin in water, globulin in salt, prolamin in ethanol, and glutelin in alkali. Therefore, the alkali extraction method is mainly practiced for isolating the rice starch (de Souzaet et al., 2016). The method for isolation of starch by alkali method is described in Figure 2.10. Choi et al. (2018) isolated starch from broken rice grains. The grains were wet, ground, and soaked in distilled water at 25°C for 1 h for enzymatic extraction. Afterward, either 1% w/v prozyme BPN (food-grade protease generated from *Bacillus amyloliquefaciens*) or prozyme PSP (food-grade endo-protease generated from *Aspergillus niger*) was added to the samples, with subsequent agitation at 45°C for 24 h. The samples were then filtered and centrifuged. These authors further studied the efficacy of the extraction from broken rice using a freeze–thaw infusion method, which involves rapid infusion with food enzymes combined with food-grade proteases. The freeze–thaw infusion method facilitates the rapid entry of enzymes into substrates with enzymatic reactions efficiently

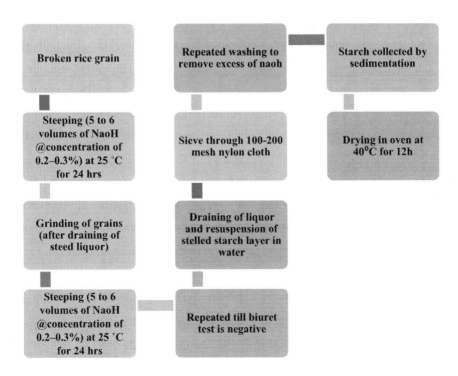

Figure 2.10 *Flow diagram for isolation of starch from broken rice: a by-product of the rice milling industry.*

occurring both inside and outside the substrate. The starch extraction efficiencies for the alkaline method, protease treatment, and freeze–thaw infusion methods were 66.8%, 61.7%, and 69.3%, respectively. The physicochemical properties of the starches, including shape, pasting, and thermal transition, were not affected by the extraction method. The characterization of extracted starch from broken parboiled rice was investigated (Raajeswari et al., 2023) by Fourier-transform infrared spectroscopy (FTIR), X-ray diffraction (XRD), thermogravimetric analysis (TGA), field emission scanning electron microscopy (FESEM). The quality of starch from broken rice estimated through physical, thermal, and physiological properties show a similar property as like starch extracted from rice. It is the colorless and odorless compound that doesn't affect the color and flavor of the product to be added that can be easily used in food industries at low cost.

Esterified starch is used as an emulsion stabilizer and for encapsulation, and can also be used to partially replace fat in emulsion-based food products (Hung et al., 2004), such as plant-based sausages. As a thickening agent in foods, etherified or cross-linked starch is used frequently. Both modifications lead to decreased solubility of starch, by adding inter- and intramolecular bonds, strengthening and stabilizing the starch polymers. Cross-linked starch is increasingly resistant to high and low temperatures and pH; however, it also comes with smaller swelling volume (Singh, 2007). Limberger et al. (2011) evaluated the applicability of chemically and physically modified broken rice starch as a fat substitute in sausages. Extruded and phosphorylated broken rice starch was used for the preparation of low-fat sausage formulation. Results suggest that modified broken rice can be used as a fat substitute in sausage formulations, yielding lower caloric value products with acceptable sensory characteristics. All low-fat sausages presented about 55% reduction in the fat content and around 28% reduction in the total caloric value. Fat replacement with phosphorylated and extruded broken rice starch increased the texture acceptability of low-fat sausages, when compared to low-fat sausages with no modified broken rice starch.

At present, study on extraction of starch and its modification from broken grains for advanced utilization in food industry is high concentrated in the rice milling industry. No studies have been conducted for the utilization of starch from brokens of other grains as the percentage of brokens as by-product in other cereal grains is comparatively less.

2.10.3 Antioxidants and Fiber Extraction from By-Products of Cereal and Millets Processing

Cereal grain by-products present a rich source of diverse phytonutrients and bioactive compounds, which can be harnessed for the development of food products or the extraction of nutraceutical ingredients. In the contemporary era, consumers exhibit heightened interest in products that offer health-related benefits. Cereal by-products manifest as abundant and cost-effective reservoirs of phytochemicals with promising potential for both nutraceutical and pharmaceutical applications. Notably, γ-oryzanols, found in rice bran and rice bran oil, exhibit antioxidant properties surpassing tocopherols by tenfold. Among cereals, corn bran emerges as an exceptional source of the potent antioxidant ferulic acid, rivaling fruits and vegetables. Uniquely, sorghum bran stands as a dietary reservoir of 3-deoxy anthocyanidin, a rare flavonoid type displaying robust cytotoxic activities. Wheat and rye bran feature arabinoxylans that contribute to blood glucose reduction, while oat bran serves as the foundational material for the extraction of dietary fibers, particularly β-glucans.

Millet, a significant drought-resistant crop, takes its place as a nourishing staple in the diets of Africa and Asia. Beyond its nutritional merits, millet boasts an array of bioactive compounds celebrated for their antioxidant prowess. The infusion of antioxidants into our diets assumes paramount importance for promoting human well-being. This review aims to evaluate the antioxidant compounds inherent in millet, shedding light on the factors that govern their antioxidant activity. Phenolic compounds, including phenolic acids, flavonoids, tannins, as well as xylo-oligosaccharides (XOs), insoluble fibers, and peptides, are among the naturally occurring constituents within millet. Additionally, lipophilic antioxidants like vitamin E and carotenoids are widely dispersed across diverse millet varieties. Noteworthy is the influence of food processing on antioxidant bioactivity. Processes like germination and fermentation hold the potential to amplify antioxidant properties due to heightened levels of antioxidants, particularly phenolic compounds. Apart from lysine and threonine, millet proteins stand as excellent sources of essential amino acids, while the grains also offer an abundance of phytochemicals and micronutrients. Beyond its nutritional contributions, millet is recognized for its potential health benefits, encompassing wound healing enhancement, cardiovascular disease prevention, and the reduction of blood glucose and cholesterol levels. Oxidative stress has been implicated in various chronic conditions, making antioxidants vital in mitigating oxidative damage. Several compounds within millet exhibit antioxidant properties, with phytochemicals such as phenolics and dietary fiber primarily concentrated in the bran layers, accompanied by antioxidant-rich micronutrients like carotenoids and tocopherols (Table 2.7). Furthermore, millet can be enriched with antioxidants, such as peptides, through fermentation and germination. In summation, both cereal grains and millet present themselves as promising sources of nutraceuticals and antioxidants. Their integration into diets holds the potential to alleviate disease risks and bolster overall health through their remarkable antioxidant attributes.

2.11 Conclusion

This chapter aimed to delve into the potential utilization of by-products derived from various cereal and millets. Through a comprehensive exploration

Table 2.7 Antioxidants in Cereal and Millet Grains

Name of compound	Major grains
Phenolic acids	Wheat, corn, kodo, finger, foxtail, pearl millets, rice
Flavonoids	Kodo, finger, foxtail, proso, pearl millets, wheat, corn, rice
Tannins	Finger millets, sorghum
Xylo-oligosaccharides	Finger millets
Insoluble fibers	Foxtail millets
Protein and peptides	Peal, foxtail millet
Carotenoid	Finger, little, foxtail, proso millets, wheat, rice, corn
Vitamin E	Finger, little, foxtail, proso millets, corn, wheat, rice

Source: (Liang & Liang, 2019)

of their nutritional, physical, functional attributes, and antioxidant capacity, this research enhances our understanding of metabolite composition and functional compounds in husks and brans, particularly in terms of their antioxidant activity. This deeper understanding aids in evaluating the viability of employing these by-products on a larger scale. In order to fully harness the potential of these processed products, it is imperative to fully comprehend their strengths and limitations, especially in contexts such as food or medicine. Subsequent research endeavors are warranted. In conclusion, cereal bran, a significant by-product of the milling industry, constitutes a reservoir of diverse components with potential health benefits that can be harnessed as food ingredients. With tailored processing techniques for each component of cereal bran in mind, processed cereal by-products can be seamlessly integrated into conventional food products, offering nutrition, bioactive compounds, shelf stability, and convenience at a cost-effective rate. However, due to the lack of available data on the quantities of cereal and millet by-products generated as bran, germ, and fractions, determining the feasibility of obtaining ample nutraceuticals from these sources remains challenging. Consequently, the incorporation of cereal by-products as ingredients in foods emerges as a promising strategy to foster the aforementioned health advantages.

References

Abdullah, W. J., and Zhang, H. (2020). Recent advances in the composition, extraction and food applications of plant-derived oleosomes. Trends in Food Science & Technology 106: 322–332.

Bayramoglu, B., Sahin, S., and Sumnu, G. (2008). Solvent-free microwave extraction of essential oil from oregano. Journal of Food Engineering 88: 535–540.

Boukid, F., Folloni, S., Ranieri, R., & Vittadini, E. (2018). A compendium of wheat germ: Separation, stabilization and food applications. *Trends in Food Science & Technology,* 78, 120–133. https://doi.org/10.1016/j.tifs.2018.06.001

Bousbia, N., Vian, M., Ferhat, M., Meklati, B., and Chemat, F. (2009). A new process for extraction of essential oil from Citrus peels: Microwave hydro diffusion and gravity. Journal of Food Engineering 90: 409–413.

Choi, J. M., Park, C. S., Baik, M. Y., Kim, H. S., Choi, Y. S., Choi, H. W., and Seo, D. H. (2018). Enzymatic extraction of starch from broken rice using freeze-thaw infusion with food-grade protease. Starch -St€arke 70 (1–2): 1700007.

Da Porto, C., Decorti, D., and Kikic, I. (2009). Flavour compounds of Lavandula angustifolia L. to use in food manufacturing: Comparison of three different extraction methods. Food Chemicals 112: 1072–1078.

Da-wei, W., Ting, L. T., Chen, S. Y., and Jun, Y. U. (2009). Supercritical fluid extraction of conjugated linoleic acid from corn germ oil. Food Science 30 (20): 64–67.

Dehghan, M., Mente, A., Zhang, X., Swaminathan, S., Li, W., Mohan, V., and Mapanga, R. (2017). Associations of fats and carbohydrate intake with cardiovascular disease and mortality in 18 countries from five continents (PURE): A prospective cohort study. The Lancet 390 (10107): 2050–2062. https://doi.org/10.1016/S0140-6736(17)32252-3.

Deng, C., Xu, X., Yao, N., Li, N., and Zhang, X. (2006). Rapid determination of essential oil compounds in Artemisia Selengensis Turcz by gas chromatography-mass spectrometry with microwave distillation and simultaneous solid-phase microextraction. Analytical Chim Acta 556: 289–294.

FAO-FNP (Food and Agricultural Organisation—Food and Nutrition Program). *Modified Starch* (FNP 40, 1990). Available from: http://www.fao.org/3/w6355e/w6355e0o.htm

Fărcaş, Anca Corina, Sonia Ancuţa Socaci, Silvia Amalia Nemeş, Liana Claudia Salanţă, Maria Simona Chiş, Carmen Rodica Pop, Andrei Borşa, Zoriţa Diaconeasa, and Dan Cristian Vodnar. 2022. "Cereal Waste Valorization through Conventional and Current Extraction Techniques—An Up-to-Date Overview" *Foods 11*(16): 2454. https://doi.org/10.3390/foods11162454

Ghafoor, K., Ozcan, M. M., AL-Juhaimi, F., Babıker, E. E., Sarker, Z. I., Ahmed, I. A. M., and Ahmed, M. A. (2017). Nutritional composition, extraction, and utilization of wheat germ oil: A review. European Journal of Lipid Science and Technology 119 (7): 1600160.

Giordano, D., Vanara, F., Reyneri, A., and Blandino, M. (2016). Effect of dry-heat treatments on the nutritional value of maize germ. International Journal of Food Science & Technology 51 (11): 2468–2473.

Hung, PV., Morita, N. (2004). Dough properties and bread quality of flours supplemented with cross-linked corn starches. *Food Research International. 37*(5): 461–467.

Karimi, A., Azizi, M. H., and Ahmadi Gavlighi, H. (2020). Fractionation of hydrolysate from corn germ protein by ultrafiltration: In vitro antidiabetic and antioxidant activity. Food Science & Nutrition 8 (5): 2395–2405. https://doi.org/10.1002/fsn3.v8.510.1002/fsn3.1529.

Limberger, V. M., Brum, F. B., Patias, L. D., Daniel, A. P., Comarela, C. G., Emanuelli, T., and Silva, L. P. D. (2011). Modified broken rice starch as fat substitute in sausages. Ciência E Tecnologia De Alimentos 31 (3): 789–792. doi: 10.1590/S0101-20612011000300037.

Liang Shan, and Liang Kehong. (2019). Millet grain as a candidate antioxidantfood resource: a review, *International Journal of Food Properties, 22*: 1, 1652-1661. (5) (PDF) Millet grain as a candidate antioxidant food resource: a review. Available from: https://www.researchgate.net/publication/336132040_Millet_grain_as_a_candidate_antioxidant_food_resource_a_review [accessed Dec 12 2023].

Ling, B., Ouyang, S., & Wang, S. (2019). Radio-frequency treatment for stabilization of wheat germ: Storage stability and physicochemical properties. *Innovative Food Science & Emerging Technologies*, 52, 158–165. https://doi.org/10.1016/j.ifset.2018.12.002.

Luithui, Y., Nisha, R. B., and Meera, M. S. (2019). Cereal by-products as an important functional ingredient: Effect of processing. Journal of Food Science and Technology 56: 1–11.

Marinho, C., Lemos, C. O. T., Arvelos, S., and Barraza, M. (2019). Extraction of corn germ oil with supercritical CO2 and cosolvents. Journal of Food Science and Technology 56 (10): 12–25.

Metherel, A. H., Taha, A. Y., Izadi, H., and Stark, K. D. (2009). The application of ultrasound energy to increase lipid extraction throughput of solid matrix samples (flaxseed). Prostaglandins: Leukotrienes and Essential Fatty Acids (PLEFA) 81 (5–6): 417–423.

Nde, D. B., and Foncha, A. C. (2020). Optimization methods for the extraction of vegetable oils: A review. Processes 8: 1–21.

Raajeswari, PA, Devatha, S M., and Pragatheeswari, R. 2023. Characterization and Property Analysis of Starch from Broken Parboiled Rice. *Sustainability, Agri, Food and Environmental Research*, (ISSN: 0719-3726), 11(X) http://dx.doi.org/10.7770/safer-V11N1-art2913

Rebolleda, S., Rubio, N., Beltrán, S., Sanz, M. T., and González-Sanjosé, M. L. (2012). Supercritical fluid extraction of corn germ oil: Study of the influence of process parameters on the extraction yield and oil quality. The Journal of Supercritical Fluids 72: 270–277.

Rondanelli, M., Miccono, A., Peroni, G., Nichetti, M., Infantino, V., Spadaccini, D., and Perna, S. (2021). Rice germ macro- and micronutrients: A new opportunity for the nutraceutics. Natural Product Research 35(9): 1532–1536.

Rondanelli, M., Miccono, A., Peroni, G., Nichetti, M., Infantino, V., Spadaccini, D., Perna, S. (2021). Rice germ macro- and micronutrients: A new opportunity for the nutraceutics. *Natural Product Research, 35*(9), 1532–1536. https://doi.org/10.1080/14786419.2019.1660329.

Saldivar, X., Wang, Y.-J., Chen, P., and Hou, A. (2011). Changes in chemical composition during soybean seed development. Food Chemistry 124 (4): 1369–1375.

Slavin, J. L., Jacobs, D., and Marquart, L. (2000). Grain processing and nutrition. Critical Reviews in Food Science and Nutrition 40 (4): 309–326.

Singh, J., Kaur, L., and OJ, M. C. (2007). Factors influencing the physico-chemical, morphological, thermal and rheological properties of some chemically modified starches for food applications—A review. Food Hydrocolloids 21 (1): 1–22.

Wang, W., Wu, X., Han, Y., Zhang, Y., Sun, T., &and Dong, F. (2011),. Investigation on ultrasoundassisted extraction of three dibenzylbutyrolactone lignins from Hemistepta Iyrata. Journal of Applied Pharmaceutical Science, 01 (09),: 24–28.

Wu, J., Lin, L., and Chau, F. (2001). Ultrasound-assisted extraction of ginseng saponins from ginseng roots and cultured ginseng cells. Elsevier Journal of Ultrasonics Sonochemistry 8: 347–352.

Yadav, R., Yadav, N., Saini, P., and Kaur, D. (2020). Potential Value Addition from Cereal and Pulse Processed By-Products: A Review. Book- Sustainable Food Waste Management, pp. 155–176. (Yadav R. et al., 2020)

Zhang, L., Ying, S., Keji, T., and Ramesh, P. (2009). Ultrasound-assisted extraction flavonoids from Lotus (Nelumbo nuficera Gaertn) leaf and evaluation of its anti-fatigue activity. International Journal of Physical Sciences 4 (8): 418–422.

Zhu, K. X., Sun, X. H., and Zhou, H. M. (2009). Optimization of ultrasound-assisted extraction of defatted wheat germ proteins by reverse micelles. Journal of Cereal Science 50: 266–271.

3

Value Addition of Oilseed Processing Industrial Waste

M. Selvamuthukumaran

3.1 Introduction

The waste derived out of oilseed processing industries can be efficiently utilized for the preparation of several value-added products. They can be successfully utilized for developing various functional foods (Usman et al., 2022). The cake, which is a by-product obtained after pressing and extracting the crushed oilseeds, are a valuable source of proteins and one of the cheapest raw material sources for the production of protein isolates, bioactive peptides and protein hydrolysates. The meal obtained from sesame and mustard contains appreciable amounts of antioxidants (Saeed et al., 2022).

The oilseed processing industries can generate huge amounts of oilseed meal, which can be incorporated into food products to make them more nutritious and will help undernourished communities by supplementing their diet with proteins and other valuable micronutrients (Ancuța and Sonia, 2020).

3.2 Production of Oilseed Cake Meal

The oilseeds were cleaned, washed, dried, flaked and subjected to extraction either by a solvent extraction process by using solvents like food-grade hexane or by mechanical extraction process by pressing it through screw press to extract crude oil. Once the crude oil is extracted, the residue which is left over, known as cake, can be powdered in a hammer mill for fortification with several food products. The obtained oilseed cake meal contains several antioxidants (Table 3.1) and proteins so they can be efficiently used for the production of protein isolates and concentrates.

DOI: 10.1201/9781003269199-3

Table 3.1 Antioxidants Present in Various Oilseed Cakes (Modified from Usman et al., 2023)

Name of the oilseed cake	Antioxidants present in specific oilseed cake	References
Sunflower	Chlorogenic acid, catechin, epicatechin, p-coumaric	Şahin and Elhussein (2018)
Rape seed	P-coumaric, catechin, caffeic acid, epicatechin, ferulic acid, sinapic acid	Teh and Bekhit (2015)
Canola	Gallic acid, catechin, luteolin, caffeic acid, quercetin, ferulic acid	Şahin and Elhussein (2018)
Sesame	p-coumaric, ferulic acid, lignans	Senanayake et al. (2019), Sarkis et al. (2014)
Peanut	p-coumaric, caffeic acid	Teh and Bekhit (2015)
Flax seed	p-coumaric, ferulic acid	Şahin and Elhussein (2018)
Palm	p-coumaric, caffeic acid, p-hydroxybenzoic acid	Senanayake et al. (2019)

3.3 Oilseed Cake Meal Fortification

The variety of bakery products like cookies, cakes and bread can be fortified with oilseed cake meal, which is a great source of protein cum fibre (Bochkarev et al., 2016). The consumption of fortified oilseed cake meal products has got immense health benefits (Figure 3.1). Martínez et al. (2021) incorporated oilseed meal obtained from the sources of flax seed, chia seed and sesame by replacing wheat flour, thereby adding innovation like replacing gluten. Norajit et al. (2011) developed energy bars by supplementing rice flour with proportions of hemp seed meal @ 0–40%. The fortification of the bar with hemp seed meal enhanced the antioxidant content of the final product. Sobczak et al. (2020) successfully prepared compressed tablets from oilseed meals obtained from flax seed, pumpkin, sunflower and coconut.

3.4 Enzyme Production from Oilseed Meal

The oilseed meal contains higher sources of dense nutrients like carbon and nitrogen, therefore they can be effectively used for the production of enzymes using a solid-state fermentation process. There are varieties of enzymes that are being produced using oilseed meal, such as mannose, tannase, phytase, glutaminase, protease, α-amylase etc. (Chatterjee et al., 2015). Gupta et al. (2018a) produced protease enzymes by using bacterial strains from various oilseed cakes like coconut, palm, olive, sesame and soyabean.

Treichel et al. (2010) produced enzyme lipase by using different oilseed meals like olive, coconut and sesame with the help of fungal strains like *Rhizomucor pusillus, P. simplicissimum, P. chrysogenum* and *Candida rugosa.*

74   Wealth out of Food Processing Waste

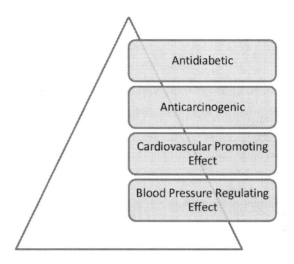

Figure 3.1 *Health benefits of consuming oilseed meal–fortified products.*
Source: **(Modified from Usman et al., 2023)**

3.5 Antibiotic and Antimicrobials Production from Oilseeds

The oil meal obtained from sesame has been used for the production of antibiotics. The cake is a rich source of antibiotics and antimicrobials production (Gupta et al., 2018b). The cake obtained from sesame, soyabean and sunflower has been used for the production of clavulanic acid and cephamycin C. Sarkar et al. (2021) reported that by using sunflower and sesame oil meal the Bacitracin, Bacillus licheniformis and Bacillus thuringiensis was produced.

3.6 Mushroom Production from Oilseed Meal

The mushroom has been produced by using a solid-state fermentation process with the help of oilseed meal. Jatuwong et al. (2020) obtained increased yield by using meal obtained from cottonseed and the resultant mushroom was found to be the rich source of both protein and fat.

3.7 Hulls

Hulls are one of the important wastes generated by oilseed processing industries. They contain rich sources of antioxidants ascribed to polyphenols (Jamwal, 2022). Currently they are utilized only as an animal feed. The hull can be powdered and fortified with food products during processing to enhance the antioxidant status of the products viz. catechins, flavonoids etc.

3.8 Wastewater Sludge

Wastewater sludge, which is one of the indirect solid waste that are being sent to the landfill (Chipasa, 2001; Welz et al., 2017). Pandey et al. (2003) found that the wastewater sludge contains organic matter of 76% with dense nutrient content, and they can be efficiently recycled. The wastewater sludge from oilseed processing industries can be successfully used to generate biofuel (Ngoie et al., 2019a, 2019b; Muanruksa and Kaewkannetra, 2020).

3.9 Conclusions

The food industries, especially oilseed processing, can efficiently utilize and convert the various waste generated by these industries into various value-added food additives. They can efficiently be utilized by fortifying with various food products. Such industries should adopt circular economies so that waste generated by these industries can be minimized.

References

Ancuţa, P., & Sonia, A. (2020). Oil press-cakes and meals valorization through circular economy approaches: A review. Applied Sciences, 10(21), 7432.

Bochkarev, M. S., Egorova, E. Y., Reznichenko, I. Y., & Poznyakovskiy, V. M. (2016). Reasons for the ways of using oilcakes in food industry. Foods and Raw Materials, 4(1), 4–12.

Chatterjee, R., Dey, T. K., Ghosh, M., & Dhar, P. (2015). Enzymatic modification of sesame seed protein, sourced from waste resource for nutraceutical application. Food and Bioproducts Processing, 94, 70–81.

Chipasa, K. B. (2001). Limits of physicochemical treatment of wastewater in the vegetable oil refining industry. Polish Journal of Environmental Studies, 10, 141–147.

Gupta, A., Sharma, A., Pathak, R., Kumar, A., & Sharma, S. (2018a). Solid state fermentation of non-edible oil seed cakes for production of proteases and cellulases and degradation of anti-nutritional factors. Journal of Food Biotechnology Research, 1, 3–8.

Gupta, A., Sharma, R., Sharma, S., & Singh, B. (2018b). Oilseed as potential functional food ingredient. In Trends & Prospects in Food Technology, Processing and Preservation (1st ed.). Today and Tomorrow's Printers and Publishers.

Jamwal, P. (2022). Oilseeds industry by-products used as functional food ingredients. The Pharma Innovation Journal 2022, SP-11(6), 2648–2658.

Jatuwong, K., Kumla, J., Suwannarach, N., Matsui, K., & Lumyong, S. (2020). Bioprocessing of agricultural residues as substrates and optimal conditions for phytase production of chestnut mushroom, Pholiota adiposa, in solid state fermentation. Journal of Fungi, 6(4), 384.

Martínez, E., García-Martínez, R., Álvarez-Ortí, M., Rabadán, A., PardoGiménez, A., & Pardo, J. E. (2021). Elaboration of gluten-free cookies with defatted seed flours: Effects on technological, nutritional, and consumer aspects. Food, 10(6), 1213.

Muanruksa, P., & Kaewkannetra, P. (2020). Combination of fatty acids extraction and enzymatic esterification for biodiesel production using sludge palm oil as a low cost substrate. Renewable Energy, 146, 901–906.

Ngoie, W. I., Oyekola, O. O., Ikhu—Omoregbe, D., & Welz, P. J. (2019a). Valorisation of edible oil wastewater sludge: Bioethanol and biodiesel production. Waste and Biomass Valorization. https://doi.org/10.1007/s12649-019-00633-w.

Ngoie, W. I., Oyekola, O. O., Ikhu—Omoregbe, D., & Welz, P. J. (2019b). Qualitative assessment of biodiesel produced from primary edible oil wastewater sludge. Waste and Biomass Valorization. https://doi.org/10.1007/s12649-019-00772-0.

Norajit, K., Gu, B.-J., & Ryu, G.-H. (2011). Effects of the addition of hemp powder on the physicochemical properties and energy bar qualities of extruded rice. Food Chemistry, 129(4), 1919–1925.

Pandey, R. A., Sanyal, P. B., Chattopadhyay, N., & Kaul, S. N. (2003). Treatment and reuse of wastes of a vegetable oil refinery. Resources Conservation and Recycling, 37, 101–117.

Saeed, K., Pasha, I., Jahangir Chughtai, M. F., Ali, Z., Bukhari, H., & Zuhair, M. (2022). Application of essential oils in food industry: Challenges and innovation. Journal of Essential Oil Research, 34(2), 97–110.

Şahin, S., & Elhussein, E. A. A. (2018). Assessment of sesame (Sesamum indicum L.) cake as a source of high-added value substances: From waste to health. Phytochemistry Reviews, 17(4), 691–700.

Sarkar, N., Chakraborty, D., Dutta, R., Agrahari, P., Sundaram, D., Singh, A., & Jacob, S. (2021). A comprehensive review on oilseed cakes and their potential as a feedstock for integrated biorefinery. Journal of Advanced Biotechnology and Experimental Therapeutics, 4(3), 376.

Sarkis, J. R., Michel, I., Tessaro, I. C., & Marczak, L. D. F. (2014). Optimization of phenolics extraction from sesame seed cake. Separation and Purification Technology, 122, 506–514.

Senanayake, C. M., Algama, C. H., Wimalasekara, R. L., Weerakoon, W., Jayathilaka, N., & Seneviratne, K. N. (2019). Improvement of oxidative stability and microbial shelf life of vanilla cake by coconut oil meal and sesame oil meal phenolic extracts. Journal of Food Quality, 2019, 1–8.

Sobczak, P., Zawiślak, K., Starek, A., Żukiewicz-Sobczak, W., Sagan, A., Zdybel, B., & Andrejko, D. (2020). Compaction process as a concept of press-cake production from organic waste. Sustainability, 12(4), 1567.

I, S.-S., & Bekhit, A. E.-D. A. (2015). Utilization of oilseed cakes for human nutrition and health benefits. In K. R. Hakeem, M. Jawaid, & O. Y. Alothman (Eds.) Agricultural Biomass based Potential Materials (pp. 191–229). Springer.

Treichel, H., de Oliveira, D., Mazutti, M. A., di Luccio, M., & Oliveira, J. V. (2010). A review on microbial lipases production. Food and Bioprocess Technology, 3(2), 182–196.

Usman, I., Hussain, M., Imran, A., Afzaal, M., Saeed, F., Javed, M., Afzal, A., Ashfaq, I., al Jbawi, E., & A. Saewan, S. (2022). Traditional and innovative approaches for the extraction of bioactive compounds. International Journal of Food Properties, 25(1), 1215–1233.

Usman, I., Saif, H., Imran, A., Afzaal, M., Saeed, F., Azam, I., Afzal, A., Ateeq, H., Islam, F., Shah, Y. A., & Shah, M. A. (2023). Innovative applications and therapeutic potential of oilseeds and their by-products: An eco-friendly and sustainable approach. Food Science & Nutrition, 11, 2599–2609.

Welz, P. J., Le Roes-Hill, M., & Swartz, C. D. (2017). NATSURV 6: Water and Wastewater Management in the Edible Oil Industry (Edition 2). WRC TT 702/16.

4

Value Addition of Pulse Processing Industrial Waste

M.Selvamuthukumaran

4.1 Introduction

Food waste generation can lead to major loss of important natural resources, like land, energy, water and labour. The post-harvest losses account for around 14% globally, which is estimated to be around 400 billion USD (FAO, 2022). Around 17% of food is being wasted at the retail level by consumers, especially in households, according to the Food Waste Index as reported by the United Nations Environment Programme. Annual food waste accounts for one-third, which can feed around 1.26 billion undernourished people (Jan et al., 2013).

The generation of such waste can lead to environmental damage. These wastes can discharge around 4.4 kilotons of carbon dioxide equivalent every year, which accounts for emission of around 8% anthropogenic greenhouse gas to the environment (Sharma et al., 2020). The processing of pulse can lead to the generation of various by-products, such as husks and germ. These can be effectively recycled to produce a variety of value-added products.

Red gram, known as pigeon pea, is one of the most important pulses produced in India. Parate and Talib (2015) utilized red gram husk for preparing activated carbon, which can remove heavy metals. They prepared char by carbonizing the dehydrated red gram husk in an airtight container at 500°C for a period of 1 hour. They activated the obtained char by soaking it in concentrated sulfuric acid at a ratio of 1:1 for one day at room temperature in order to obtain activated carbon. They found that the prepared activated carbon from red gram had showed significant metal adsorption capacity. They further recommended that the waste generated from pulse processing industries could be significantly utilized in an economical and eco-friendly manner.

4.2 Bioactive Constituents from By-Products of Pulses

The major edible pulses were pigeon pea, black gram, kidney bean, chickpea, moth bean, field pea, lima bean, rice bean, moth bean, kidney bean and lentils.

78   DOI: 10.1201/9781003269199-4

The processing of such edible pulses can generate waste, which contains ample amounts of various bioactive constituents like phytosterols, saponins, lectins, phytates, phenolic compounds etc. They are low glycemic index (GI) foods, having GI values between 28 and 52, so incorporation of such pulse-based foods can reduce the risk of heart diseases (Atkinson et al., 2008; Rizkalla et al., 2002). Rebello et al. (2014) reported that obesity can be drastically reduced by incorporating pulse-based products, which can replace energy-rich carbohydrate foods. Pulse by-products also contain antioxidants like polyphenols, which can contribute to several biological properties and help to enhance the product's sensory characteristics, especially colour (Balasundram et al., 2006). These components are highly present in seed coat, which are removed during the milling process. The processing of black gram can remove around 25% of black gram's constituents, which are the richest source of polyphenols (Girish et al., 2012). The various phenolic compounds detected from dry kidney bean seed coat was found to be cyanidin 3-glucoside (0–0.12 mg/g), petunidin 3-glucoside (0–0.17 mg/g), delphinidin 3-glucoside (0–2.61 mg/g) and pelargonidin 3-glucoside (0–0.59 mg/g) (Choung et al., 2003).

The defatted lentil sample contains total phenolic content in esterified form of 2.32–21.54 mg GAE/g, insoluble form of 2.55–17.51 mg GAE/g and free form of 1.37–5.53 mg GAE/g (Alshikh et al., 2015).

4.3 Heath Benefits of Consuming Pulse-Based By-Products

It has been reported that the incorporation of pulse-based by-products can reduce the risk of cancer because of the various antioxidants, micronutrients and fibre that are present (Avune et al., 2009; Mudryj et al., 2014). Risk of heart disease can also be significantly reduced as a result of the regular consumption of a pulse-based diet because it helps with weight loss (Mc Crory et al., 2010; Papanikolaou and Fulgoni, 2008; Singh et al., 2017). The various bioactive constituents present in pulses can contribute to a variety of health benefits because of their anti-inflammatory, anti-diabetic, anti-ulcer, anti-carcinogenic, anti-thrombotic and antioxidative effects (Daglia, 2012; Jin et al., 2012; Ozcan et al., 2014; He and Giusti, 2010) (Figure 4.1). Bazzano et al. (2001) justified that the regular consumption of pulse, i.e. four times a week, can reduce the risk of coronary heart disease by 22%.

During harvesting, damaged pulse seeds are discarded and considered a by-product, and during commercial scale processing the pods and other seed residues are separated, especially during cleaning and splitting operations. Canning, freezing and drying processes can also generate waste like stems, leaves and empty pods (Tassoni et al., 2020). Around 5–25% of cultivated leguminous crops are generated as waste. The milling processing industries, especially pulse, can generate around 25% by-products, which includes powder, husks and shriveled, broken and unprocessed seeds (Pojic et al., 2018).

Figure 4.1 Health properties of consuming value-added products obtained from pulse processing industrial wastes.

The waste obtained during pulse processing can be converted into several value-added products. They are a rich source of constituents like dietary fibres, proteins and phenolic compounds.

4.4 Extraction of Proteins from Wastes of Pulse Processing Industry

The waste obtained from the pulse processing industry can be effectively utilized for recovering proteins by means of both wet and dry fractionation methods (Table 4.1). The various wet fractionation methods used for the successful extraction of pulse are organic solvent extraction, isoelectric precipitation, membrane separation and enzyme-assisted extraction (Tassoni et al., 2020). The most popular technique used for protein extraction is alkaline extraction, followed by isoelectric precipitation. In this type, protein is solubilized at a pH ranging from 8 to 11, followed by precipitating the solubilized protein at a pH close to isoelectric point, i.e. 4.5–4.6. Higher protein yield recovery can also be obtained through the membrane separation process. Some cost-effective and environmentally friendly techniques, like aqueous two-phase system extraction and subcritical water extraction, have gained more attention in recent years compared to traditional techniques for protein extraction (Pojic et al., 2018). Enzyme-assisted extraction techniques involve higher costs for extracting proteins. Air classification techniques and sieving can be adopted as dry methods for recovering proteins from pulse processing industrial wastes. Pulse flour fractionation can be achieved by electrostatic separation (Pojic et al., 2018).

Table 4.1 Recovery of Protein Ingredients from Pulse Processing Industrial Wastes

Source	Methods used for protein recovery	Reference
Disease-infected beans	Alkaline extraction followed by isoelectric precipitation	Hernandez-Alvarez et al. (2013)
Chickpea and pea processing feedstocks	Aqueous extraction and enzyme-assisted extraction	Prandi et al. (2021)
Black gram milled by-products	Dry fractionation	Girish et al. (2012)
Moth bean milled by-products	Dry fractionation	Kamani et al. (2020)

4.5 Fortification

The recovered proteins from pulse processing industrial wastes can be effectively fortified for developing various snacks and bakery products (Tiwari et al., 2011a, 2011b). Disease-infected and frost-affected pulses are also effectively used for recovering functional components for developing novel food products. Hernandez-Alvarez et al. (2013) also explored the possibility of producing angiotensin-converting enzyme inhibitory (ACE-I) peptides from disease-damaged beans of *P. vulgaris*. They extracted proteins successfully from anthracnose-infected beans by alkaline extraction followed by an isoelectric precipitation method. Portman et al. (2020) had also successfully developed extruded foods by using downgraded frost-affected lentils as one of the ingredients by effectively utilizing the wastes for further value addition.

4.6 Conclusions

Pulse processing industrial waste can be effectively utilized for developing various value-added products like proteins, fibres, starch and several antioxidative constituents, especially polyphenols. The recovered bioactive constituents can be effectively utilized for developing various health-oriented food products.

References

Alshikh N, de Camargo AC, Shahidi F. "Phenolics of selected lentil cultivars: Antioxidant activities and inhibition of low-density lipoprotein and DNA damage." Journal of Functional Foods. 2015, 18, 1022–1038.

Atkinson FS, Foster-Powell K, Brand-Miller JC. "International tables of glycemic index and glycemic load values." Diabetes Care. 2008, 31, 2281–2283.

Avune D, De Stefani E, Ronco A, Boffetta P, Deneo-Pellegrini H, Acosta G, et al. "Legume intake and the risk of cancer: A multisite case-control study in Uruguay." Cancer Causes & Control. 2009, 20, 1605–1615.

Balasundram N, Sundaram K, Samman S. "Phenolic compounds in plants and agri-industrial by-products: Antioxidant activity, occurrence, and potential uses." Food Chemistry. 2006, 99, 191–203.

Bazzano LA, He J, Ogden LG, Loria C, Vupputuri S, Myers L, et al. "Legume consumption and risk of coronary heart disease in US men and women: NHANES I epidemiologic follow-up study." Archives in International Medicine. 2001, 161, 2573–2578.

Choung G, Choi BR, An YN, Chu YH, Cho YS. "Anthocyanidin profile of Korean cultivated kidney bean (Phaseolus vulgaris L.)." Journal of Agricultural and Food Chemistry. 2003, 51, 7040–7043.

Daglia M. "Polyphenols as antimicrobial agents." Current Opinion in Biotechnology. 2012, 23, 174–181.

FAO. FAO Cereal Supply and Demand Brief I World Food Situation; Food and Agriculture Organization of the United Nations: Roma, Italy, 2022.

Girish TK, Pratape VM, Rao UJSP. "Nutrient distribution, phenolic acid composition, antioxidant and alpha-glucosidase inhibitory potentials of black gram (Vigna mungo L.) and its milled by-products." Food Research International. 2012, 46, 370–377.

He JA, Giusti MM. "Anthocyanins: Natural colorants with health-promoting properties." Annual Review in Food Science and Technology. 2010, 163–186.

Hernandez-Alvarez AJ, Carrasco-Castilla J, Davila-Ortiz G, Alaiz M, Giron-Calle J, et al. "Angiotensin-converting enzymeinhibitory activity in protein hydrolysates from normal and anthracnose disease-damaged Phaseolus vulgaris seeds." Journal of Science of Food and Agriculture. 2013, 93, 961–966.

Jan O, Tostivint C, Turbé A, O'Connor C, Lavelle P. Food Wastage Footprint: Impacts on Natural Resources; Food and Agriculture Organization of the United Nations: Roma, Italy, 2013.

Jin L, Ozga JA, Lopes-Lutz D, Schieber A, Reinecke DM. "Characterization of proanthocyanidins in pea (Pisum sativum L.), lentil (Lens culinaris L.), and faba bean (Vicia faba L.) seeds." Food Research International. 2012, 46, 528–535.

Kamani MH, Martin A, Meera MS. "Valorization of byproducts derived from milled moth bean: Evaluation of chemical composition, nutritional profile and functional characteristics." Waste Biomass Valorization. 2020, 11, 4895–4906.

Mc Crory, Hamaker BR, Lovejoy JC, Eichelsdoerfer PE. "Pulse consumption, satiety, and weight management." Advances in Nutrition. 2010, 1, 17–30.

Mudryj N, Yu N, Aukema HM. "Nutritional and health benefits of pulses." Applied Physiology, Nutrition and Metabolism. 2014, 39, 1197–1204.

Ozcan T, Akpinar-Bayizit A, Yilmaz-Ersan L, Delikanli B. "Phenolics in human health." International Journal of Chemical Engineering and Applications. 2014, 5, 393.

Papanikolaou Y, Fulgoni VL. "Bean consumption is associated with greater nutrient intake, reduced systolic blood pressure, lower body weight, and a smaller waist circumference in adults: Results from the National Health and Nutrition Examination Survey 1999–2002." Journal of American College and Nutrition. 2008, 27, 569–576.

Parate VR, Talib MI. "Utilization of pulse processing waste (Cajanus cajan Husk) for developing metal adsorbent: A value-added exploitation of food industry waste." American Journal of Food Science and Technology. 2015, 3(1), 1–9. doi: 10.12691/ajfst-3-1-1.

Pojic M, Misan A, Tiwari B. "Eco-innovative technologies for extraction of proteins for human consumption from renewable protein sources of plant origin." Trends in Food Science and Technology. 2018, 75, 93–104.

Portman D, Dolgow C, Maharjan P, Cork S, Blanchard C, Naiker M, Panozzo JF. "Frost-affected lentil (Lens culinaris M.) compositional changes through extrusion: Potential application for the food industry." Cereal Chemistry. 2020, 97, 818–826.

Prandi B, Zurlini C, Maria CI, Cutroneo S, Di Massimo M, et al. "Targeting the nutritional value of proteins from legumes byproducts through mild extraction technologies." Frontier in Nutrition. 2021, 8, 695793.

Rebello J, Greenway FL, Finley JW. "A review of the nutritional value of legumes and their effects on obesity and its related co-morbidities." Obesity Reviews. 2014, 15, 392–407.

Rizkalla SW, Bellisle F, Slama G. "Health benefits of low glycaemic index foods, such as pulses, in diabetic patients and healthy individuals." British Journal of Nutrition. 2002, 88, S255–S262.

Sharma P, Gaur VK, Sirohi R, Larroche C, Kim SH, Pandey A. "Valorization of cashew nut processing residues for industrial applications." Industrial Crop Production. 2020, 152, 112550.

Singh B, Singh JP, Shevkani K, Singh N, Kaur A. "Bioactive constituents in pulses and their health benefits." Journal of Food Science and Technology. 2017, 54, 858–870.

Tassoni A, Tedeschi T, Zurlini C, Cigognini IM, Petrusan JI, et al. "State-of-the-art production chains for peas, beans and chickpeas valorization of agro-industrial residues and applications of derived extracts." Molecules 2020, 25, 1383.

Tiwari BK, Brennan CS, Jaganmohan R, Surabi A, Alagusundaram K. "Utilisation of pigeon pea (Cajanus cajan L) byproducts in biscuit manufacture." LWT-Food Science and Technology. 2011a, 44, 1533–1537.

Tiwari U, Gunasekaran M, Jaganmohan R, Alagusundaram K, Tiwari BK. "Quality characteristic and shelf life studies of deepfried snack prepared from rice brokens and legumes by-product." Food Bioprocess Technology. 2011b, 4, 1172–1178.

5

Value Addition of Fruit Processing Industrial Waste

Merve Aydin, Vildan Eyiz, and İsmail Tontul

5.1 Introduction

Fruit waste is a significant concern both environmentally and economically, resulting from fruit production, distribution, and consumption. It includes discarded peels, cores, seeds, and unsold or spoiled fruits (Kumar and Manimegalai, 2004). This global issue has wide-ranging implications for sustainability, resource management, and food security. The production and disposal of fruit waste contribute to environmental degradation and greenhouse gas emissions. When fruit waste ends up in landfills, it decomposes and releases methane, a potent greenhouse gas that significantly contributes to climate change (Nordin et al., 2020). Moreover, the extensive use of land, water, and energy resources in fruit production becomes wasted when fruits are discarded instead of being utilized. From an economic perspective, fruit waste represents lost investments in agricultural practices, labor, and transportation costs. Additionally, managing and disposing of fruit waste incurs extra expenses for waste management systems and municipalities (Ishangulyyev et al., 2019).

Addressing fruit waste is crucial for several reasons. Firstly, reducing fruit waste can relieve pressure on landfills and minimize greenhouse gas emissions, thereby mitigating the environmental impact. Secondly, effective utilization of fruit waste allows for the recovery of valuable resources, such as organic matter for composting or energy generation. Finally, reducing fruit waste aligns with sustainable agriculture and resource conservation principles (Couth and Trois, 2010). In recent years, there has been a growing awareness and focus on finding innovative solutions to tackle fruit waste. Various initiatives, organizations, and businesses are working toward waste reduction, improved post-harvest handling, and the development of value-added products from fruit by-products. These efforts aim to minimize waste, create economic opportunities, and promote a more sustainable food system (Grech et al., 2020). In this chapter, we first identified fruit by-products and their valuable components. We then discussed the valorization steps for these by-products based on recent literature.

5.2 The Wastes of the Fruit Processing Industry

Fruits play a crucial role in our diet since they are good sources of energy, dietary fiber, and bioactive compounds. The demand for fruit and fruit-based products has increased significantly because of the increasing world population and changing dietary habits (Schieber et al., 2001; Vilariño et al., 2017). Fruits, such as berries, contain bioactive compounds that can affect physiological processes, reduce the risks of certain diseases, and improve overall health (Molet-Rodríguez et al., 2018). These bioactive compounds are mainly vitamins, minerals, carotenoids, polyphenols, pigments, fibers, and phytosterols (Ignat et al., 2011; Saini et al., 2021).

Fruits are consumed fresh or after processing, such as dried, juiced, or canned. As a result of fruit processing, 10–35% of the raw mass is discarded in the form of by-products, including pomace, seeds, peel, pulp, and oilseed meals (Majerska et al., 2019; Rao and Rathod, 2019). Such by-products are usually discarded as residues and used as feed, fertilizer, or compost (Dilucia et al., 2020). Therefore, this waste has economic significance for fruit processors and environmental impact. Thus, evaluating these wastes is important in reducing their environmental impact and encouraging producers to establishm a sustainable economy (Wadhwa and Bakshi, 2013). Moreover, these by-products contain significant amounts of pigments, phenolic compounds, vitamins, dietary fibers, sugar derivatives, organic acids, and mineral components (Sagar et al., 2018), which can be used in value-added products.

With an increasing demand for healthy and functional foods, the food industry has become more aware of the importance of natural food additives to obtain clean-label products. Therefore, it has become important to identify bioactive compounds to get value-added products from fruit waste/by-products (Jara-Palacios et al., 2015; Kandemir et al., 2022). Research is being conducted on using agricultural wastes as a cheap and sustainable source for producing valuable products. To extract and purify phytochemical compounds from by-products efficiently and at low costs, environmentally friendly advanced techniques need to be explored (Doria et al., 2021).

The fruit juice industry is one of the biggest agricultural industries in the world. Different fruits (such as citrus, apples, peaches, and grapes) used in producing fruit juices generate large amounts of waste (Sagar et al., 2018). As seen in Table 5.1, these wastes contain valuable bioactive components and have important health benefits, such as antioxidant, anti-cancer, and anti-inflammatory activity (Dilucia et al., 2020; Rodríguez García and Raghavan, 2021). This section presents the compositions of the main fruit wastes, their importance in the industry, and their usage areas.

5.2.1 Citrus Wastes

According to FAO data, total citrus fruit production is higher than 110 million tons, and citrus fruits arc the most traded fruit in the world. Citrus has a

Table 5.1 Percentages of Some Fruit By-Products Rich in Antioxidants and Therapeutic Properties

Fruit	By-products	(%)	Valuable compounds	Therapeutic effect	Reference
Apple	Peel, pomace, seed	25–40	Chlorogenic acid, caffeic acid, flavonoids catechins, epicatechins, oleic and linoleic acids, dietary fiber, pectin	Antioxidant, anti-inflammatory, anti-obesity, anti-cancer, antibacterial, antiviral	(Gumul et al., 2021; Sharma et al., 2021)
Banana	Peel	35	Rutin, ferulic acid, quercetin, caffeic acid, dietary fiber	Antioxidant, antimicrobial, anti-inflammatory, neuroprotective	(Vu et al., 2018, 2019; Sharma et al., 2021)
Apricot	Skin, pulp	40	Carotenoids, catechin, chlorogenic acid, cyanidin-3-galactoside, cyanidin-3-glucoside, oleic acid, linoleic acid, dietary fiber	Antioxidant, antimicrobial, anticarcinogenic	(Femenia et al., 1995; Soong and Barlow, 2004; Yiğit et al., 2009)
Citrus	Peel, seed, fibrous middle part of citrus	50	Hesperidin, naringin, neohesperidin, narirutin, carotenoids, dietary fiber, pectin	Antioxidant, anti-inflammatory, antimicrobial, anti-obesity	(Marín et al., 2007; Sharma et al., 2021)
Grape	Pomace, peels, seeds, stems	15–20	Resveratrol, flavonol glycosides, anthocyanins, procyanidins, lycopen, dietary fiber, stilbenes	Antioxidant, antimicrobial, anti-cancer, anti-inflammatory, antiaging, antiviral	(Deng et al., 2012; Gordillo et al., 2018; Schieber et al., 2001)
Mango	Peel, kernel, leaves	45	Gallotannins, gallates, gallic acid, ellagic acid, carotenoids, enzyme, vitamins (C and E), dietary fiber	Antioxidant, anti-cancer, anti-obesity, anti-inflammatory, antitumor, antibacterial, antiviral, immunomodulatory	(Ben-Othman et al., 2020; Sharma et al., 2021)
Pineapple	Skin, core	33	Ferulic acid, vanillic acid, dietary fiber, gallotannins, gallates, gallic acid, ellagic acid, myriceti, salicylic acid, tannic acid	Antioxidant, antimicrobial, anti-cancer, anti-diabetic, anti-inflammatory, immunomodulatory, cytotoxic, antihyperlipidemic	(Schieber et al., 2001; Priefert et al., 2001; Sagar et al., 2018; Rico et al., 2020; Kumar, 2021)
Pomegranate	Peel, pomace	40–49	Punicalagin, ellagic acid, gallic acid, caffeic acid, chlorogenic acid	Antioxidant, antimicrobial, anti-hypertension, anti-cancer, anti-inflammatory, antiatherosclerotic, heart disease	(Viladomiu et al., 2013; Andrade et al., 2019)
Blueberries	Pomace	20–30	Anthocyanins, cinnamic acid derivatives, flavonol-glycosides	Antioxidant, antimicrobial, anti-cancer, anti-inflammatory, neuroprotective	(Avram et al., 2017; Ma et al., 2018)
Watermelon	Peel, seed	25	Gallic acid, catechin, ellagic acid, kaempferol, cinnamic acid, ferulic acid, chlorogenic acid, rutin, lycopene, citrulline, arginin	Antioxidant, anti-apoptosis, antiaging, anti-carcinogen, anti-inflammatory, anti-atherosclerosis, anti-cardiovascular	(Rimando and Perkins-Veazie, 2005; Kolawole et al., 2017)

juice yield of only 50% of its weight; the rest is considered by-product (peel, pulp, seeds, and whole fruits that do not meet quality requirements). The increased production capacity of the food industry has led to the generation of significant by-products with high levels of bioactive components from the citrus industry.

Although there are many species, the most important citrus fruits cultivated are *Citrus lemon* (lemon), *C. aurantifolia* (lime), *C. aurantium* (bitter orange), *C. sinensis* (sweet orange), *C. reticulata* (mandarin, satsuma), *C. grandis* or *C. maxima* (pummelo) and *C. paradisi (*grapefruit) (Ledesma-Escobar and de Castro, 2014; Suri et al., 2021). The residue after producing citrus juice, called pomace, represents 50–70% of total fruit weight. Citrus waste contains 60–65% peel, 30–35% internal tissues, and about 10% seeds (Rezzadori et al., 2012). Industrial citrus waste is estimated to be over 40 million tons worldwide (Marín et al., 2007). Citrus waste has a wide range of applications, such as a source of dietary fiber, pectin, essential oils and flavonoids, a binder in food technology, and a medium for producing single-cell protein (Van Heerden et al., 2002).

Citrus waste is a rich source of biologically active compounds, including natural antioxidants such as phenolic acids and flavonoids (Sharma et al., 2017). Glycosylated flavones (luteolin, apigenin, and diosmin glucosides) and polymethoxylated flavones are two flavonoid groups in citrus fruits. In addition, the main flavonoids found in citrus species are hesperidin, narirutin, naringin, and eriocitrin (Schieber et al., 2001). The skins and seeds are an important source of phenolic compounds, including phenolic acids and flavonoids. The content is richer in the shells of seeds than in the seeds themselves. The composition of seeds and peel are not always the same for a given species. For example, lemon seeds contain mainly eriocitrin and hesperidin, while the peel is rich in neoeriocitrin, naringin, and neohesperidin (Bocco et al., 1998).

Citrus by-products are functional ingredients in developing healthy foods (functional foods) containing indigestible carbohydrates and bioactive compounds, such as dietary fibers, ascorbic acid, and flavonoids(Marín et al., 2007).

The traditional method of extracting essential oils from citrus peels is cold pressing. Cold pressing mechanically removes peel and cuticle oils to produce an aqueous emulsion, which is centrifuged to separate the essential oil. Essential oils derived from citrus waste are known for their potent antimicrobial, antioxidant, and anti-inflammatory properties. They have many potential applications, including food additives, spoilage inhibitors, pharmaceuticals, and cosmetics. Beverages, ice cream, and other foods can be flavored with these essential oils. Also, they are used in cosmetics and cleaning products, such as soaps, perfumes, and other home care products (Raeissi et al., 2008).

Citrus processing produces a large amount of waste in the form of press cakes composed of flavedo, albedo, seeds, and the remains of juice sachets. One innovative use of citrus waste is to produce anti-melanogenic extracts. This citrus by-product is a promising candidate for treating skin pigmentation disorders, according to an in vivo evaluation of the biological activity of citrus cake extracts (Costin and Hearing, 2007).

5.2.2 Dark-Colored Fruit Wastes

Grape pomace, seeds, and stalks are the by-products of juice and wine processing (Javier et al., 2018). Grape pomace represents approximately 20% of the total fruit weight and is a rich source of phenolic components (catechins, proanthocyanidins, and glycosylated flavonols) (Gülcü et al., 2019; Javier et al., 2018). The phenolic content of the grape stalk is approximately twice that of the pulp (Monagas et al., 2003). This variability in phenolic content has been reported to be due to the preservation of phenolic content as the bioactive compounds in the pomace are transferred to the juice while the stems are not processed (Javier et al., 2018). The total phenolic content of grape seeds (11.6 g GAE/100 g dw) was higher than that of grape pulp (2.36 g GAE/100 g dw) (Llobera and Cañellas, 2007).

By-products of strawberry processing (achene, stalk) are rich in phenolic compounds such as tannins, phenolic acids, and flavonoids (Misran et al., 2015; Cubero-Cardoso et al., 2020). Strawberry achenes make up 1% of the fruit weight. Still, they constitute 11% of the total phenolics and are responsible for 14% of antioxidant activity (Aaby et al., 2005).

Due to the limited fresh consumption of the phenolic-rich blackcurrant, it is processed into juices, jams, and alcoholic beverages, resulting in a by-product rich in bioactive compounds (anthocyanins and phenolics) (Nawirska and Kwaśniewska, 2005; Jurgoński et al., 2014). More than 50% of blackcurrants become pomace after processing (Sójka and Król, 2009; Plainfossé et al., 2020).

Bioactive compounds such as (+)-catechin, epicatechin, phenolic acids (vanillic, caftaric, and chlorogenic), rutin, naringenin, and anthocyanins have been shown to be the main components in berry pomaces (Gülcü et al., 2019; Vorobyova and Skiba, 2021). Blackberries are rich in phenolic acids (gallic acid and caffeic acid), flavonols (quercetin), and total anthocyanins (cyanidin-3-glucoside), and about 20% of by-products (peel, stem, seeds) are released during processing into products such as juice, wine, or jam (Rohm et al., 2015; Zafra-Rojas et al., 2020). Similarly, red berries (blueberries, raspberries, redcurrants, cranberries, currants), rich in phenolic compounds containing mainly anthocyanins, are processed into various products and a valuable by-product is obtained. Studies have shown that berry peels are rich in compounds such as anthocyanins, (+)-catechin, epicatechin, gallic acid, and caftaric acid (Gülcü et al., 2019; Bilawal et al., 2021), considering the bioactive components of berry seeds have been found to be rich in (+)-catechin, epicatechin, gallic acid, vanillic acid, and procyanidins (Gülcü et al., 2019; Bilawal et al., 2021).

5.2.3 Other Fruit Waste Sources Rich in Valuable Functional Compounds

Fruits such as apples, stone fruits, and tropical fruits used in fruit juice production generate waste containing significant amounts of valuable components (Sagar et al., 2018; Kandemir et al., 2022). Therefore, the assessment of these wastes is critical. A significant amount of produced apples are processed into apple juice concentrate (Bhushan et al., 2008). The production of apple juice, the second most consumed fruit juice after orange juice, results in 25–40% waste (pulp, peel, and seeds), depending on the method used (USDA/FAS, 2014).

Stone fruits, including apricots, nectarines, peaches, plums, and cherries, belong to the genus Prunus. The skin of peaches and apricots has a fine hairy structure, while other fruits have a smooth surface. These fruits are also classified according to whether they are endocarps or mesocarps (Walker et al., 2020). The by-products (e.g., skin, pulp, seeds) of these fruits are good sources of bioactive compounds (Maragò et al., 2015). By-products such as peels and seeds are generated during the production of plum juice (Dwivedi et al., 2014). Like other stone fruits, plums are a rich source of polyphenols (flavonols, phenolic acids, anthocyanins, etc.) (Milala et al., 2013). Similarly, sour cherries produce a large amount of by-products after processing. In the case of sour cherry juice, these by-products are the skin and pulp (which make up the pomace) and the seeds (Yılmaz et al., 2018).

Melon (also known as cantaloupe), watermelon, and honeydew are members of the Cucurbitaceae family. The juice production from these fruits results in by-products composed of seeds, peels, and pomace (Gómez-García et al., 2020). Peels and seeds are reported to provide approximately 50% of the phenolic content and antioxidant capacity of these fruits.

Tropical fruits (mango, pineapple, kiwi, and pomegranate) generate important by-products and are rich in phenolic compounds such as gallotannins, gallates, gallic acid, and ellagic acid (Arogba, 2000; Schieber et al., 2001; Kandemir et al., 2022). After processing the pineapple, the core and rind of the fruit are exposed as by-products (Freitas et al., 2015). Pomegranate by-products are also an important source of bioactive compounds (flavonoids, ellagitannins, and proanthocyanidin) and approximately 40–50% of the total fruit weight is separated as by-products as a result of processing (Andrade et al., 2019).

5.2.3.1 Fruit Pomace

Stone fruit pomace contains bioactive components such as β-carotene, neochlorogenic acid, and catechin (Rohm et al., 2015). Apple pomace is rich in phytochemicals and bioactive constituents (polyphenols, dietary fibers, triterpenoids, and volatile compounds) (Barreira et al., 2019). The main polyphenolic constituents of apple pomace are epicatechin, caffeic acid, phloridzin, and quercetin derivatives (Lu and Foo, 1997). It is considered a good source of polyphenols due to the phenolic content of the fruit peel. In response to the increasing demand for natural antioxidant compounds, apple pulp has been reported as a potential source (Ćetković et al., 2008; Lu and Yeap Foo, 2000). The skin and flesh of the peach contain considerable amounts of rutin, quercetin, and anthocyanin glycosides, chlorogenic acid, neochlorogenic acid, catechin, and epicatechin (Andreotti et al., 2008). The main bioactive constituents of plum pulp are anthocyanins, hydroxycinnamic acid, and quercetin glycosides (Milala et al., 2013). Sour cherry pomace is rich in phenolic compounds; the major phenolic constituents are neochlorogenic acid, catechin, and cyanidin-3-glucosyl-rutinoside (Yılmaz et al., 2015). In a study on microencapsulation of the phenolics of sour cherry pomace, the total phenolic content of this pomace powder was 91.29 mg GAE/g dw (Cilek et al., 2012). In another study on plum pomace extract, the total phenolic content, total anthocyanin contents, and antioxidant activity were reported as 453.27–493.84 mg/L, 35.08–38.20 mg/L, and 59.61–106.8 mM Trolox/mL, respectively (Demirdöven et al., 2015). The main kiwi juice by-product is pomace (skin, pulp, seeds, and calyx), which averages 20–40% of the fruit. The main phenolics found in kiwi pomace are catechin, chlorogenic acid, p-coumaric acid, protocatechuic acid, quinic acid, caffeic acid (and its derivatives), kaempferol derivatives, and low amounts of quercetin derivatives (Kheirkhah et al., 2019; Zhu et al., 2019). In one study, the total phenolic content of kiwifruit pulp, excluding its skin, was 421 mg GAE/g dw (Zhu et al., 2019).

5.2.3.2 Fruit Skin/Peel

Apple peel is the apple by-product richest in phenolics (anthocyanins, flavan-3-ols, dihydrochalcones, chlorogenic acid, procyanidin B2, epicatechin, caffeic acid, p-coumaric acid, ferulic acid) (Leontowicz et al., 2007; Lončarić et al., 2020) and protects the fruit from UV radiation due to its physiological properties (D'Abrosca et al., 2007). One study found that the phenolic content of apples with peel (290 mg/100 g) was higher than that of apples without peel (219.9 mg/100 g) (Eberhardt et al., 2000). The total phenolic content of the peel of Idared and Northern Spy apple cultivars was reported to be higher (sevenfold) compared to their pomace (Rupasinghe and Kean, 2011). During apricot juice production, up to 40% of the total weight (pulp and skin) are released (Kasapoğlu et al., 2021). These parts are rich in β-carotene, and the content in the peel is two to three times higher than the pulp (Ruiz et al., 2005). A similar situation has been reported for phenolic compounds. The phenolic content (catechin, chlorogenic acid, cyanidin-3-galactoside, cyanidin-3-glucoside) of the peel was found to be two to four times higher than that of the flesh (Ruiz et al., 2005; Fan et al., 2018). The catechin and chlorogenic acid content of the white-fleshed peach peel was determined to be 1342–4578 mg/kg. The total phenolic content of the peach peel was found to be about three times higher than the other parts (Montevecchi et al., 2012). Furthermore, one study (Muradoğlu and Küçük, 2018) found that the total phenolic content of peach peel was higher (two or three times) than that of flesh and whole extracts. It must be noted that Loizzo et al. (2015) observed the opposite result. Anthocyanins are also concentrated in the skin. The phenolic content of the skin of the plum was found to be about four times higher than that of its flesh (Cevallos-Casals et al., 2006). Melon and watermelon rinds contain gallic acid, catechin, ellagic acid, and kaempferol, as well as phenolic constituents such as cinnamic acid, ferulic acid, chlorogenic acid, and rutin (Saad et al., 2021). In a study comparing the phenolic content of different fruit peels (such as pomegranate, apple, orange, banana, mango, and pineapple), watermelon rind had the highest phenolic content, which was 335 mg catechin/100 g (Duda-Chodak and Tarko, 2007). The fact that mango peels contain important bioactive components (phytochemicals, polyphenols, carotenoids, enzymes, and vitamins [C and E]) has recently increased demand for them (Ajila et al., 2007). The content of gallotannin, which has antioxidant, anti-cancer, and antiproliferative properties, has been determined to be 4 mg/g dw in mango peel (Berardini et al., 2004). Sogi et al. (2013) showed that mango peel powder had total phenolic, ascorbic acid, and carotenoid contents of 2032–3185 mg/100 g, 68.49–84.74 mg/100 g, and 1880–4050 µg/100 g, respectively. It has been reported that the dietary fiber in the pineapple peel is rich in myricetin, salicylic acid, tannic acid, and trans-cinnamic acid components, thereby having strong antioxidant activity (Larrauri et al., 1997). Phenolic compounds in pineapple peel have been reported to be 2.01 mmol/100 g (Guo et al., 2003). The pomegranate peel, which makes up 50% of the pomegranate, is rich in ellagic acid derivatives, such as ellagitannins, punicalagin, and punicalin. Punicalagin, a phenolic compound unique to pomegranate, accounts for 85% of the tannins found in the pomegranate peel (Viladomiu et al., 2013; Andrade et al., 2019). Pomegranate peel has the highest antioxidant capacity compared to other parts of the fruit due to its high polyphenol content (Hajimahmoodi et al., 2008).

5.2.3.3 Fruit Seeds

Apple seeds are rich in oil, containing mainly oleic and linoleic acids (Yu et al., 2007). They are also reported as a rich source of polyphenols (chlorogenic acid, protocatechuic acid, coumaric acid, caffeic acid, and ferulic acid) (Lu and Foo, 1997; Xu et al., 2016). The antioxidant content of apple seeds was higher than that of the skin or flesh (Xu et al., 2016). They also accumulate cyanogenic glycosides, which are considered toxic because they release cyanide during digestion (Senica et al., 2019). The most abundant cyanogenic glycoside is amygdalin; its concentration was 1–4 mg/g (Bolarinwa et al., 2014). It has been reported that consuming a single apple seed has no adverse effects, and excessive consumption may cause poisoning. The high levels of amygdalin in apple seed extract may require its removal. The usage of amygdalin as a treatment for some diseases is unclear and more research is needed (Kandemir et al., 2022). The seeds of stone fruits are rich in polyphenolic compounds such as gallic acid, vanillic acid, benzoic acid, phloridzin, quercetin derivatives, catechin, and epicatechin (Savić et al., 2016). The antioxidant, antimicrobial, and anticarcinogenic effects of apricot seeds have made it possible to use them in many industries, such as food, medicine, pharmacy, and cosmetics (Yiğit et al., 2009). However, apricot seeds also contain amygdalin, and therefore their usage is limited in the food industry (Senica et al., 2017). Studies have shown that the antioxidant capacity and phenolic content of apricot seeds are higher than that of the fleshy part (Soong and Barlow, 2004). Apricot seed oil is rich in oleic and linoleic acids (Femenia et al., 1995). The main constituents of plum seeds are oil (about 30%), protein, lipids, and polyphenolic compounds (Savić et al., 2016). Cherry kernels have important bioactive components, including essential fatty acids, carotenoids, sterols, and tocopherols. Its oil content ranges from 17% to 36% and is rich in unsaturated oil acids (primarily oleic and linoleic acids), which have beneficial properties due to their antioxidant, antimicrobial, and anti-inflammatory effects (Yılmaz et al., 2015; Yılmaz, 2018; Kandemir et al., 2022). Phenolic compounds that are abundant in melon seeds have been identified as gallic acid, caffeic acid, rosmarinic acid, protocatechuic acid, quercetin-3-rutinoside, and ellagitannins (Mallek-Ayadi et al., 2018; Kandemir et al., 2022). Total phenolic compounds in melon seeds were determined to be 285 mg/100 g for cantaloupe, 304 mg/100 g for Maazoun cultivar, 29.93 mg/100 g for golden melon, and 57.2 mg/100 g for Galia melon. The total flavonoid content was found to be 162 µg/100 g for the Maazoun cultivar and 20.67 mg/100 g for golden melon (Duda-Chodak and Tarko, 2007; Ismail et al., 2010; Mallek-Ayadi et al., 2018). Furthermore, watermelon seeds were rich in phenolics, with approximately 969.3 mg catechin/100 g (Duda-Chodak and Tarko, 2007). When examined in terms of their oil content, it is similar to soybean and sunflower oils (Dubois et al., 2007; Foster et al., 2009). The mango seed, about 20% of the fruit, has been reported to contain phenolic compounds with high antioxidant capacity and tyrosinase inhibitory activity (Jahurul et al., 2015). The gallotannin content of the mango seeds was 15.5 mg/g dw (Berardini et al., 2004). The seeds were reported to contain higher amounts of phenolics than the peel (Luo et al., 2014; Ribeiro et al., 2008). Total phenolic, ascorbic acid, and carotenoid contents of mango seed powder were reported as 11228–20034 mg/100 g, 61.22–74.48 mg/100 g, and 370–790 µg/100 g, respectively (Sogi et al., 2013). The molecular size of the bromelain (protease) extract found in many parts of pineapple was found to be 26 kDa, while it was 23.8 kDa for seed extract (Arshad et al., 2014; Umesh Hebbar et al., 2008). Wang et al. (2020) identified the major phenolic compounds in kiwi seeds as caffeic acid, quercetin, catechin, p-hydroxybenzoic acid, and p-cumaric acid. In one study, inhibition concentration (IC50) according to DPPH radical activity of kiwi seeds ranged from 25.7 to 35.1 mg/mL (Deng et al., 2018; Guo et al., 2003). In another study investigating the oil composition of kiwi seeds, the main fatty acids were palmitic, oleic, and linoleic acids. Although the phenolic concentration and antioxidant capacity of the pomegranate seed is low compared to other parts of the fruit, it is known to be rich in punicic acid, a linolenic acid derivative (Guo et al., 2003; Viladomiu et al., 2013).

5.3 Valorization Approaches for Fruit Wastes

Fruit and vegetable wastes and by-products from processing industries mainly comprise peels, seeds, and pomace. As outlined in the previous section, these by-products are excellent sources of bioactive components such as polyphenols, antioxidants, and natural pigments, along with macro components such as protein, polysaccharides, and dietary fibers (Ayala-Zavala et al., 2010; Sharma et al., 2021). Increasing health awareness and changing dietary trends have increased the demand for these ingredients in fruits and encouraged their transformation into more value-added products. Evaluation of these by-products is, therefore, utterly important for food producers.

Unless these by-products are evaluated, they may cause environmental problems. For example, the decomposition of these by-products over time releases greenhouse gases, which is a reason for global warming. Therefore, producers must also be focused on reducing the environmental impact of fruit by-products (Vilariño et al., 2017; Cano-Lamadrid and Artés-Hernández, 2021).

The bioactive components of these by-products should be extracted and added to other products to increase their value. In addition, these wastes can be processed or used directly as a functional ingredient. Due to the insufficient consumption of fruit and vegetables by certain segments of the population, their by-products are important to enrich foods or turn them into supplementary food (Mateos-Aparicio and Matias, 2019). Furthermore, fruit by-products are also used as an alternative to reduce the production cost of edible films/coatings and to improve their antioxidant, antimicrobial, physical, and mechanical properties (Otoni et al., 2017; Dilucia et al., 2020).

Today, there is an increasing interest in using bioactive compounds extracted from fruit by-products as antimicrobials, antioxidants, colorants, emulsifiers, thickeners, and gelling agents. While orange peel essential oil is used as an antimicrobial in kashar cheese (Kavas and Kavas, 2016), mango seeds, coconut peel, pomegranate waste (Ayala-Zavala et al., 2010), and grape waste (Mattos et al., 2017) can also be used as antimicrobial components in different food formulations. Vitamin E extracted from grape seed can be used as an antioxidant to prevent lipid oxidation (dos Santos Freitas et al., 2008). Similarly, the presence of phenolic components in grape by-products also limited lipid oxidation (Mattos et al., 2017). Its efficiency to prevent lipid oxidation in oil-in-water (fish) emulsions was as prominent as the effectiveness of propyl gallate (Pazos et al., 2005). As an alternative to butylated hydroxytoluene (BHT E321), apple by-products also show antioxidant activity and prevent lipid peroxidation (Peschel et al., 2006). By-products off anthocyanin-rich fruits, such as cherries, strawberries, and grapes, can be evaluated for anthocyanin production (Ayala-Zavala et al., 2010), and mango peels as a source of carotenoids (Ayala-Zavala et al., 2010). Citrus peels and apple pulp are important sources of pectin (Burey et al., 2008). To this end, the addition of blueberry waste (Luchese et al., 2018), citrus peel and leaves (Kasaai and Moosavi, 2017), and mango (Adilah et al., 2018) and pomegranate peel (Mushtaq et al., 2018) to edible films increased the film's swelling index, tensile strength, water vapor, and oxygen barriers. Apricot kernel (Priyadarshi et al., 2018), grape seed (Sogut and Seydim, 2018), and pomegranate peel (Yuan et al., 2016) extracts used as antimicrobial agents in chitosan-containing films were found to be effective against *Escherichia coli, Bacillus subtilis, Listeria monocytogenes, Staphylococcus aureus, Pseudomonas aeruginosa*, and total aerobic bacteria.

5.4 Green Extraction and Recovery Techniques

The extraction process is the most important stage in obtaining bioactive components from fruit by-products (Khoddami et al., 2013). Extraction of bioactive compounds is usually accomplished by conventional extraction techniques. The plant matrix, temperature, time, pressure, solvent, and method affect the extraction efficiency and yield of bioactive compounds. For an efficient extraction, it is necessary to optimize the processing conditions, such as the solid-to-liquid ratio, the physicochemical characteristics of the solvent, particle size, as well as the extraction time and temperature (Rodríguez García and Raghavan, 2021). Conventional techniques usually require large amounts of organic solvents, high energy usage, and longer durations. Therefore, in recent years, researchers and the food industry focused on developing and optimizing suitable novel "green" extraction techniques to reduce energy needs and organic solvent usage (Table 5.2) (Azmir et al., 2013).

Novel extraction techniques provide higher efficiency, reduce energy consumption and solvent usage, increase yield, improve process control, and result in higher-quality extracts than conventional extraction (Mena-García et al., 2019). Ultrasound-assisted extraction, microwave-assisted extraction (MAE), supercritical fluid extraction (SFE), enzyme-assisted extraction (EAE), pulsed electric field extraction, pressurized liquid extraction, high-voltage electrical discharge, deep eutectic solvent-assisted extraction, and cold plasma-assisted extraction are some examples of these novel techniques (Picot-Allain et al., 2021). The green extraction methods use water, ionic liquids, ethanol, fatty acid/oils esters, and glycerol as solvents (Mohapatra and Kate, 2019). Additionally, they require less energy and time.

5.4.1 Basics of Novel Extraction Techniques

5.4.1.1 Ultrasound-Assisted Extraction (UAE)

The desirable extracts can be obtained from fruit by-products using ultrasound with improved productivity and reduced operating costs. In ultrasound-assisted

Table 5.2 Benefits and Drawbacks of Green Solvents Variety and Extraction Techniques

VARIETIES of GREEN SOLVENTS	Benefits	Drawbacks	Can be used to extract	Reference
Water	Cost ↓ Ecological impact ↓ No special equipment	Low solvency potential	Anthocyanin, betalains, phenolic acids, phenolic acid glycosides, saponins, terpenoids, aromas, pectins	(Chemat et al., 2012; Mena-García et al., 2019; Rodriguez Garcia and Raghavan, 2021)
Ethanol	Medium cost Ecological impact ↓ Reusable ↑	Polarity ↓ Emission ↑ Corrosive Flammable and explosive	Flavonoids, phenolic compounds, flavonoid glycosides, tannins, terpenoids, alkaloids, pigments, pectins	(Chemat et al., 2012; Sarris and Papanikolaou, 2016; Mena-García et al., 2019; Rodriguez Garcia and Raghavan, 2021)
Methanol	Medium cost Ecological impact ↓ Reusable ↑	Corrosive Flammable and explosive Toxic	Flavonoids, phenolic acids, anthocyanin, flavonoid glycosides, tannins, terpenoids, saponins	(Naczk and Shahidi, 2004)
Glycerol	Cost ↓ Ecological impact ↓ Odorless Biodegradable Chemical stability ↑ Sweet tasting Boiling ↑	Polarity ↓ Capital cost ↑ High boiling point	Polyphenols	(Chemat et al., 2012; Mena-García et al., 2019; Makris and Lalas, 2020; Rodriguez Garcia and Raghavan, 2021)
CO₂	Cost ↓ Risk ↓ Renewable ↑ Non-toxic Odorless Ecological impact ↓	Polar and non-polar ↓ Required pressure ↑ Equipment maintenance prices ↑	Phenolics	(Chemat et al., 2012; Jesus and Meireles, 2014; Mena-García et al., 2019; Rodriguez Garcia and Raghavan, 2021)
Supercritical water	Renewable ↑ No toxicity problems	Energy consumption ↑ Equipment oxidation problems		(Demirbas and Akdeniz, 2002; Jesus and Meireles, 2014; Carpentieri et al., 2021)

(Continued)

Table 5.2 (*Continued*)

VARIETIES of GREEN SOLVENTS

	Benefits	Drawbacks	Can be used to extract	Reference
Deep Eutectic Solvent (NADESs)	Solubilizing capacity ↑ Cost ↓ Biodegradability ↑ Liquid (even below 0°C) Volatility ↓ Toxicity ↓ Viscosity control Easy preparation Eco-friendly	Viscosity ↑ Not entirely handled toxicity issues	Phenolics, flavonoids, isoflavonoids, terpenoids, alkaloids, anthocyanins, aromas	(Radošević et al., 2016; Wu et al., 2018; Gullón et al., 2020; Dheyab et al., 2021)
Ionic liquids (ILs)	Insignificant vapor pressure Thermally stable (at temperatures >200°C) Excellent solubility (organic, inorganic, and organometallic materials)	Cost ↑ Viscosity ↑ Not entirely handled toxicity issues	Polyphenols, natural colorants, aromas	(Carpentieri et al., 2021; McDaniel and Yethiraj, 2019)
Supramolecular solvents (SUPRAs)	Bioactive components, alkaloids	Extracting amphiphilic components ↑	Not entirely handled extracting solutes from solid materials	(Ballesteros-Gómez et al., 2010; Carpentieri et al., 2021)
GREEN EXTRACTION TECHNIQUES *Ultrasound-assisted extraction (UAE)*	Energy and power consumption ↓ Extraction yield ↑ Extraction time ↓ Extraction temperature↓ Use of solvent ↓ Maintenance savings ↑ Equipment price ↓ Suitable for heat-sensitive ingredients	Free radicals ↑ at high power levels Solvent-based process Difficult to industrial scale-up Need for separation and purification steps Non-uniform ultrasound energy distribution Change of macromolecules ↑	Phenolics, carotenoids	(Azmir et al., 2013; Ameer et al., 2017; Carpentieri et al., 2021; Rodriguez Garcia and Raghavan, 2021)
Microwave-assisted extraction (MAE)	Extract quality ↑ Selectivity ↑ Extraction yield ↑ Extraction time ↓ Use of solvent ↓	Equipment cost ↑ Less operable than UAE Volatile components efficiency ↓ Use of organic solvent Non-selective extraction	Flavonoids, phenolics, carotenoids, essential oils, volatile oils, terpenes	(Chen and Chen, 2013; Sagar et al., 2018; Carpentieri et al., 2021; Rodriguez Garcia and Raghavan, 2021)

	Power usage ↓ Cost ↓ More operable than SFE Equipment dimensions ↓ Solvent-free process possible Economically viable compared to SFE Reproducibility ↑ Escalation in industry ↑	Unsuitable for non-polar components Need for purification step Not suitable for heat-sensitive materials		
Supercritical fluid extraction (SFE)	Mass transfer ↑ Extraction time↓ Use of solvent ↓ (nontoxic) Waste amount ↓ Solvent recovery No filtration needed Extraction in room temperature Suitable for volatile components Automated system Polarity adjustment possible Eco-friendly	Cost of system ↑ Required pressure Risk of loss of volatiles	Flavonoids, phenolics, anthocyanin, volatile compounds, carotenoids	(Sagar et al., 2018; Carpentieri et al., 2021; Rodríguez García and Raghavan, 2021)
Enzyme-assisted extraction (EAE)	Extraction yield ↑ Extraction of bound components Eco-friendly	Enzyme susceptibility Industrial scale-up difficult Enzyme cost ↑	Anthocyanin, carotenoids, essential oils, bounded phytochemicals	(Puri et al., 2012; Sagar et al., 2018; Rodríguez García and Raghavan, 2021)
Pulsed electric field extraction (PEFE)	Extraction time ↓ Extract effective ↑ Energy consumption ↓ Extraction yield ↑ Selectivity ↑ Scalability ↑ No energy-intensive drying required Eco-friendly Non-destruction No thermal impact Continuous process	Maintenance ↑ Equipment prices ↑ Depending on medium conductivity	Polyphenols, phytosterols	(Abbas et al., 2008; Pourzaki et al., 2013; Sagar et al., 2018; Carpentieri et al., 2021; Rodríguez García and Raghavan, 2021)

(Continued)

Table 5.2 (*Continued*)

VARIETIES of GREEN SOLVENTS		Benefits	Drawbacks	Can be used to extract	Reference
	Pressurized liquid extraction (PLE)	Suitable for solid materials for biomolecule isolation Better than SFE for polar compounds Extraction time ↓ Use of solvent ↓ Extraction yield ↑ Use of fully GRAS solvents (water or other solvents)	Equipment cost ↑ Not suitable for samples with very low target analytes (max. 10 g)	Phenolics, flavonoids, anthocynanin, carotenoids, essential oils	(Suchan et al., 2004; Sagar et al., 2018; Žlabur et al., 2018)
	Cold plasma-assisted extraction (CPAE)	Nontoxic Extraction temperature ↓ Pressure ↓ or atmospheric Synergistic impact ↑ (when used with other extraction methods) Cost of reaction chamber ↓ (for pressure and temperature) No chemical residue	Difficult to industrial scale-up Investment cost ↑ Bioactive compound penetration ↓	Phenolic compounds	(Gavahian and Cullen, 2019; Bao et al., 2020; Rashid et al., 2020)
	High hydrostatic pressure extraction	Mass transfer ↑ Suitable polar/non-polar for compounds No heating Energy consumption ↓ Extraction yield ↑	Equipment cost ↑ Maintenance ↑ Required pressure ↑	Polyphenols	(Shinwari and Rao, 2018; Rodríguez García and Raghavan, 2021)

extraction, high-power sound waves (20 kHz and higher) are used in the medium. As a result of compression and rarefaction cycles of sound waves, cavitation phenomena occur under ultrasonication. Cavitation is the process by which tiny gas bubbles expand, oscillate, and burst in a fluid under the influence of ultrasound waves (Vernes et al., 2019). The burst of bubbles is responsible for the formation of local high-temperature and high-pressure spots. Therefore, cavitation leads to better cell disruption of plant tissues and improved solvent penetration into the matrix (Chemat et al., 2017). It has been shown that extraction efficiency is higher at low frequencies (20–40 kHz). Moreover, higher cavitation occurs at low temperatures and in less viscous solvents (Tiwari, 2015). Thus, power, frequency, ultrasonic intensity, temperature, and solvent type are among the main parameters that affect ultrasound-assisted extraction. In addition, a combination of ultrasonication with other extraction techniques can increase extraction efficiency (Vernes et al., 2019).

5.4.1.2 Microwave-Assisted Extraction (MAE)

Microwave-assisted extraction is a novel and green extraction technique that provides dielectric heating, offers high reproducibility in a shorter time, reduces solvent consumption, and requires lower energy input. It provides extraction without decreasing the extraction yield, uses radiation to heat the solvent, and facilitates the location of the target compounds in the solvent (Boukroufa et al., 2015; Chuyen et al., 2018). Microwave (300 MHz–300 GHz) has been described as a promising alternative technology for conventional heating. The mechanism of microwave heating is based on the continuous reorientation of polar molecules, especially water, to face opposite poles when exposed to the microwave alternating electric field (Soquetta et al., 2018). The extraction starts as the solvent penetrates in the plant matrix. Electromagnetic waves then break up the components, and the soluble components pass from the insoluble matrix to the solution, resulting in liquid and residual solid phase separation. Extraction can also be carried out without solvent. The cell lysis can be facilitated by water in the sample acting as a solvent (Costa, 2016). Many bioactive compounds can be obtained through the MAE, such as phenolics, carotenoids, essential oils, and antioxidants (Soquetta et al., 2018; Elik et al., 2020). The extraction efficiency is affected by power, temperature, solid-solvent ratio, time, and solvent in MAE (Soquetta et al., 2018; Panzella et al., 2020).

5.4.1.3 Supercritical Fluid Extraction (SFE)

The solvent in supercritical fluid extraction gains supercritical fluid properties by bringing it to critical temperature and pressure. Thus, it exhibits a gas-liquid property with high density and solubility like liquids and low viscosity and zero surface tension like gases. Low temperature, short time, high efficiency, easiness, high selectivity, and not causing environmental pollution are some of the advantages of SFE. In addition, this technique provides an extract that is not subject to oxidation and is completely solvent-free (Uwineza and Waśkiewicz, 2020). Carbon dioxide (CO_2) is the most widely used supercritical fluid, and solvent efficiency increases with co-solvents such as ethanol, methanol, acetone, and hexane (Picot-Allain et al., 2021).

5.4.1.4 Enzyme-Assisted Extraction (EAE)

In enzyme-assisted extraction, enzymatic pretreatment is seen as an effective way to release bound compounds or increase yields in general. During extraction, adding specific enzymes, such as cellulase, α-amylase, and pectinase, ensures the degradation of the cell wall and increases the hydrolysis of polysaccharides and lipid components. EAE is affected by several factors, such as pH, temperature, time, and enzyme type (Gligor et al., 2019; Akyüz and Ersus, 2021). Notably, the optimal conditions for enzyme activity are achieved at moderate pH and temperature (Picot-Allain et al., 2021).

5.4.1.5 Pulsed Electric Field Extraction (PEFE)

The pulsed electric field is a technology that is based on the application of a short-term (1–10 microsecond) high-intensity electric field (15–80 kV/cm) and can be applied in a wide frequency range (Syed et al., 2017; Käferböck et al., 2020). It is a low-cost tissue disintegration technique that causes pore formation in cell membranes without causing a temperature increase (Fincan et al., 2021). Depending on the applied electric field intensity, structure and size of the cells, the PEFE method reversibly or irreversibly disrupts the cell membrane. This effect allows the extraction solvent to easily enter the cells, allowing selective uptake of bioactive compounds. Therefore, it is more effective than traditional extraction (Puértolas and Barba, 2016).

5.4.1.6 Pressurized Liquid Extraction (PLE)

Pressurized liquid extraction is also known as accelerated solvent extraction, enhanced solvent extraction, pressure hot solvent extraction, high-pressure solvent extraction, or subcritical solvent extraction. PLE is an automated extraction technique that combines high temperatures (20–200°C) and high pressures (up to 20 MPa) for rapid and efficient extraction of bioactive compounds (Jentzer et al., 2015; Alvarez-Rivera et al., 2020). Organic solvents are liquid when attached to supercritical regions at high pressure. Thus, as the surface tension and viscosity of the solvent decrease, the mass transfer rate and diffusion increase while the extraction time and solvent usage reduce. As a result of the denaturation of the cell membrane and proteins under high pressure, extracted target components are reported as more accessible (Cascaes Teles et al., 2021).

5.4.1.7 Cold Plasma-Assisted Extraction (CPAE)

Cold plasma technology is created by energizing a gas. While some molecules in the gas remain neutral, others transform into free electrons, radicals, ions, heat, and electromagnetic radiation by breaking down the covalent bonds. CPAE is a green and clean method that uses very low energy and leaves no toxic residue. It is a non-thermal technology applied at a low temperature (Segat et al., 2016; Bao et al., 2020). Since it can be operated at atmospheric or low pressures, it does not require an expensive reaction chamber for pressure and temperature regulation (Rashid et al., 2020).

5.4.1.8 Deep Eutectic Solvent-Assisted Extraction (DESAE)

Natural deep eutectic solvents show low or no toxicity, high biodegradability and biocompatibility, high solvency capacity, and environmental friendliness. Therefore, they are used as "green solvents" as an alternative to conventional solvents (Liu et al., 2018; Trusheva et al., 2019). Extracts obtained using natural deep eutectic solvents as green alternatives to ethanol and glycol are more suitable for direct consumption, while similar compositions are obtained with alcoholic extracts (Funari et al., 2019).

5.4.2 Extraction of Essential Oil from Citrus Peels

Citrus essential oils are highly valued for their distinctive aroma and a wide range of therapeutic properties. Derived from the peels of citrus fruits such as oranges, lemons, grapefruits, and limes, these oils have been used for centuries in various traditional medicines and aromatherapy practices. In recent years, scientific research has shed light on the numerous benefits and applications of citrus essential oils, making them an important subject of study. From their antimicrobial and anti-inflammatory effects to their antioxidant activity and mood-enhancing qualities, the scientific evidence supports their importance in various fields.

Therefore, the extraction of citrus essential oils from citrus peels has been a major subject for the food science and technology industry. Hydrodistillation, cold pressing, microwave-assisted extraction, ultrasound assisted extraction, supercritical CO_2 extraction, and other techniques are frequently used for extracting industrially significant value chemicals from citrus peels (Mahato et al., 2019).

Traditionally, cold pressing and steam distillation are used for the extraction of essential oils from citrus peels. Cold pressing physically removes the oils from the peel and albedo, resulting in an emulsion. The essential oils are then recovered through centrifugation (Ferhat et al., 2007). Steam distillation is a relatively straightforward technique for extracting oil components from peels. In comparison to cold pressing (yield of 0.05%), distillation provides a higher yield of 0.21%, making it a more cost-effective method for oil recovery

(Ferhat et al., 2007). On the other hand, high temperatures can cause oxidation of essential oil components and thus change the sensorial properties of the oil. Therefore, modern techniques, such as microwave-assisted extraction, ultrasound assisted extraction, and supercritical CO_2 extraction, are employed to increase the quality and decrease the time and environmental effects of the process (Mahato et al., 2019). Some studies on extraction of citrus essential oil using different techniques are outlined in Table 5.3.

5.4.3 Extraction of the Colorants from Fruit Varieties

Synthetic pigments are widely used in food applications due to their ease of application, high stability, low production costs, and high coloring strength (Sharma et al., 2021). However, in recent years, the negative effects of these pigments on human health, changing diet styles, and health concerns of consumers have increased the demand for natural pigments. Therefore, food producers and researchers have focused on studies on natural pigments to meet these demands (González-Montelongo et al., 2010; Rodriguez et al., 2016).

Fruits have a wide range of colors and are rich sources of pigment, which can be regarded as secondary metabolites. Fruit wastes and by-products (peels, seeds, and pomace) are an important source of natural pigments, such as anthocyanins, betalains, carotenoids, and chlorophylls. These pigments significantly contribute to the functional properties of foods with their rich therapeutic potential (Ayala-Zavala et al., 2010; Sharma et al., 2021).

The market size has increased due to the widespread use of natural pigments. Thus, renewable biological resources are needed for sustainable pigment production. For this reason, using green processing and extraction technologies to obtain natural pigments from food processing by-products is important (Melgar et al., 2019). Innovative green technologies such as supercritical fluid extraction (Andrade et al., 2019), ultrasound-assisted extraction (Milea et al., 2019; Saldaña et al., 2021), and microwave-assisted extraction (Pap et al., 2013; Cunha et al., 2018) have been studied in the extraction of pigments from fruit wastes and by-products. Green extraction methods use water, ethanol, ionic liquids, fruit-vegetable oils (such as cocoa, soybean, and rapeseed), fatty acid esters, and glycerol as solvents (Rajha et al., 2015).

Pigments can be categorized according to their source (natural/synthetic), solubility (water/oil soluble), structure and colors. Generally, pigments are examined under four groups: anthocyanins, carotenoids, betalains, and chlorophyll (Sharma et al., 2021).

Anthocyanins: Anthocyanins (more than 700 different compounds) are water-soluble and nontoxic pigments in glycosylated form. They are responsible for the red, blue, and purple colors of plants. The wide color range of anthocyanins arises from structural differences, such as anthocyanidin type, hydroxyl

Table 5.3 Extraction of Essential Oil from Citrus By-Product

Method of extraction	Citrus by-product	Extraction condition	Findings	Reference
Steam distillation	Orange, sweet lime, and lemon peels	88–98°C, 15–75 min, 1:1,6–1:2,4 solid to solvent ratio	The optimum conditions were found to be 96°C, 60 min and 1:2 solid to solvent ratio	(Sikdar and Baruah, 2017)
Hydrodistillation, cold pressing and supercritical CO_2 extraction	_Citrus medica_ L. cv. Diamante peels	Hydrodistillation: boiling 3 h Supercritical CO_2 extraction: 40°C, 100bar, 6h	Yield was much higher in supercritical CO_2 extraction, Hydrodistillation and cold pressing provides more components in essential oil, Hydrodistilled sample had high anticholinesterase activity while SFE extract had no activity	(Menichini et al., 2011)
Hydrodistillation, solvent extraction with pentane, and supercritical CO_2 extraction	Kumquat peels	Supercritical CO_2 extraction: 40°C, 80–250 bar, 4h Hydrodistillation: boiling 60 min Solvent extraction: refluxed 5 h in Soxhlet apparatus	All extraction techniques provide similar limonene, myrcene, and germacrene-D ratio but supercritical CO_2 extraction provided higher content in esters and sesquiterpenes	(Sicari and Poiana, 2017)
Cold pressing, hydrodistillation, and microwave dry distillation	Eureka lemon peels	Hydrodistillation: boiling 3 h Microwave dry distillation: 1000W, 30 min	Microwave dry distillation and hydrodistillation provided higher yield than cold pressing. Microwave dry distillation was advantageous in terms of environmental effect, antimicrobial activity, and essential oil composition	(Ferhat et al., 2007)

(Continued)

Table 5.3 (*Continued*)

Method of extraction	Citrus by-product	Extraction condition	Findings	Reference
Microwave steam distillation, steam distillation	Orange peel	Microwave steam distillation: 200–700W, 2–18 g/min	Microwave steam distillation provided shorter extraction time and better sensorial properties. Optimum conditions were 500 W and 6 min.	(Sahraoui et al., 2011)
Supercritical CO_2 extraction, cold pressing	Kabosu citrus peels	Supercritical CO_2 extraction: 20 MPa and 80°C	Supercritical CO_2 extraction yielded 13 times higher essential content than cold pressing	(Roy et al., 2011)
Hydrodistillation with enzymatic pretreatment	Orange, lime, grapefruit peels	Enzymatic treatment: Powercell© at 50°C for 3–12 h	Enzymatic treatment increased the essential oil yield two- to six-fold in orange and grapefruit samples	(Chávez-González et al., 2016)
Hydrodistillation with ultrasonic pretreatment	Orange peel	Ultrasonic treatment: 500 W, 20 kHz, 5–25 min	Yield was increased by 33% using ultrasounds	(Sandhu et al., 2021)

groups, and the position of the sugars. Blackcurrants, cherries, and berries (such as blueberries, elderberries, black goji berries, raspberries, blackberries, chokeberries, and strawberries) are some examples of anthocyanin-rich fruits (Andersen and Jordheim, 2014; Farooque et al., 2018). By-products of the fruit juice and wine industries especially contain considerable amounts of anthocyanins. In addition, red stone fruits, except for apricots, are sources of anthocyanins (especially proanthocyanidins). Therefore, their by-products can also be used in anthocyanin extraction (Kandemir et al., 2022). Blackberry by-products (4.31 mg Cy3GlE/g) and apple peel (1.69 mg Cy3GlE/g) are important sources of anthocyanins and nutraceuticals (da Fonseca Machado et al., 2018). In a previous study, applying pressurized liquid or accelerated solvent extraction to blackberry by-products in anthocyanin production increased the yield and decreased extraction time and solvent use (Machado et al., 2015). Ferreira et al. (2020) studied the extraction of anthocyanins from blueberry pomace using water and citric acid (1%) at 525 W for 3 min. Anthocyanin yield from grape by-products was found to be 11.21 mg/g with a high hydrostatic pressure extraction process (100–1000 MPa, 0°C–100°C), while it was determined as 7.93 with the conventional method (Corrales et al., 2008). Zhang et al. (2020) obtained an extract with 5.98 mg cyanidin-3-glucoside equivalent/g dw anthocyanin content from mulberry wine residues with a process carried out by ultrasonic-assisted enzymatic extraction at 52°C, for 94 min, 315 W, 0.22% enzyme.

Betalains: Betalains are also water-soluble pigments divided into betacyanins and betaxanthins. Betalains are present in a limited number of plants, such as beet (*Beta vulgaris*), prickly pear (*Opuntia*), *Hylocereus* and *Mammillaria* species (Ochoa-Velasco and Guerrero-Beltrán, 2016; Stintzing and Carle, 2007). De Mello et al. (2014) applied ultrasonic extraction with 80% acetone for 15 min to obtain betalain extract from pitaya or dragon fruit peel. They determined the betalain content of peels to be 101.4 mg/100 g. Similarly, in another study, microwave-assisted extraction was applied to obtain betalain from pitaya peels (Cunha et al., 2018).

Carotenoids: Carotenoids, another natural pigment, are lipophilic and give the red, yellow, or orange color to various fruits and vegetables (Rivera and Canela-Garayoa, 2012). They are divided into two classes: carotenes (α-carotene, β-carotene, and lycopene) and xanthophylls (lutein, zeaxanthin, and β-cryptoxanthin) (Bian et al., 2012). Carotenoids can be generally chemically synthesized. However, promoting green consumerism from fruit processing by-products has gained importance (Rodriguez-Amaya, 2015). Studies on fruit by-products as a source of carotenoids such as gac fruit peel and pulp (Chuyen et al., 2018), apricot waste (Koutsoukos et al., 2019), orange peel (Montero-Calderon et al., 2019), citrus fruit waste (Boukroufa et al., 2015), and pomegranate waste (Goula et al., 2017) were carried out.

Several studies have been carried out on the production of color components from the aforementioned fruit wastes by green extraction methods (Table 5.4). Goula et al. (2017) performed ultrasound-assisted extraction of carotenoids

Table 5.4 Obtaining the Color Components Using Novel Green Extraction Methods from Fruit By-Products

Target pigment	Fruit waste source	Green extraction method	Processing conditions	Yield	Benefits	Reference
Anthocyanin	Cranberry pomace	Pressurized ethanol extraction	120°C/50 bar, 5 mL/min, 100% ethanol solvent	8.42 mg cyanidin-3-glucoside dw	Extraction yield ↑, mass of solvent ↓, high-temperature extraction	(Saldaña et al., 2021)
	Grape pomace	Ultrasound-assisted extraction	55°C, 40% amplitude, 6 min	3.74 cyanidin-3-glucoside	Antioxidant capacity ↑, total anthocyanins ↑	(Bruno Romanini et al., 2021)
	Sweet cherry skin	Ultrasound-assisted extraction	70% ethanol, 40°C, 40 kHz, 100 W	14.48 mg cyanidin 3-glucoside/100 g dw	Antioxidant capacity ↑	(Milea et al., 2019)
	Blackcurrant marc	Microwave-assisted extraction	10 min, 700 W, 1:20 liquid-solid ratio, pH 2	20.4 mg/g	Extraction yield ↑, extraction time ↓, solvent usage↓	(Pap et al., 2013)
	Grape by-product	Pulsed electric field	100—300 V/cm to 20—80 kV/cm, room temperature	14.05 mg/g	Extraction yield ↑, extraction time ↓, energy consumption ↑	(Corrales et al., 2008)
	Raspberry pomace	Ultrasound-assisted extraction	1:5 liquid-solid ratio, 80% methanol and 1% formic acid, 120 min, 50°C, 50 kHz	7.13 mg cyanidin-3-glucoside/L	Antioxidant capacity ↑, total phenolics, flavonoids and anthocyanins ↑	(Krivokapić et al., 2021)
	Strawberry pomace	Ultrasound and homogenizer-assisted extraction	0.8 g: 25 mL (%70 acetone, %29.5 water, %0.5 acetic acid), 1 min (ultraturrax), 15 min (sonication)	20.2—47.4 mg/100 g fw	Extraction time ↓, extraction temperature ↓	(Buendía et al., 2010)
	Pomegranate peel and seed	Homogenizer-assisted extraction	1:30 liquid: solid ratio (lyophilized extract: de-ionized water), 1 h	1.8 mg/g (peel), 0.007 mg/g (seed)	Extraction time ↓	(Kupnik et al., 2021)

Carotenoid	Orange peel	Ultrasound-assisted extraction	40 kHz, 150 W 35 min, 42°C, 15 mL/g liquid-solid ratio, olive oil as solvent	1.85 mg/100 g	Extraction time ↓, liquid-to-solid ratio ↓, extraction temperature ↓	(Gajic et al., 2021)
	Mandarin epicarp	Ultrasound-assisted extraction	42 kHz, 240 W, 60°C, 60 min, a 0.0004 g/mL solid-liquid ratio, sunflower oil as solvent	140.7 mg/100 g	Effectiveness ↑	(Ordóñez-Santos et al., 2021)
	Apricot waste	Natural deep eutectic solvents, microwave-assisted extraction	80:20 (choline chloride:tartaric acid), 20 min	76.11 mg/100 g (β-Carotene)	Extraction yield ↑, extraction time ↓	(Koutsoukos et al., 2019)
	Citrus by-product	Supercritical CO_2 extraction	44.88°C, 25.196 MPa	1.91 mg/g		(Ndayishimiye and Chun, 2017)
	Rowanberry pomace	Supercritical extraction		512.4–1913.9 mg/100 g (total carotenoid), 282.4–976.8 mg/100 g (β-caroten)		(Bobinaitė et al., 2020)
Betalain	Dragon fruit peel	Microwave-assisted extraction	Distilled water, 8 min, 100 W, 35°C	9 mg/L	Extraction time ↓, extraction effectiveness ↑	(Thirugnanasambandham and Sivakumar, 2017)
	Prickly pear peel	Microwave-assisted extraction	Water, 5 g/L (solid:liquid), 1.4 min, 36.6 °C, 400 W	132.9 mg of betalains/g	Antioxidant capacity ↑, liquid-to-solid ratio ↓, extraction effectiveness ↑, extraction temperature ↓, extraction temperature ↓	(Melgar et al., 2019)
	Prickly pear peel	Ultrasound-assisted extraction	Methanol, 5 g/L (solid:liquid), 2.5 min, (34.6%), 30°C, 40 kHz	201.6 mg of betalains/g	Antioxidant capacity ↑, liquid-to-solid ratio ↓, extraction effectiveness ↑, extraction temperature ↑, extraction temperature ↓	(Melgar et al., 2019)

(Continued)

Table 5.4 *(Continued)*

Target pigment	Fruit waste source	Green extraction method	Processing conditions	Yield	Benefits	Reference
	Red prickly pear peel and pulp	Ultrasound-assisted extraction	Water, 3 g/30 mL (fresh sample/solvent), 10 min, 24 kHz, 400 W	89.29 mg betalain/100 g fw for peels; 28.25 mg betalain/100 g fw for pulps	Liquid-to-solid ratio ↓, extraction effectiveness ↑, extraction time↓, energy consumption ↓	(Koubaa et al., 2016)
	Red prickly pear peel and pulp	Pulse electric field assisted extraction	Water, 3 g/30 mL (fresh sample/solvent), 50 pulses, 20 kV/cm	81.3 mg betalain/100 g fw for peels; 34.25 mg betalain/100 g fw for pulps	Liquid-to-solid ratio ↓, extraction effectiveness ↑, extraction time↓, energy consumption ↓	(Koubaa et al., 2016)
	Pitaya fruits peels	Pressurized hot water assisted extraction	1:6 (solid/liquid ratio), 56.9°C, 6.7 MPa, 9 min	2.18 mg betanin equivalents/g dry extract	Extraction temperature ↓, extraction time↓	(Shen et al., 2019)

from pomegranate peels using vegetable oils. They obtained a carotenoid content of 0.613 and 0.671 mg carotenoids/100 g dry weight using sunflower and soybean oil, respectively. Carotenoid content (32.08 µg/g) of orange peel extracts obtained using supercritical ionic liquids ([BMIM][PF6], [BMIM][Cl], [HMIM][Cl], [BMIM][BF4]) was approximately four times higher than conventional extraction using acetone (7.88 µg/g) (Murador et al., 2019). Saini et al. (2021) studied the extraction of lutein from *Citrus reticulata* (kinnow) peel using UAE. They determined that the optimum values of the extraction parameter were 6.16 mL/g (solvent: solid), 43.14°C, 32.88% amplitude, and 33.71 min. Under these conditions, the lutein yield was 29.70 µg/g. Priyadarsani et al. (2021) optimized the supercritical CO_2 extraction process conditions for obtaining lycopene from ripe grapefruit (*Citrus paradise*) endocarp. The optimum conditions were 305 bar, 35 g/min CO_2, 135 min, and 70°C. These conditions allowed the extraction of 93% of lycopene in the raw material.

5.4.4 Extraction of Flavoring and Aroma Ingredients

Fruit and vegetable wastes are an important source of a variety of bioproducts. They can be used as a source of flavor and aroma. Increased consumer demand for natural, biocompatible, and safe sources has driven the market for flavors, fragrances, and aromas.

Vanillin, which is produced from vanillic acid, is the main component of vanilla flavor, which is the most important and most widely used flavor in the food, cosmetic, pharmaceutical, and detergent industries (Tilay et al., 2008). Vanilla is extracted from the fermented pods of the vanilla orchid (*Vanilla planifolia*). The increasing demand for natural flavors has led to research into the production of natural vanillin from natural raw materials by means of microbial biotransformation (Priefert et al., 2001). In addition to vanilla, other aromatic compounds can be produced from plant by-products.

Commercially, several aromatic compounds are obtained from rhamnose, rutin, or fruit by-products by chemical hydrolysis (Sagar et al., 2018). 1-rhamnose is the main component of cell wall pectins and is the raw material used in the production of the strawberry aroma "furanol" (2,5-dimethyl-4-hydroxy-3(2H)-furanone) (Haleva-Toledo et al., 1999). The pineapple aroma (ethyl butyrate) was made from apple pulp, and the coconut aroma was made from olive paste cake (Sagar et al., 2018).

5.4.5 Extraction of Functional Ingredients (Phenolics, Antioxidants, Pectin, Dietary Fiber, etc.) from Fruit Pit, Seed, Peel, and Pomace

The by-products (such as skin, pulp, and seeds) released during the processing of fruits can be an excellent source of carbohydrates, dietary fibers, proteins, tannins, flavonoids, vitamins (A and E), minerals, oils, and pigments (Ben-Othman et al., 2020; Samsuri et al., 2020). Fruit by-products contain bioactive compounds in varying concentrations depending on the type of fruit, the climate, the region, maturity, and extraction conditions (Trigo et al., 2020). The seeds are rich in polyphenols and bioactive lipids, and the husks are rich in dietary fiber (Banerjee et al., 2017; Ran et al., 2018).

Among different compounds in these by-products, phenolic compounds are a large group of secondary metabolites that are commonly found in fruits and play an important role in human health (namely, their antibacterial, antitumor, antiviral, and antimutagenic properties) and nutrition (Heim et al., 2002; Ignat et al., 2011).

Phenolics are water-soluble substances (more than 8000 different molecules) mainly synthesized by the shikimic acid, pentose phosphate, and phenylpropanoid pathways. They contain one or more hydroxyl groups with one or more aromatic rings in their basic structure (Balasundram et al., 2006). Structurally more complex phenolics with high molecular weight are often called polyphenols (Singh et al., 2016).

The skin, outer surface, and seeds of the fruits contain high amounts of phenolic compounds (Friedman, 1997). The main classes of phenols found in fruit by-products are flavonoids (subclasses: flavonols, flavanones, flavones, flavanonols, isoflavones, flavanols, and anthocyanidins), phenolic acids, and tannins (Babbar et al., 2014; Robbins, 2003). These phenolic compounds have high antioxidant activity, preventing the formation of free radicals (Borges et al., 2010). The phenolic compounds are divided into extractable (free) and non-extractable (bound) phenols according to their solubility. Bound phenolics often covalently bind to plant cell wall components, such as pectin, cellulose, and other food matrices, making their extraction typically very difficult (Rohilla and Mahanta, 2021; Singh et al., 2016).

Table 5.5 sums up the green extraction methods to obtain phenolic compounds, antioxidants, etc., from fruit pit, seed, and peel, including their process parameters, type of extraction device, and functional compounds determined. Rohilla and Mahanta (2021) optimized the UAE conditions of phenolic compounds from red and yellow tamarillo fruit pulp. The optimum conditions were 73% and 78% acetone concentration and 43% and 46% amplitude for red and yellow pulp, respectively. Additionally, the optimum extraction time was 12 min for maximizing the phenolic content, flavonoid content, and DPPH radical scavenging activity. Egüés et al. (2021) obtained the highest efficacy of phenolic compounds from apple pomace at 20 min, 90°C, and 50% amplitude in a similar study. The antioxidant activity of the seed, skin, and flesh of various apples was determined using ultrasound-assisted extraction. The apple seed showed higher antioxidant capacity than other parts (Xu et al., 2016). In another study, the determination of the total phenolic and flavonoid content of blackcurrant pomace was studied using the UAE method. Among the phenolic components, the chlorogenic acid content of the pulp was found to be high (517.02 µg/g) (Vorobyova and Skiba, 2021). Bao et al. (2020) used the

Table 5.5 Phenolic Compounds and Polyphenols Extraction from the Fruit By-Product Using Green Extraction Processes

Method of extraction	Fruit waste	Functional ingredients	Extraction condition	Reference
Enzyme-assisted extraction	Sweet cherries (*Prunus avium* L.) pomace	Non-extractable polyphenols	70°C, 10 pH, 18.4 min, Pectinase (2 ml of Pectinase/g of sample)	(Domínguez-Rodríguez et al., 2021)
Pressurized ethanol extraction	Cranberry pomace	Phenolic compounds	140°C/50 bar, 5 mL/min, 30% ethanol as solvent	(Saldaña et al., 2021)
Ultrasound-assisted extraction	Kinnow mandarin (*Citrus reticulata*) peel	Phenolic compounds	31% amplitude, 30:1 (v/w), 41°C, 15 min	(Kaur et al., 2021)
	Apricot pomace	Total phenolic and flavonoid contents	50°C, 90 min, 50% ethanol as solvent, 1:10 (w/v)	(Kasapoğlu et al., 2021)
	Peach waste	Total phenolic, flavonoid, and anthocyanin contents	23%, 120 s	(Plazzotta et al., 2020)
	Apple (Idared and Northern Spy) skin	Phenolic compounds	1:50, 100% methanol, 30 kHz, 30°C, 45 min	(Rupasinghe and Kean, 2011)
	Grape pomace	Phenolic compounds	55°C, 40% amplitude, 6 min	(Bruno Romanini et al., 2021)
	Blackcurrant pomace	Phenolic compounds	1:10, 27 kHz, 6 W/cm2, 2 h	(Vorobyova et al., 2021)
	Orange pomace	Phenolic compounds	1:10 (70% ethanol: su), 26.43 min, 350°C, 65.66%	(Ghasempour et al., 2019)
Supercritical fluid extraction	Grape peel	Radical scavenging activity	45–46°C, 160–165 kg/cm2 pressure, 6–7% ethanol	(Ghafoor et al., 2010)
Microwave-assisted extraction	Kiwi pomace	Polyphenols	1:15 (50% ethanol:water), 75°C, 15 min	(Carbone et al., 2020)
	Apple peel	Phenolic compounds	60W, 110°C, 60 min	(Casazza et al., 2015)
	Peach waste	Total phenolic, flavonoid, and anthocyanin contents	540 W, 50 s	(Plazzotta et al., 2020)
High-voltage electrical discharge	Grape stem	Phenolic compounds	pH 2.5, 4.0 ms, 50% ethanol	(Brianceau et al., 2016)
	Pomegranate peel	Phenolic compounds	9 kV, 100 Hz, 2 µs	(Xi et al., 2017)
Pulsed electric field	Orange peel	Naringin and hesperidin	Water, 5 kV/cm, 60 µs	(Redondo et al., 2018)

(Continued)

Table 5.5 *(Continued)*

Method of extraction	Fruit waste	Functional ingredients	Extraction condition	Reference
	Apple peel	Phenolic compounds	Pure water, 1200 V/cm	(Wang et al., 2020)
	Peach by-product	Biocompounds	Ethanol %70, 0.0014 kJ/kg, 16 μs	(Plazzotta et al., 2021)
Combined technologies				
Enzyme-assisted extraction + High hydrostatic pressure	Alicante Bouschet grape pomace (skin, seeds, and stalks)	Phenolic compounds	32°C, 100 rpm, 1:8 (w/v), pH 5 Polygalacturonase (0.25% p/v, 200 MPa/10 min), Carboxymethylcellulase (4% p/v, 50 MPa/10min), β-glycosidase (4% p/v, 50 MPa/5 min)	(Cascaes Teles et al., 2021)
Supercritical extraction +Subcritical water extraction	Mandarin (*Citrus unshia*) peel	Hesperidin- narirutin	300 bar, 175°C, 5 min, solid:liquid ratio 30 mL/g– 300 bar, 175°C, 10 min, solid:liquid ratio 20 mL/g	(Šafranko et al., 2021)
Microwave- and ultrasound-assisted extraction	Grape pomace	Phenolic compounds	Distilled water (2% citric acid), 1:3 1000 W, 10 min (MAE), 450 W, 15 min (UAE)	(Da Rocha and Noreña, 2020)
Natural deep eutectic solvent +Ultrasound-assisted extraction	Apricot pomace	Phenolic compounds	Choline chloride:lactic acid (molar ration 1:2 with 30% of water v/v), solid:liquid ratio 1:12.5, 65°C, 50 min, 27 kHz, 2 h	(Vorobyova and Skiba, 2021)
Deep eutectic solvents +Microwave-assisted extraction	Mango peel	Polyphenols	59.82 mL/g liquid:solid ratio (lactic acid/sodium acetate), 436.45 W, 19.66 min	(Pal and Jadeja, 2020)
Natural deep eutectic solvent + High-voltage electrical discharge	Grape peel	Polyphenols	Lactic acid:glucose, 40 kV, 10 kA, 0.5Hz	(El Kantar et al., 2019)

high-voltage atmospheric cold plasma method to enhance the extraction of phenolic compounds, particularly anthocyanins, from grape marc. As a result of this application, an increase in the yield of phenolic compounds, antioxidant capacity, and diversity of phenolic compounds was observed with the degradation of grape pomace cell walls.

Today, there is a growing interest in the recovery and evaluation of dietary fibers (non-starch polysaccharides and lignin) from food by-products. Fruit by-products (such as pulp and peel) contain mainly pectin, cellulose, hemicellulose, lignin, and gums (Banerjee et al., 2017; Garcia-Amezquita et al., 2018). The water-holding and intestinal regulating properties of cellulose, hemicellulose, and lignin and the cholesterol-lowering and glucose-regulating effects of pectin and gums have increased the interest in dietary fiber (Maurya et al., 2015). Fruit by-products contain a more balanced ratio of soluble and insoluble dietary fiber than cereals.

The dietary fiber content of apples, apricot, grape, passion fruit, and orange by-products was reported to be 75.8, 72.3, 67.2, 64.2, and 58.2 g/100 dw, respectively (Ayar et al., 2018; Casarotti et al., 2018). Apple peel and pomace, the by-product of apple juice, contain significant amounts of dietary fiber. However, it was reported that the pomace (0.91% fw) contains more dietary fiber than the peel (Gorinstein et al., 2001). The total dietary fiber content of apple pomace was determined as 442–495 g/kg and 480 g/kg, respectively, as a result of vacuum and freeze-drying (Yan and Kerr, 2013). It has been reported that the pomace among the grape by-products has the highest dietary fiber content (González-Montelongo et al., 2010). The total dietary fiber content of red grape pomace (77.20% dm) (Llobera and Cañellas, 2007) and white grape pomace (71.56% dm) was found to be similar (Llobera and Cañellas, 2008). On the other hand, Deng et al. (2011) reported that dried red grape pomace (51–56%) had higher dietary fiber content than white grape pomace (17–28%). Mango by-products are also rich in dietary fiber. It has been reported that mango peels contain total dietary fiber of 40.6–72.5% (Ajila et al., 2007, 2008; Ajila and Prasada Rao, 2013). Crude fiber has been reported to be between 1.75% and 3.96% in mango seeds (Dhingra and Kapoor, 1985; Kittiphoom, 2012). The total dietary fiber contents of orange (Liucheng) and lemon peels were reported to be 57% (9.41% soluble) and 53.02% (18.48% soluble), respectively (Gorinstein et al., 2001; Chau and Huang, 2003). However, the dietary fiber content of the lemon pulp was higher than that of the peel (77.93%, of which 27.91% was soluble) (Russo et al., 2021). The results of the most recently preferred innovative extraction methods for the extraction of dietary fibers from the aforementioned fruit by-products are summarized in Table 5.6.

In a study conducted by Dominiak et al. (2014), enzymes such as cellulase, Laminex C2K, Validase TRL, Multifect B, GC220, and GC880 were utilized to extract pectin from lime peel through enzymatic processes. The most effective enzyme was found to be Laminex C2K, which exhibited optimal performance

Table 5.6 Pectin and Dietary Fiber Extraction from Fruit By-Product Using Green Extraction Processes

Method of extraction	Fruit waste	Functional ingredients	Extraction condition	Reference
Enzymatic-assisted extraction	Yellow passion fruit peel	Soluble dietary fibers (GalpA and high methyl-esterified homogalacturonan)	Defatted with hexane (1:7 solid/liquid); suspended with water (1:20 solid/liquid); 90°C, α-Amylase (5 units/g of sample), 3 h; 60°C, amyloglucosidase (1.3 units/g of sample), 1 h	(Abboud et al., 2019)
	Apple pomace	Pectin	1:20 (apple pomace:water), 20°C, 20 min (prior to extraction), Celluclast (18.95%) and Viscoferm (17.86%), 40°C, 3 h, pH 2	(Wikiera et al., 2015)
	Kiwi pomace	Pectin	Celluclast (1.05 mL/kg), 25°C, 30 min	(Yuliarti et al., 2015)
High-pressure assisted extraction	Mango peel	Total soluble dietary fiber	600 MPa, 10 min, 55°C	(Tejada-Ortigoza et al., 2017)
Microwave-assisted extraction	Orange peel	Pectin	1:16.9 (g/ml), 422 W, 169 s, pH 1.4	(Prakash Maran et al., 2013)
Moderate electric field assisted extraction	Yellow passion fruit peel	Galacturonic acid	1:30 (solid:solvent), 30 min, 50°C, pH 2, 60 Hz, 50 V	(De Oliveira et al., 2015)
Ultrasound-assisted extraction	Grapefruit peel	Pectin	12.56 W/cm², 66.71°C, 27.95 min	(Wang et al., 2015)

114   Wealth out of Food Processing Waste

at 50°C, pH 3.5, and a duration of four hours. The extracted pectin underwent enzymatic de-esterification, leading to improved stability against calcium sensitivity and high-acidity milk beverages. Additionally, the study concluded that the obtained pectin had the ability to form a gel under low-pH conditions in a sugar solution.

In another study by Moorthy et al. (2015), ultrasound-assisted extraction was employed to obtain pectin from pomegranate peel. The process conditions that resulted in the highest pectin yield were determined to be 62°C, 29 minutes, and a solid-liquid ratio of 1:18 g/ml, utilizing an ultrasound frequency of 20 kHz, power of 130 W, and a flat-tip probe with a diameter of 2 cm.

Li et al. (2013) compared ultrasound- and microwave-assisted extraction methods for obtaining soluble dietary fiber from apple pomace. The findings revealed that the ultrasound method yielded a higher amount of dietary fiber compared to the microwave-assisted extraction technique.

The techno-functional properties of by-products, such as swelling, gelling, water/oil retention, and thickening, have mainly been attributed to dietary fibers, together with the contribution of proteins and other (non-nutritive) components. Compounds exhibiting these properties may provide an opportunity and solution for low-cost and value-added bakery product formulation (Mateos-Aparicio and Matias, 2019).

5.5 Efficient Utilization of Pomace for Effective Fortification

While processing fruit juices, purees, or concentrates, huge volumes of pomace waste are released. It has been known for a long time that pomace has been used as animal feed, fertilizer, compost, and beverage (alcoholic/non-alcoholic) components (Majerska et al., 2019). Although pomace has a high potential for use in food, pharmaceuticals, and medicine, its use has not yet become widespread. It has been determined that fruit pomace is a good source of hemicellulose, cellulose, and lignin and is rich in polyphenols. This situation has increased the interest of food researchers and the food industry that pomace can be transferred into new food products or that its bioactive components can be obtained (Banerjee et al., 2017; Majerska et al., 2019). It has been reported that the physical and chemical properties of food products enriched with pomace's bioactive components may also improve. Along with the developing technology, some new methods have been developed to evaluate pomace (Majerska et al., 2019). Some of these technologies are outlined in Table 5.7.

Aronia, blueberry, and elderberry pomace powders are reported to be good, stable, and inexpensive natural food colorants that can be used in different foods (Nemetz et al., 2021). In another study, grape pomace was used as a substitute for flour, sugar, and fat in cake (Quiles et al., 2018). The final product obtained by adding apple and grape pomace to various foods, such as bakery

Table 5.7 Efficient Use of Various Fruit Pomaces/Pulps in Food Products

Product group	Fruit material	Added amount/ratio	Results	Reference
Cereal products				
Extruded products	Apple pomace	30%	— Increase in structure, antioxidant capacity, functional properties in extruded foods — Gluten-free	(Lohani and Muthukumarappan, 2017)
Bread	Grape pomace (powder)	5% and 10%	— Increase in fiber and polyphenol contents	(Yu and Smith, 2015)
Muffin	Grape pomace (powder)	5%, 10%, and 15%	— Decrease in volume index, springiness, lightness, aroma preference — Increase in firmness and color	(Walker et al., 2014)
Sponge cake	Mango pulp (powder)	5%, 10%, 20%, and 30%	— Increase in dietary fiber content, firmness, gumminess, chewiness, springiness, redness, and chroma (crumb) — Decrease in fat content, caloric value and glycemic index, brightness, redness, yellowness, hue angle, and chroma (crust)	(Noor Aziah et al., 2011)
Cookies	Blackcurrant pomace (powder)	20%	— No significant change in physical properties — Increase in product's hardness, redness, and yellowness values — Decrease in brightness value — No significant effect on sensory properties	(Tańska et al., 2016)
	Orange pulp (powder)	5%, 15%, and 25%	— Increase in technological quality and level of acceptance — Decrease in caloric value	(Larrea et al., 2005)

Biscuits	Grape pomace (powder)	10%, 20%, and 30%	— Increase in dietary fiber and phenolic content and antioxidant capacity — Decrease in water absorption of dough and hardness — Deterioration in brightness and yellowness values	(Mildner-Szkudlarz et al., 2013)
	Apple pomace (powder)	10%, 20%, and 30%	— Decrease in glycemic index — Increase in technological quality and acceptance	(Alongi et al., 2019)
Rice cracker	Apple pomace (powder)	3,% 6%, and 9%	— Increase in dietary fiber and mineral contents	(Mir et al., 2017)
Pasta	Grape pomace (Extract)	30%	— Increase in phenolic content	(Marinelli et al., 2015)
Dairy products				
Fermented skim milk	Winery pomace (Pinot Noir) (extract)	3.6 mL (to increase the total phenolic to 80 mg/L)	— High lactic acid bacteria amount — Extended shelf life — Functional food	(de Souza de Azevedo et al., 2018)
Yoghurt	Apple pomace (powder)	0.1%, 0.5%, and 1%	— Increase in gelatinization pH — Shortening in fermentation time	(Wang et al., 2019)
	Grape pomace (fiber)	1%, 2%, and 3%	— Improved product quality — Enriched polyphenol content — Extended shelf life	(Tseng and Zhao, 2013)
Salad dressing	Grape pomace (fiber)	0.5%, 1%, and 2%	— Improving product quality — Enriched polyphenol content — Extended shelf life	(Tseng and Zhao, 2013)
Meat products				
Chicken patty	Plum pomace (extract)	1.6%	— Preventing lipid oxidation — Extended shelf life — Increase in sensory properties of the product	(Basanta et al., 2018)

(Continued)

Table 5.7 *(Continued)*

Product group	Fruit material	Added amount/ratio	Results	Reference
Beef frankfurter	Grape seed flour	0.5%, 1%, 2%, 3%, 4%, and 5%	— Decrease in oxidation range — Increase in the content of protein and total dietary fiber as well as water-holding capacity	(Özvural and Vural, 2011)
Functional foods				
Edible film	Grape pomace	20%	— Increased phenolic content and antioxidant activity — Alternative of synthetic edible film — Carrier of valuable components	(Ferreira et al., 2014)
Natural colorant	Grape pomace		— Industrial production — Increase in purity of anthocyanin	(Zhao et al., 2020)
Dietary supplement	Jabuticaba pomace		— Functional food and additive — Dietary fiber and phenolic source	(Gurak et al., 2014)
Cellulose (bacterial)	Citrus fruit pomace		— Low cost and eco-friendly production	(Fan et al., 2016)
Pectin	Apple pomace		— Functional-dietetic additive	(Rana et al., 2015)
Polyphenol preparations (high concentration)	Plum pomace (extract)		— Increase in bacteriostatic activity — Functional additive	(Sójka et al., 2015)
Nutrient carrier				
Probiotic carriers for fermented/ non-fermented beverages	Passion pulp-pomace	20%	— Extended shelf life	(Santos et al., 2017)

products, fruit juice, yogurt, jam, and meat, is rich in fiber and phenolic compounds, is naturally balanced and textured, has a long shelf life, and is less susceptible to lipid oxidation (for meat products) (Bordiga et al., 2019; Lyu et al., 2020). Tumbas Šaponjac et al. (2016) added bioactive compounds obtained from cherry pulp to whey and soy protein to replace flour in cookies. While no loss of the polyphenol content was observed in cookies stored for four months, a slight decrease in antioxidant activity and a large decrease in anthocyanin content were observed. Moreover, the cookies had an acceptable sensorial property (Tumbas Šaponjac et al., 2016). The addition of 20% elderberry, rosehip, and rowanberry pomace to the shortbread cookies did not have a significant effect on the physical properties of the products. While the change of textural properties of the cookies with elderberry pomace was insignificant, the addition of rosehip and rowanberry pomace increased the hardness of the cookies. Furthermore, adding pomace to the samples reduced their brightness and increased their redness (Tańska et al., 2016). Adding grape pomace powder to breadsticks negatively influenced the textural properties, but the sensory properties were acceptable (Rainero et al., 2022). Bender et al. (2016) added 5%, 7.5%, and 10% Tannat grape pomace skin to muffins. The acceptable ratio was found to be 10%. While the extensibility of the obtained dough and the elastic resistance-extensibility balance of the flour dough increased, the deformation energy and the dough resistance to the deformation (tenacity) decreased. With regard to the textural properties of the final product, hardness and chewiness increased while cohesiveness and resilience decreased. In addition, there was an increase in crumb redness and a decrease in crumb brightness and yellowness. Similarly, Bchir et al. (2014) found that the water absorption, stability, and tenacity of dough prepared with apple, pear, and date pomace increased while extensibility, softening, breakdown, and setback decreased. In addition, the specific volume and crumb-crust texture of bread with added dietary fiber did not show a significant difference compared to the control sample. On the other hand, the inner crust color showed a small difference. Among these fibers, apple gave the best viscoelastic structure to the dough.

The chemical changes that may occur due to the high water content and low pH of the pomace may cause the degradation of the bioactive components in its content and reduce its quality. In addition, not processing the pomace immediately and improper storage conditions can easily lead to microbiological contamination and serious adverse effects on human health. The suitability of storage or drying conditions is important in preserving the quality of fresh pomace and prolonging its durability. In addition, the safety of raw materials is also significant in terms of pomace quality, as they may contain fertilizers or pesticides (Reißner et al., 2019). It is, therefore, essential to take the necessary precautions to maintain the quality of the pomace and protect its bioactive components.

Food businesses that invest in pomace processing to produce value-added products will increase their profitability. At the same time, choosing the right pomace processing methods will help minimize the loss of pomace's bioactive components and maintain final product quality (Majerska et al., 2019).

5.6 Production of Enzymes from By-Products of Fruit Processing Industries

In several studies, amylases, cellulases, invertases, pectinases, and other enzymes are purified from different fruit wastes.

Pectinases: Pectinases are the general term for a group of enzymes that catalyze the degradation, depolymerizing, or de-esterifying pectins. The main role of pectinases is to break down the pectic substances found in the cell walls and the intermediate lamellae of plants. They are classified into various groups and subgroups (Yadav et al., 2009), such as protopectinases, pectinesterases, and depolymerises, based on the type of pectic substances they use as substrates and the mechanisms. Pectinase enzymes are obtained from cashew, banana, and pineapple wastes (Venkatesh et al., 2009), grape pomace (Botella et al., 2005) and lemon peel, orange peel, banana peel, wheat bran, rice bran, and sugar cane pulp (Mrudula and Anitharaj, 2011) in different studies.

Cellulases: Cellulases consist of exo-1,4-beta-glucanase, endo-1,4-beta-d-glucanase, and beta-d-glucosidase. They are used in the food industry to recover valuable compounds from plants and release flavoring chemicals (Meyer et al., 1998). In previous studies, the cellulase enzyme is obtained from potato peels (Dos Santos et al., 2012), banana waste (Dabhi et al., 2014), and palm kernel (Tengku Norsalwani and Nik Norulaini, 2012).

Invertase: The hydrolysis of sucrose to glucose and fructose is catalyzed by the invertase. Invertase is particularly useful in producing sweets, jam, confectionery, and pharmaceuticals. The invertase enzyme is preferred because its product is sweeter and hard to crystallize compared to sucrose (Aranda et al., 2006). Uma et al. (2010) produced invertase by *Aspergillus flavus* using fruit peel waste as substrate. They used orange, pineapple, and pomegranate wastes in their studies.

Amylases: This group consists of three enzymes: α-amylase, β-amylase, and glucoamylase. Solid-state fermentation is often used to produce amylase because environmental factors such as temperature and pH can easily be controlled (Gangadharan et al., 2008). Many fruit wastes, such as citrus waste, date waste, banana waste, potato skins, loquat seeds, and mango seeds, are used to produce amylase (Sagar et al., 2018).

Other enzymes: In addition to the aforementioned enzymes, proteases, xylanases, tannases, and laccases are other enzymes that can be obtained from fruit waste. Solid-state fermentation is used to produce these enzymes (Kumar, 2020). Proteases, also known as proteolytic enzymes or peptidases, play a fundamental role in numerous biological processes by catalyzing the hydrolysis of peptide bonds in proteins. Proteases are ubiquitous enzymes found in all living organisms. They are involved in critical physiological processes such as digestion, protein transformation, blood clotting, immune response, and cell signaling. By selectively cleaving peptide bonds, proteases regulate protein structure, function, and degradation, thereby exerting control over cellular processes (Jamrath et al., 2012). Xylanases are versatile enzymes with significant industrial applications. Their ability to hydrolyze xylan into valuable products has made them valuable in various biotechnological processes, ranging from biofuel production to food processing. The ongoing research and development efforts in this field continue to unravel new xylanases and improve their properties, further enhancing their potential in sustainable and eco-friendly bioprocessing (da Silva et al., 2019). Tannase is an enzyme that catalyzes the hydrolysis of tannins, which are complex polyphenolic compounds found in various plants. Tannins contribute to the astringency and bitterness of certain foods and beverages. Tannase finds applications in the food, pharmaceutical, and beverage industries, where it is used to modify the sensorial properties of products (Dhiman et al., 2022). Laccases are enzymes belonging to the class of oxidoreductases. They are widely distributed in nature and play a significant role in the breakdown and modification of various organic compounds. Laccases are capable of oxidizing a wide range of substrates, including phenolic compounds, aromatic amines, and even non-phenolic compounds (Selvam et al., 2022).

5.7 Circular Bio-Economy and Techno-Economic Prospects

Evaluating suitable by-products and waste is important for the development and sustainability of the food industry (Mateos-Aparicio and Matias, 2019). Fruit loss does not only cause wastage of food products, it also indirectly involves the wastage of critical resources, such as soil, water, fertilizer, chemicals, energy, and labor. The utilization of these wastes is therefore critical from an environmental, economic, and social point of view. These large amounts of loss and waste also cause huge environmental problems as they decompose in landfills and release harmful greenhouse gases (Vilariño et al., 2017). The amount of by-products that have the potential to be used after processing is estimated to be millions of tons per year. The economic development of the food industry can be promoted by using fruit wastes and by-products as raw materials for new food production or by obtaining food additives from these products. It can also contribute to eliminating nutritional problems (Mateos-Aparicio and Matias, 2019). Therefore, the food industry and researchers make great efforts to evaluate and reuse these valuable resources because of their low prices, high available quantities, and high content of bioactive components (Dilucia et al., 2020).

According to one study, food loss and waste are between 194 and 389 kg/person/year worldwide and 158–298 kg/person/year in Europe (Corrado and Sala, 2018). It has been observed that the amount of waste per capita increases with income level (Xue et al., 2017). Bread and bakery products and fruits and vegetables account for the largest proportion of food losses and waste in the retail sector (Katajajuuri et al., 2014). According to a 2020 report, fruit and vegetable wastage in Denmark was 11%, which may correspond to an average of €425 per household/year (Teuber and Jensen, 2020). In this sense, the management of food waste is a very important issue for global food security. The European Commission has started to work on the reduction of fruit waste in order to promote the transition from a linear to a circular economy (Campos et al., 2020). Preventing food waste, consuming food for as long as possible, and reusing it as a raw material by evaluating it at the end of its shelf life are included in the circular economy (Wilts et al., 2020). The combined use of a circular economy and good chemistry not only preserves the value of the food, but also adds value to its by-products and can reduce waste by keeping them in the industrial cycle. As a result, such new methods contribute to the growth of global competitiveness and sustainability by creating new jobs and environments (Campos et al., 2020).

The valorization of fruit by-products not only solves environmental problems, it also enables the creation of multi-million-dollar businesses or the development of closed-loop systems for the development of new food ingredients, food additives, and value-added food products, supporting the circular economy (Campos et al., 2020).

5.8 Legislation and Regulations

Fruit wastes exhibit high bioinstability, auto-oxidation, and enzymatic (pathogenic) properties due to their composition and high water content. For this reason, assessing these wastes must depend on specific legislation and regulations (Anal, 2017; Schanes et al., 2018). The aim of the European Union's Waste Directive that all waste, including biodegradable food, should not be transported to landfills has been one of the key factors affecting waste assessment policy (European Commission, 2014). The use of these wastes/by-products as a food component or additive (natural) is done in the EU by considering the European Community Regulation 178/2002 (Article 2) and Codex Alimentarius Directives. Some additives approved by the European Commission, such as anthocyanins (E136, grape by-products) and chlorophylls (E140, green leaves, mango peels), already originate from fruit by-products (Ajila et al., 2007; Faustino et al., 2019). Natural additives derived from such by-products must fulfill the requirements of the current directive (EC Regulation No. 258/97). Novel foods must be assessed for safety before being placed on the market (Mateos-Aparicio and Matias, 2019).

5.9 Future Trends

Valuable components derived from fruit by-products can enhance the antioxidant capacity, fiber content, vitamins, minerals, flavor, and rheological properties of food products when added to them (Majerska et al., 2019). By-products rich in color pigments can serve as a future source of natural colorants, enhancing the visual appeal of food products. They are considered safe alternatives to artificial colorants (Sharma et al., 2021). The use of encapsulation methods to stabilize and prolong the shelf life of these components shows potential for practical applications (Kandemir et al., 2022).

Further research is needed to develop safe, efficient, solvent-free, environmentally friendly, and cost-effective extraction methods. Additionally, investment in new infrastructure for the safe processing of by-products in industry is necessary. The cost involved in the entire process poses a challenge for researchers. In this regard, government-funded projects have gained importance, serving as incentives and investments in the future (Sharma et al., 2021; Kandemir et al., 2022). Future studies should also focus on products incorporating fruit by-products, as there is limited research in this area. Sensory properties and bioavailability of these products should be included in future studies (Kandemir et al., 2022). Furthermore, the potential of these new products to be better tailored to individual needs has started gaining recognition in the global market as "personalized functional" foods (Cai, 2019).

Recent studies have primarily focused on the extraction of valuable components from by-products and their utilization in products, particularly in the following areas:

- Identifying highly innovative and economically viable resources,
- Developing innovative, convenient, and environmentally friendly extraction processes,
- Technological advancements in bioprocessing systems,
- Innovations in color stabilization,
- Exploring the availability of different types of components in a single process, and
- Producing functional foods enriched with valuable components to create high-quality products with significant bioactivity (Sharma et al., 2021).

It was emphasized that future research should concentrate on optimizing the quantity and quality of bioactive components of fruits, through biochemical, physiological, and molecular biological studies. There is a possibility that the development of fruit-derived components with curative effects for certain diseases may be achievable in the near future (Cai, 2019).

5.10 Conclusions

This book chapter examines the various types of fruit by-products; their bioactive components, such as phenolic compounds, pigments, enzymes, and dietary fibers; as well as novel and environmentally friendly extraction methods. It explores the application of these extracted ingredients as raw materials or additives in food products. Furthermore, the chapter assesses the economic and technological prospects of these products, along with the relevant regulations. During fruit processing, by-products (pulp, seeds, and peels) are released in amounts up to 50% of the raw mass. The fact that these by-products contain significant amounts of valuable components (phenolic compounds, pigments, essential oils, dietary fibers, flavorings, vitamins, and minerals) has increased the demand for their evaluation and use in various food products. Conventional extraction of these components usually requires large amounts of solvent, high energy consumption, and longer extraction times, leading to green extraction, which provides high efficiency, reduced energy and solvent consumption, increased yields, improved process control, and high-quality extracts. As a result, efforts to increase, valorize, and enrich existing food resources have intensified in response to the increase in food consumption that will accompany the growth of the world's population in the coming years. Valuable ingredients such as antioxidants, colorings, fiber, and essential oils derived from existing fruit by-products have become an alternative to artificial additives and have improved the functional properties and quality of the foods to which they are added. It has also provided a resource for people on gluten-free, vegan, and similar special diets. Today, the conversion of fruit by-products into new and upcycled food products is being evaluated as a solution to reduce food waste. As today shapes the future, there is a need for more research into the food technology, health, economic, and regulatory aspects of fruit wastes to provide more sustainable and bioavailable food products in the future.

References

Aaby, K., Skrede, G., & Wrolstad, R. E. (2005). Phenolic composition and antioxidant activities in flesh and achenes of strawberries (Fragaria ananassa). *Journal of Agricultural and Food Chemistry, 53*(10), 4032–4040. https://doi.org/10.1021/JF0480010

Abbas, K. A., Mohamed, A., Abdulamir, A. S., & Abas, H. A. (2008). A review on supercritical fluid extraction as new analytical method. *American Journal of Biochemistry and Biotechnology, 4*(4), 345–353. https://doi.org/10.3844/AJBBSP.2008.345.353

Abboud, K. Y., da Luz, B. B., Dallazen, J. L., Werner, M. F. de P., Cazarin, C. B. B., Maróstica Junior, M. R., Iacomini, M., & Cordeiro, L. M. C. (2019). Gastroprotective effect of soluble dietary fibres from yellow passion fruit (Passiflora edulis f. flavicarpa) peel against ethanol-induced ulcer in rats. *Journal of Functional Foods, 54*, 552–558. https://doi.org/10.1016/J.JFF.2019.02.003

Adilah, A. N., Jamilah, B., Noranizan, M. A., & Hanani, Z. A. N. (2018). Utilization of mango peel extracts on the biodegradable films for active packaging. *Food Packaging and Shelf Life, 16*, 1–7. https://doi.org/10.1016/J.FPSL.2018.01.006

Ajila, C. M., Bhat, S. G., & Prasada Rao, U. J. S. (2007). Valuable components of raw and ripe peels from two Indian mango varieties. *Food Chemistry, 102*(4), 1006–1011. https://doi.org/10.1016/J.FOODCHEM.2006.06.036

Ajila, C. M., Leelavathi, K., & Prasada Rao, U. J. S. (2008). Improvement of dietary fiber content and antioxidant properties in soft dough biscuits with the incorporation of mango peel powder. *Journal of Cereal Science, 48*(2), 319–326. https://doi.org/10.1016/J.JCS.2007.10.001

Ajila, C. M., & Prasada Rao, U. J. S. (2013). Mango peel dietary fibre: Composition and associated bound phenolics. *Journal of Functional Foods, 5*(1), 444–450. https://doi.org/10.1016/J.JFF.2012.11.017

Akyüz, A., & Ersus, S. (2021). Optimization of enzyme assisted extraction of protein from the sugar beet (Beta vulgaris L.) leaves for alternative plant protein concentrate production. *Food Chemistry, 335*, 127673. https://doi.org/10.1016/J.FOODCHEM.2020.127673

Alongi, M., Melchior, S., & Anese, M. (2019). Reducing the glycemic index of short dough biscuits by using apple pomace as a functional ingredient. *LWT, 100*, 300–305. https://doi.org/10.1016/J.LWT.2018.10.068

Alvarez-Rivera, G., Bueno, M., Ballesteros-Vivas, D., Mendiola, J. A., & Ibanez, E. (2020). Pressurized liquid extraction. In C. F. Poole (Ed.), *In liquid-phase extraction* (pp. 375–398). Elsevier.

Ameer, K., Shahbaz, H. M., & Kwon, J. H. (2017). Green extraction methods for polyphenols from plant matrices and their byproducts: A review. *Comprehensive Reviews in Food Science and Food Safety, 16*(2), 295–315. https://doi.org/10.1111/1541-4337.12253

Anal, A. K. (2017). Food processing by-products and their utilization: Introduction. In *Food processing by-products and their utilization* (pp. 1–10). https://doi.org/10.1002/9781118432921.CH1

Andersen, Ø. M., & Jordheim, M. (2014). Basic anthocyanin chem-istry and dietary sources. In C. Taylor, T. C. Wallace, & M. M. Giusti (Eds.), *Anthocyanins in health and disease* (pp. 13–90). CRC Press.

Andrade, M. A., Lima, V., Sanches Silva, A., Vilarinho, F., Castilho, M. C., Khwaldia, K., & Ramos, F. (2019). Pomegranate and grape by-products and their active compounds: Are they a valuable source for food applications? *Trends in Food Science & Technology, 86*, 68–84. https://doi.org/10.1016/J.TIFS.2019.02.010

Andreotti, C., Ravaglia, D., Ragaini, A., & Costa, G. (2008). Phenolic compounds in peach (Prunus persica) cultivars at harvest and during fruit maturation. *Annals of Applied Biology, 153*(1), 11–23. https://doi.org/10.1111/J.1744-7348.2008.00234.X

Aranda, C., Robledo, A., Loera, O., Contreras-Esquivel, J. C., Rodríguez, R., & Aguilar, C. N. (2006). Fungal invertase expression in solid-state fermentation. *Food Technology & Biotechnology, 44*(2), 229–233.

Arogba, S. S. (2000). Mango (Mangifera indica) kernel: Chromatographic analysis of the Tannin, and stability study of the associated polyphenol oxidase activity. *Journal of Food Composition and Analysis, 13*(2), 149–156. https://doi.org/10.1006/JFCA.1999.0838

Arshad, Z. I. M., Amid, A., Yusof, F., Jaswir, I., Ahmad, K., & Loke, S. P. (2014). Bromelain: An overview of industrial application and purification strategies. *Applied Microbiology and Biotechnology, 98*(17), 7283–7297. https://doi.org/10.1007/S00253-014-5889-Y/TABLES/5

Avram, A. M., Morin, P., Brownmiller, C., Howard, L. R., Sengupta, A., & Wickramasinghe, S. R. (2017). Concentrations of polyphenols from blueberry pomace extract using nanofiltration. *Food and Bioproducts Processing, 106*, 91–101. https://doi.org/10.1016/J.FBP.2017.07.006

Ayala-Zavala, J. F., Rosas-Domínguez, C., Vega-Vega, V., & González-Aguilar, G. A. (2010). Antioxidant enrichment and antimicrobial protection of fresh-cut fruits using their own byproducts: Looking for integral exploitation. *Journal of Food Science, 75*(8), R175–R181. https://doi.org/10.1111/J.1750-3841.2010.01792.X

Ayar, A., Siçramaz, H., Öztürk, S., & Öztürk Yilmaz, E. (2018). Probiotic properties of ice creams produced with dietary fibres from by-products of the food industry. *International Journal of Dairy Technology, 71*(1), 174–182. https://doi.org/10.1111/1471-0307.12387

Azmir, J., Zaidul, I. S. M., Rahman, M. M., Sharif, K. M., Mohamed, A., Sahena, F., Jahurul, M. H. A., Ghafoor, K., Norulaini, N. A. N., & Omar, A. K. M. (2013). Techniques for extraction of bioactive compounds from plant materials: A review. *Journal of Food Engineering, 117*(4), 426–436. https://doi.org/10.1016/J.JFOODENG.2013.01.014

Babbar, N., Oberoi, H. S., & Sandhu, S. K. (2014). Therapeutic and nutraceutical potential of bioactive compounds extracted from fruit residues. *Critical Reviews in Food Science and Nutrition, 55*(3), 319–337. https://doi.org/10.1080/10408398.2011.653734

Balasundram, N., Sundram, K., & Samman, S. (2006). Phenolic compounds in plants and agri-industrial by-products: Antioxidant activity, occurrence, and potential uses. *Food Chemistry, 99*(1), 191–203. https://doi.org/10.1016/J.FOODCHEM.2005.07.042

Ballesteros-Gómez, A., Sicilia, M. D., & Rubio, S. (2010). Supramolecular solvents in the extraction of organic compounds: A review. *Analytica Chimica Acta, 677*(2), 108–130. https://doi.org/10.1016/J.ACA.2010.07.027

Banerjee, J., Singh, R., Vijayaraghavan, R., MacFarlane, D., Patti, A. F., & Arora, A. (2017). Bioactives from fruit processing wastes: Green approaches to valuable chemicals. *Food Chemistry, 225*, 10–22. https://doi.org/10.1016/J.FOODCHEM.2016.12.093

Bao, Y., Reddivari, L., & Huang, J. Y. (2020). Development of cold plasma pretreatment for improving phenolics extractability from tomato pomace. *Innovative Food Science & Emerging Technologies, 65*, 102445. https://doi.org/10.1016/J.IFSET.2020.102445

Barreira, J. C. M., Arraibi, A. A., & Ferreira, I. C. F. R. (2019). Bioactive and functional compounds in apple pomace from juice and cider manufacturing: Potential use in dermal formulations. *Trends in Food Science & Technology, 90*, 76–87. https://doi.org/10.1016/J.TIFS.2019.05.014

Basanta, M. F., Rizzo, S. A., Szerman, N., Vaudagna, S. R., Descalzo, A. M., Gerschenson, L. N., Pérez, C. D., & Rojas, A. M. (2018). Plum (Prunus salicina) peel and pulp microparticles as natural antioxidant additives in breast chicken patties. *Food Research International, 106*, 1086–1094. https://doi.org/10.1016/J.FOODRES.2017.12.011

Bchir, B., Rabetafika, H. N., Paquot, M., & Blecker, C. (2014). Effect of pear, apple and date fibres from cooked fruit by-products on dough performance and bread quality. *Food and Bioprocess Technology, 7*(4), 1114–1127. https://doi.org/10.1007/S11947-013-1148-Y/FIGURES/3

Bender, A. B. B., Speroni, C. S., Salvador, P. R., Loureiro, B. B., Lovatto, N. M., Goulart, F. R., Lovatto, M. T., Miranda, M. Z., Silva, L. P., & Penna, N. G. (2016). Grape pomace skins and the effects of its inclusion in the technological properties of muffins. *Journal of Culinary Science & Technology, 15*(2), 143–157. https://doi.org/10.1080/15428052.2016.1225535

Ben-Othman, S., Jõudu, I., Bhat, R., Beatriz, M., Oliveira, P., & Alves, R. C. (2020). Bioactives from agri-food wastes: Present insights and future challenges. *Molecules, 25*(3), 510. https://doi.org/10.3390/MOLECULES25030510

Berardini, N., Carle, R., & Schieber, A. (2004). Characterization of gallotannins and benzophenone derivatives from mango (Mangifera indica L. cv. 'Tommy Atkins') peels, pulp and kernels by high-performance liquid chromatography/electrospray ionization mass spectrometry. *Rapid Communications in Mass Spectrometry, 18*(19), 2208–2216. https://doi.org/10.1002/RCM.1611

Bhushan, S., Kalia, K., Sharma, M., Singh, B., & Ahuja, P. S. (2008). Processing of apple pomace for bioactive molecules. *Critical Reviews in Biotechnology, 28*(4), 285–296. https://doi.org/10.1080/07388550802368895

Bian, Q., Gao, S., Zhou, J., Qin, J., Taylor, A., Johnson, E. J., Tang, G., Sparrow, J. R., Gierhart, D., & Shang, F. (2012). Lutein and zeaxanthin supplementation reduces photooxidative damage and modulates the expression of inflammation-related genes in retinal pigment epithelial cells. *Free Radical Biology and Medicine, 53*(6), 1298–1307. https://doi.org/10.1016/J.FREERADBIOMED.2012.06.024

Bilawal, A., Ishfaq, M., Gantumur, M. A., Qayum, A., Shi, R., Fazilani, S. A., Anwar, A., Jiang, Z., & Hou, J. (2021). A review of the bioactive ingredients of berries and their applications in curing diseases. *Food Bioscience, 44*, 101407. https://doi.org/10.1016/J.FBIO.2021.101407

Bobinaitė, R., Kraujalis, P., Tamkutė, L., Urbonavičienė, D., Viškelis, P., & Venskutonis, P. R. (2020). Recovery of bioactive substances from rowanberry pomace by consecutive extraction with supercritical carbon dioxide and pressurized solvents. *Journal of Industrial and Engineering Chemistry, 85*, 152–160. https://doi.org/10.1016/J.JIEC.2020.01.036

126   Wealth out of Food Processing Waste

Bocco, A., Cuvelier, M. E., Richard, H., & Berset, C. (1998). Antioxidant activity and phenolic composition of citrus peel and seed extracts. *Journal of Agricultural and Food Chemistry, 46*(6), 2123–2129. https://doi.org/10.1021/JF9709562

Bolarinwa, I. F., Orfila, C., & Morgan, M. R. A. (2014). Amygdalin content of seeds, kernels and food products commercially-available in the UK. *Food Chemistry, 152*, 133–139. https://doi.org/10.1016/J.FOODCHEM.2013.11.002

Bordiga, M., Travaglia, F., & Locatelli, M. (2019). Valorisation of grape pomace: An approach that is increasingly reaching its maturity—A review. *International Journal of Food Science & Technology, 54*(4), 933–942. https://doi.org/10.1111/IJFS.14118

Borges, G., Degeneve, A., Mullen, W., & Crozier, A. (2010). Identification of flavonoid and phenolic antioxidants in black currants, blueberries, raspberries, red currants, and cranberries. *Journal of Agricultural and Food Chemistry, 58*(7), 3901–3909. https://doi.org/10.1021/JF902263N/SUPPL_FILE/JF902263N_SI_001.PDF

Botella, C., De Ory, I., Webb, C., Cantero, D., & Blandino, A. (2005). Hydrolytic enzyme production by Aspergillus awamori on grape pomace. *Biochemical Engineering Journal, 26*(2–3), 100–106. https://doi.org/10.1016/J.BEJ.2005.04.020

Boukroufa, M., Boutekedjiret, C., Petigny, L., Rakotomanomana, N., & Chemat, F. (2015). Bio-refinery of orange peels waste: A new concept based on integrated green and solvent free extraction processes using ultrasound and microwave techniques to obtain essential oil, polyphenols and pectin. *Ultrasonics Sonochemistry, 24*, 72–79. https://doi.org/10.1016/J.ULTSONCH.2014.11.015

Brianceau, S., Turk, M., Vitrac, X., & Vorobiev, E. (2016). High voltage electric discharges assisted extraction of phenolic compounds from grape stems: Effect of processing parameters on flavan-3-ols, flavonols and stilbenes recovery. *Innovative Food Science & Emerging Technologies, 35*, 67–74. https://doi.org/10.1016/J.IFSET.2016.04.006

Bruno Romanini, E., Misturini Rodrigues, L., Finger, A., Perez Cantuaria Chierrito, T., Regina da Silva Scapim, M., & Scaramal Madrona, G. (2021). Ultrasound assisted extraction of bioactive compounds from BRS Violet grape pomace followed by alginate-Ca2+ encapsulation. *Food Chemistry, 338*, 128101. https://doi.org/10.1016/J.FOODCHEM.2020.128101

Buendía, B., Gil, M. I., Tudela, J. A., Gady, A. L., Medina, J. J., Soria, C., López, J. M., & Tomás-Barberán, F. A. (2010). HPLC-MS analysis of proanthocyanidin oligomers and other phenolics in 15 strawberry cultivars. *Journal of Agricultural and Food Chemistry, 58*(7), 3916–3926. https://doi.org/10.1021/jf9030597

Burey, P., Bhandari, B. R., Howes, T., & Gidley, M. J. (2008). Hydrocolloid gel particles: Formation, characterization, and application. *Critical Reviews in Food Science and Nutrition, 48*(5), 361–377. https://doi.org/10.1080/10408390701347801

Cai, M. (2019). Fruit-based functional food. In *The role of alternative and innovative food ingredients and products in consumer wellness* (pp. 35–72). https://doi.org/10.1016/B978-0-12-816453-2.00002-4

Campos, D. A., Gómez-García, R., Vilas-Boas, A. A., Madureira, A. R., & Pintado, M. M. (2020). Management of fruit industrial by-products—A case study on circular economy approach. *Molecules, 25*(2), 320. https://doi.org/10.3390/MOLECULES25020320

Cano-Lamadrid, M., & Artés-Hernández, F. (2021). By-products revalorization with nonthermal treatments to enhance phytochemical compounds of fruit and vegetables derived products: A review. *Foods 2022, 11*(1), 59. https://doi.org/10.3390/FOODS11010059

Carbone, K., Amoriello, T., & Iadecola, R. (2020). Exploitation of kiwi juice pomace for the recovery of natural antioxidants through microwave-assisted extraction. *Agriculture, 10*(10), 435. https://doi.org/10.3390/AGRICULTURE10100435

Carpentieri, S., Soltanipour, F., Ferrari, G., Pataro, G., & Donsì, F. (2021). Emerging green techniques for the extraction of antioxidants from agri-food by-products as promising ingredients for the food industry. *Antioxidants, 10*(9), 1417. https://doi.org/10.3390/ANTIOX10091417

Casarotti, S. N., Borgonovi, T. F., Batista, C. L. F. M., & Penna, A. L. B. (2018). Guava, orange and passion fruit by-products: Characterization and its impacts on kinetics of acidification and properties of probiotic fermented products. *LWT*, *98*, 69–76. https://doi.org/10.1016/J.LWT.2018.08.010

Casazza, A. A., Aliakbarian, B., Mura, M., Chasseur, M., Freguglia, M., Valentini, S., Palombo, D., & Perego, P. (2015). Polyphenols from grape and apple skin: A study on nonconventional extractions and biological activity on endothelial cell cultures. *Chemical Engineering Transactions*, *44*, 205–210. https://doi.org/10.3303/CET1544035

Cascaes Teles, A. S., Hidalgo Chávez, D. W., Zarur Coelho, M. A., Rosenthal, A., Fortes Gottschalk, L. M., & Tonon, R. V. (2021). Combination of enzyme-assisted extraction and high hydrostatic pressure for phenolic compounds recovery from grape pomace. *Journal of Food Engineering*, *288*, 110128. https://doi.org/10.1016/J.JFOODENG.2020.110128

Ćetković, G., Čanadanović-Brunet, J., Djilas, S., Savatović, S., Mandić, A., & Tumbas, V. (2008). Assessment of polyphenolic content and in vitro antiradical characteristics of apple pomace. *Food Chemistry*, *109*(2), 340–347. https://doi.org/10.1016/J.FOODCHEM.2007.12.046

Cevallos-Casals, B. A., Byrne, D., Okie, W. R., & Cisneros-Zevallos, L. (2006). Selecting new peach and plum genotypes rich in phenolic compounds and enhanced functional properties. *Food Chemistry*, *96*(2), 273–280. https://doi.org/10.1016/J.FOODCHEM.2005.02.032

Chau, C. F., & Huang, Y. L. (2003). Comparison of the chemical composition and physicochemical properties of different fibers prepared from the peel of citrus sinensis L. Cv. liucheng. *Journal of Agricultural and Food Chemistry*, *51*(9), 2615–2618. https://doi.org/10.1021/JF025919B

Chávez-González, M. L., López-López, L. I., Rodríguez-Herrera, R., Contreras-Esquivel, J. C., & Aguilar, C. N. (2016). Enzyme-assisted extraction of citrus essential oil. *Chemical Papers*, *70*(4), 412–417. https://doi.org/10.1515/CHEMPAP-2015-0234/MACHINEREADABLECITATION/RIS

Chemat, F., Rombaut, N., Sicaire, A. G., Meullemiestre, A., Fabiano-Tixier, A. S., & Abert-Vian, M. (2017). Ultrasound assisted extraction of food and natural products: Mechanisms, techniques, combinations, protocols and applications: A review. *Ultrasonics Sonochemistry*, *34*, 540–560. https://doi.org/10.1016/J.ULTSONCH.2016.06.035

Chemat, F., Vian, M. A., & Cravotto, G. (2012). Green extraction of natural products: Concept and principles. *International Journal of Molecular Sciences*, *13*(7), 8615–8627. https://doi.org/10.3390/IJMS13078615

Chen, A. Y., & Chen, Y. C. (2013). A review of the dietary flavonoid, kaempferol on human health and cancer chemoprevention. *Food Chemistry*, *138*(4), 2099–2107. https://doi.org/10.1016/J.FOODCHEM.2012.11.139

Chuyen, H. V., Nguyen, M. H., Roach, P. D., Golding, J. B., & Parks, S. E. (2018). Microwaveassisted extraction and ultrasound-assisted extraction for recovering carotenoids from Gac peel and their effects on antioxidant capacity of the extracts. *Food Science & Nutrition*, *6*(1), 189–196. https://doi.org/10.1002/FSN3.546

Cilek, B., Luca, A., Hasirci, V., Sahin, S., & Sumnu, G. (2012). Microencapsulation of phenolic compounds extracted from sour cherry pomace: Effect of formulation, ultrasonication time and core to coating ratio. *European Food Research and Technology*, *235*(4), 587–596. https://doi.org/10.1007/S00217-012-1786-8/FIGURES/5

Corrado, S., & Sala, S. (2018). Food waste accounting along global and European food supply chains: State of the art and outlook. *Waste Management*, *79*, 120–131. https://doi.org/10.1016/J.WASMAN.2018.07.032

Corrales, M., Toepfl, S., Butz, P., Knorr, D., & Tauscher, B. (2008). Extraction of anthocyanins from grape by-products assisted by ultrasonics, high hydrostatic pressure or pulsed electric fields: A comparison. *Innovative Food Science & Emerging Technologies*, *9*(1), 85–91. https://doi.org/10.1016/J.IFSET.2007.06.002

Costa, R. (2016). The chemistry of mushrooms: A survey of novel extraction techniques targeted to chromatographic and spectroscopic screening. *Studies in Natural Products Chemistry*, *49*, 279–306. https://doi.org/10.1016/B978-0-444-63601-0.00009-0

Costin, G.-E., & Hearing, V. J. (2007). Human skin pigmentation: Melanocytes modulate skin color in response to stress. *The FASEB Journal*, *21*(4), 976–994. https://doi.org/10.1096/FJ.06-6649REV

Couth, R., & Trois, C. (2010). Carbon emissions reduction strategies in Africa from improved waste management: A review. *Waste Management*, *30*(11), 2336–2346. https://doi.org/10.1016/J.WASMAN.2010.04.013

Cubero -Cardoso, J., Serrano, A., Trujillo-Reyes, A., Villa-Gómez, D. K., Borja, R., & Fermoso, F. G. (2020). Valorization options of strawberry extrudate agro-waste: A review. In A. N. de Barros & I. Gouvinhas (Eds.), *Innovation in the food sector through the valorization of food and agro-food by-product* (pp. 271–282). IntechOpen.

Cunha, L. C. M., Monteiro, M. L. G., Costa-Lima, B. R. C., Guedes-Oliveira, J. M., Alves, V. H. M., Almeida, A. L., Tonon, R. V., Rosenthal, A., & Conte-Junior, C. A. (2018). Effect of microencapsulated extract of pitaya (Hylocereus costaricensis) peel on color, texture and oxidative stability of refrigerated ground pork patties submitted to high pressure processing. *Innovative Food Science & Emerging Technologies*, *49*, 136–145. https://doi.org/10.1016/J.IFSET.2018.08.009

Dabhi, B. K., Vyas, R. V., & Shelat, H. N. (2014). Use of banana waste for the production of cellulolytic enzymes under solid substrate fermentation using bacterial consortium. *Int. J. Curr. Microbiol. App. Sci.*, *3*(1). www.ijcmas.com

D'Abrosca, B., Pacifico, S., Cefarelli, G., Mastellone, C., & Fiorentino, A. (2007). 'Limoncella' apple, an Italian apple cultivar: Phenolic and flavonoid contents and antioxidant activity. *Food Chemistry*, *104*(4), 1333–1337. https://doi.org/10.1016/J.FOODCHEM.2007.01.073

da Fonseca Machado, A. P., Alves Rezende, C., Alexandre Rodrigues, R., Fernández Barbero, G., de Tarso Vieira e Rosa, P., & Martínez, J. (2018). Encapsulation of anthocyanin-rich extract from blackberry residues by spray-drying, freeze-drying and supercritical antisolvent. *Powder Technology*, *340*, 553–562. https://doi.org/10.1016/J.POWTEC.2018.09.063

Da Rocha, C. B., & Noreña, C. P. Z. (2020). Microwave-assisted extraction and ultrasound-assisted extraction of bioactive compounds from grape pomace. *International Journal of Food Engineering*, *16*(1–2). https://doi.org/10.1515/IJFE-2019-0191/MACHINEREADABLECITATION/RIS

da Silva, P. O., de Alencar Guimarães, N. C., Serpa, J. D. M., Masui, D. C., Marchetti, C. R., Verbisck, N. V., Zanoelo, F. F., Ruller, R., & Giannesi, G. C. (2019). Application of an endo-xylanase from Aspergillus japonicus in the fruit juice clarification and fruit peel waste hydrolysis. *Biocatalysis and Agricultural Biotechnology*, *21*, 101312. https://doi.org/10.1016/J.BCAB.2019.101312

de Mello, F. R., Bernardo, C., Dias, C. O., Gonzaga, L., Amante, E. R., Fett, R., & Candido, L. M. B. (2014). Antioxidant properties, quantification and stability of betalains from pitaya (Hylocereus undatus) peel. *Ciência Rural*, *45*(2), 323–328. https://doi.org/10.1590/0103-8478CR20140548

Demirbaş, A., & Akdeniz, F. (2002). Fuel analyses of selected oilseed shells and supercritical fluid extraction in alkali medium. *Energy conversion and management*, *43*(15), 1977–1984.

Demirdöven, A., Karabiyikli, Ş., Tokatli, K., & Öncül, N. (2015). Inhibitory effects of red cabbage and sour cherry pomace anthocyanin extracts on food borne pathogens and their antioxidant properties. *LWT—Food Science and Technology*, *63*(1), 8–13. https://doi.org/10.1016/J.LWT.2015.03.101

Deng, G. F., Shen, C., Xu, X. R., Kuang, R. D., Guo, Y. J., Zeng, L. S., Gao, L. L., Lin, X., Xie, J. F., Xia, E. Q., Li, S., Wu, S., Chen, F., Ling, W. H., & Li, H. B. (2012). Potential of fruit wastes as natural resources of bioactive compounds. *International Journal of Molecular Sciences*, *13*(7), 8308–8323. https://doi.org/10.3390/IJMS13078308

Deng, J., Liu, Q., Zhang, Q., Zhang, C., Liu, D., Fan, D., & Yang, H. (2018). Comparative study on composition, physicochemical and antioxidant characteristics of different varieties of kiwifruit seed oil in China. *Food Chemistry*, *264*, 411–418. https://doi.org/10.1016/J.FOODCHEM.2018.05.063

Deng, Q., Penner, M. H., & Zhao, Y. (2011). Chemical composition of dietary fiber and polyphenols of five different varieties of wine grape pomace skins. *Food Research International*, *44*(9), 2712–2720. https://doi.org/10.1016/J.FOODRES.2011.05.026

De Oliveira, C. F., Giordani, D., Gurak, P. D., Cladera-Olivera, F., & Marczak, L. D. F. (2015). Extraction of pectin from passion fruit peel using moderate electric field and conventional heating extraction methods. *Innovative Food Science & Emerging Technologies*, *29*, 201–208. https://doi.org/10.1016/J.IFSET.2015.02.005

de Souza de Azevedo, P. O., Aliakbarian, B., Casazza, A. A., LeBlanc, J. G., Perego, P., & de Souza Oliveira, R. P. (2018). Production of fermented skim milk supplemented with different grape pomace extracts: Effect on viability and acidification performance of probiotic cultures. *PharmaNutrition*, *6*(2), 64–68. https://doi.org/10.1016/J.PHANU.2018.03.001

Dheyab, A. S., Bakar, M. F. A., Alomar, M., Sabran, S. F., Hanafi, A. F. M., & Mohamad, A. (2021). Deep Eutectic Solvents (DESs) as green extraction media of beneficial bioactive phytochemicals. *Separations*, *8*(10), 176. https://doi.org/10.3390/SEPARATIONS8100176

Dhiman, S., Mukherjee, G., Kumar, A., & Majumdar, R. S. (2022). Purification, characterization and application study of bacterial tannase for optimization of gallic acid synthesis from fruit waste. *Journal of Scientific & Industrial Research*, *81*(10), 1029–1036. https://doi.org/10.56042/JSIR.V81I10.55236

Dhingra, S., & Kapoor, A. C. (1985). Nutritive value of mango seed kernel. *Journal of the Science of Food and Agriculture*, *36*(8), 752–756. https://doi.org/10.1002/JSFA.2740360817

Dilucia, F., Lacivita, V., Conte, A., & Del Nobile, M. A. (2020). Sustainable use of fruit and vegetable by-products to enhance food packaging performance. *Foods*, *9*(7), 857. https://doi.org/10.3390/FOODS9070857

Domínguez-Rodríguez, G., Marina, M. L., & Plaza, M. (2021). Enzyme-assisted extraction of bioactive non-extractable polyphenols from sweet cherry (Prunus avium L.) pomace. *Food Chemistry*, *339*, 128086. https://doi.org/10.1016/J.FOODCHEM.2020.128086

Dominiak, M., Søndergaard, K. M., Wichmann, J., Vidal-Melgosa, S., Willats, W. G. T., Meyer, A. S., & Mikkelsen, J. D. (2014). Application of enzymes for efficient extraction, modification, and development of functional properties of lime pectin. *Food Hydrocolloids*, *40*, 273–282. https://doi.org/10.1016/J.FOODHYD.2014.03.009

Doria, E., Boncompagni, E., Marra, A., Dossena, M., Verri, M., & Buonocore, D. (2021). Polyphenols extraction from vegetable wastes using a green and sustainable method. *Frontiers in Sustainable Food Systems*, *5*, 690399. https://doi.org/10.3389/FSUFS.2021.690399/BIBTEX

Dos Santos, T. C., Gomes, D. P. P., Bonomo, R. C. F., & Franco, M. (2012). Optimisation of solid state fermentation of potato peel for the production of cellulolytic enzymes. *Food Chemistry*, *133*(4), 1299–1304. https://doi.org/10.1016/J.FOODCHEM.2011.11.115

dos Santos Freitas, L., Jacques, R. A., Richter, M. F., Silva, A. L. da, & Caramão, E. B. (2008). Pressurized liquid extraction of vitamin E from Brazilian grape seed oil. *Journal of Chromatography A*, *1200*(1), 80–83. https://doi.org/10.1016/J.CHROMA.2008.02.067

Dubois, V., Breton, S., Linder, M., Fanni, J., & Parmentier, M. (2007). Fatty acid profiles of 80 vegetable oils with regard to their nutritional potential. *European Journal of Lipid Science and Technology*, *109*(7), 710–732. https://doi.org/10.1002/EJLT.200700040

Duda-Chodak, A., & Tarko, T. (2007). Antioxidant properties of different fruit seeds and peels. *Acta Scientiarum Polonorum Technologia Alimentaria*, *6*(3), 29–36.

Dwivedi, S. K., Joshi, V. K., & Mishra, V. (2014, January). Extraction of anthocyanins from plum pomace using XAD-16 and determination of their thermal stability. *JSIR*, *73*(1), 230. http://nopr.niscpr.res.in/handle/123456789/25428

Eberhardt, M. V., Lee, C. Y., & Liu, R. H. (2000). Antioxidant activity of fresh apples. *Nature*, *405*(6789), 903–904. https://doi.org/10.1038/35016151

Egüés, I., Hernandez-Ramos, F., Rivilla, I., & Labidi, J. (2021). Optimization of ultrasound assisted extraction of bioactive compounds from apple pomace. *Molecules*, *26*(13), 3783. https://doi.org/10.3390/MOLECULES26133783

Elik, A., Yanık, D. K., & Göğüş, F. (2020). Microwave-assisted extraction of carotenoids from carrot juice processing waste using flaxseed oil as a solvent. *LWT, 123*, 109100. https://doi.org/10.1016/J.LWT.2020.109100

El Kantar, S., Rajha, H. N., Boussetta, N., Vorobiev, E., Maroun, R. G., & Louka, N. (2019). Green extraction of polyphenols from grapefruit peels using high voltage electrical discharges, deep eutectic solvents and aqueous glycerol. *Food Chemistry, 295*, 165–171. https://doi.org/10.1016/J.FOODCHEM.2019.05.111

European Commission. (2014). *Landfill waste.* Https://Environment.Ec.Europa.Eu/Topics/Waste-and-Recycling/Landfill-Waste_en.

Fan, X., Gao, Y., He, W., Hu, H., Tian, M., Wang, K., & Pan, S. (2016). Production of nano bacterial cellulose from beverage industrial waste of citrus peel and pomace using Komagataeibacter xylinus. *Carbohydrate Polymers, 151*, 1068–1072. https://doi.org/10.1016/J.CARBPOL.2016.06.062

Fan, X., Jiao, W., Wang, X., Cao, J., & Jiang, W. (2018). Polyphenol composition and antioxidant capacity in pulp and peel of apricot fruits of various varieties and maturity stages at harvest. *International Journal of Food Science & Technology, 53*(2), 327–336. https://doi.org/10.1111/IJFS.13589

Farooque, S., Rose, P. M., Benohoud, M., Blackburn, R. S., & Rayner, C. M. (2018). Enhancing the potential exploitation of food waste: Extraction, purification, and characterization of renewable specialty chemicals from blackcurrants (Ribes nigrum L.). *Journal of Agricultural and Food Chemistry, 66*(46), 12265–12273. https://doi.org/10.1021/ACS.JAFC.8B04373/SUPPL_FILE/JF8B04373_SI_001.PDF

Faustino, M., Veiga, M., Sousa, P., Costa, E. M., Silva, S., & Pintado, M. (2019). Agro-food byproducts as a new source of natural food additives. *Molecules, 24*(6), 1056. https://doi.org/10.3390/MOLECULES24061056

Femenia, A., Rosselló, C., Mulet, A., & Cañellas, J. (1995). Chemical composition of bitter and sweet apricot kernels. *Journal of Agricultural and Food Chemistry, 43*(2), 356–361. https://doi.org/10.1021/JF00050A018/ASSET/JF00050A018.FP.PNG_V03

Ferhat, M. A., Meklati, B. Y., & Chemat, F. (2007). Comparison of different isolation methods of essential oil from Citrus fruits: Cold pressing, hydrodistillation and microwave 'dry' distillation. *Flavour and Fragrance Journal, 22*(6), 494–504. https://doi.org/10.1002/FFJ.1829

Ferreira, A. S., Nunes, C., Castro, A., Ferreira, P., & Coimbra, M. A. (2014). Influence of grape pomace extract incorporation on chitosan films properties. *Carbohydrate Polymers, 113*, 490–499. https://doi.org/10.1016/J.CARBPOL.2014.07.032

Ferreira, L. F., Minuzzi, N. M., Rodrigues, R. F., Pauletto, R., Rodrigues, E., Emanuelli, T., & Bochi, V. C. (2020). Citric acid water-based solution for blueberry bagasse anthocyanins recovery: Optimization and comparisons with microwave-assisted extraction (MAE). *LWT, 133*, 110064. https://doi.org/10.1016/J.LWT.2020.110064

Fincan, M., Çiftci, Y., Üniversitesi, E., Fakültesi, M., Mühendisliği Bölümü, G., & Geliş, T. (2021). VURGULU ELEKTRİK ALAN ÖN İŞLEMİ İLE DEREOTUNDAN FENOLİKLERİN EKSTRAKSİYONU: DONDURUP ÇÖZÜNDÜRME, ISIL İŞLEM, MİKRODALGA ÖN İŞLEMLERİ VE SOLVENT EKSTRAKSİYONU İLE KARŞILAŞTIRILMASI. *Gida the Journal of Food, 46*(6), 1343–1357. https://doi.org/10.15237/gida.GD21092

Foster, R., Williamson, C. S., & Lunn, J. (2009). Briefing paper: Culinary oils and their health effects. *Nutrition Bulletin, 34*(1), 4–47. https://doi.org/10.1111/J.1467-3010.2008.01738.X

Freitas, A., Moldão-Martins, M., Costa, H. S., Albuquerque, T. G., Valente, A., & Sanches-Silva, A. (2015). Effect of UV-C radiation on bioactive compounds of pineapple (Ananas comosus L. Merr.) by-products. *Journal of the Science of Food and Agriculture, 95*(1), 44–52. https://doi.org/10.1002/JSFA.6751

Friedman, M. (1997). Chemistry, biochemistry, and dietary role of potato polyphenols: A review. *Journal of Agricultural and Food Chemistry, 45*(5), 1523–1540. https://doi.org/10.1021/JF960900S

Funari, C. S., Sutton, A. T., Carneiro, R. L., Fraige, K., Cavalheiro, A. J., da Silva Bolzani, V., Hilder, E. F., & Arrua, R. D. (2019). Natural deep eutectic solvents and aqueous solutions as an alternative extraction media for propolis. *Food Research International, 125,* 108559. https://doi.org/10.1016/J.FOODRES.2019.108559

Gajic, I. M. S., Savic, I. M., Gajic, D. G., & Dosic, A. (2021). Ultrasound-assisted extraction of carotenoids from orange peel using olive oil and its encapsulation in ca-alginate beads. *Biomolecules, 11*(2), 225. https://doi.org/10.3390/BIOM11020225

Gangadharan, D., Sivaramakrishnan, S., Nampoothiri, K. M., Sukumaran, R. K., & Pandey, A. (2008). Response surface methodology for the optimization of alpha amylase production by Bacillus amyloliquefaciens. *Bioresource Technology, 99*(11), 4597–4602. https://doi.org/10.1016/J.BIORTECH.2007.07.028

Garcia-Amezquita, L. E., Tejada-Ortigoza, V., Serna-Saldivar, S. O., & Welti-Chanes, J. (2018). Dietary fiber concentrates from fruit and vegetable by-products: Processing, modification, and application as functional ingredients. *Food and Bioprocess Technology, 11*(8), 1439–1463. https://doi.org/10.1007/S11947-018-2117-2

Gavahian, M., & Cullen, P. J. (2019). Cold plasma as an emerging technique for mycotoxin-free food: Efficacy, mechanisms, and trends. *Food Reviews International, 36*(2), 193–214. https://doi.org/10.1080/87559129.2019.1630638

Ghafoor, K., Park, J., & Choi, Y. H. (2010). Optimization of supercritical fluid extraction of bioactive compounds from grape (Vitis labrusca B.) peel by using response surface methodology. *Innovative Food Science & Emerging Technologies, 11*(3), 485–490. https://doi.org/10.1016/J.IFSET.2010.01.013

Ghasempour, N., Rad, A. H. E., Javanmard, M., Azarpazhouh, E., & Armin, M. (2019). Optimization of conditions of ultrasound-assisted extraction of phenolic compounds from orange pomace (Citrus sinensis). *International Journal of Biology and Chemistry, 12*(2), 10–19. https://doi.org/10.26577/IJBCH-2019-V2-2

Gligor, O., Mocan, A., Moldovan, C., Locatelli, M., Crişan, G., & Ferreira, I. C. F. R. (2019). Enzyme-assisted extractions of polyphenols—A comprehensive review. *Trends in Food Science & Technology, 88,* 302–315. https://doi.org/10.1016/J.TIFS.2019.03.029

Gómez-García, R., Campos, D. A., Aguilar, C. N., Madureira, A. R., & Pintado, M. (2020). Valorization of melon fruit (Cucumis melo L.) by-products: Phytochemical and biofunctional properties with emphasis on recent trends and advances. *Trends in Food Science & Technology, 99,* 507–519. https://doi.org/10.1016/J.TIFS.2020.03.033

González-Montelongo, R., Gloria Lobo, M., & González, M. (2010). Antioxidant activity in banana peel extracts: Testing extraction conditions and related bioactive compounds. *Food Chemistry, 119*(3), 1030–1039. https://doi.org/10.1016/J.FOODCHEM.2009.08.012

Gordillo, B., Sigurdson, G. T., Lao, F., González-Miret, M. L., Heredia, F. J., & Giusti, M. M. (2018). Assessment of the color modulation and stability of naturally copigmented anthocyanin-grape colorants with different levels of purification. *Food Research International, 106,* 791–799. https://doi.org/10.1016/J.FOODRES.2018.01.057

Gorinstein, S., Zachwieja, Z., Folta, M., Barton, H., Piotrowicz, J., Zemser, M., Weisz, M., Trakhtenberg, S., & Màrtín-Belloso, O. (2001). Comparative contents of dietary fiber, total phenolics, and minerals in persimmons and apples. *Journal of Agricultural and Food Chemistry, 49*(2), 952–957. https://doi.org/10.1021/JF000947K

Goula, A. M., Ververi, M., Adamopoulou, A., & Kaderides, K. (2017). Green ultrasound-assisted extraction of carotenoids from pomegranate wastes using vegetable oils. *Ultrasonics Sonochemistry, 34,* 821–830. https://doi.org/10.1016/J.ULTSONCH.2016.07.022

Grech, A., Howse, E., & Boylan, S. (2020). A scoping review of policies promoting and supporting sustainable food systems in the university setting. *Nutrition Journal, 19*(1), 1–13. https://doi.org/10.1186/S12937-020-00617-W/TABLES/2

Gülcü, M., Uslu, N., Özcan, M. M., Gökmen, F., Özcan, M. M., Banjanin, T., Gezgin, S., Dursun, N., Geçgel, Ü., Ceylan, D. A., & Lemiasheuski, V. (2019). The investigation of bioactive compounds of wine, grape juice and boiled grape juice wastes. *Journal of Food Processing and Preservation, 43*(1), e13850. https://doi.org/10.1111/JFPP.13850

Gullón, P., Gullón, B., Romaní, A., Rocchetti, G., & Lorenzo, J. M. (2020). Smart advanced solvents for bioactive compounds recovery from agri-food by-products: A review. *Trends in Food Science & Technology, 101*, 182–197. https://doi.org/10.1016/J.TIFS.2020.05.007

Gumul, D., Ziobro, R., Korus, J., & Kruczek, M. (2021). Apple pomace as a source of bioactive polyphenol compounds in gluten-free breads. *Antioxidants, 10*(5), 807. https://doi.org/10.3390/ANTIOX10050807

Guo, C., Yang, J., Wei, J., Li, Y., Xu, J., & Jiang, Y. (2003). Antioxidant activities of peel, pulp and seed fractions of common fruits as determined by FRAP assay. *Nutrition Research, 23*(12), 1719–1726. https://doi.org/10.1016/J.NUTRES.2003.08.005

Gurak, P. D., De Bona, G. S., Tessaro, I. C., & Marczak, L. D. F. (2014). Jaboticaba pomace powder obtained as a co-product of juice extraction: A comparative study of powder obtained from peel and whole fruit. *Food Research International, 62*, 786–792. https://doi.org/10.1016/J.FOODRES.2014.04.042

Hajimahmoodi, M., Oveisi, M. R., Sadeghi, N., Jannat, B., Hadjibabaie, M., Farahani, E., Akrami, M. R., & Namdar, R. (2008). Antioxidant properties of peel and pulp hydro extract in ten Persian pomegranate cultivars. *Pakistan Journal of Biological Sciences: PJBS, 11*(12), 1600–1604. https://doi.org/10.3923/PJBS.2008.1600.1604

Haleva-Toledo, E., Naim, M., Zehavi, U., & Rouseff, R. L. (1999). Effects of l-cysteine and N-acetyl-l-cysteine on 4-hydroxy-2,5-dimethyl-3(2H)-furanone (Furaneol), 5-(Hydroxymethyl) furfural, and 5-methylfurfural formation and browning in buffer solutions containing either rhamnose or glucose and arginine. *Journal of Agricultural and Food Chemistry, 47*(10), 4140–4145. https://doi.org/10.1021/JF9813788

Heim, K. E., Tagliaferro, A. R., & Bobilya, D. J. (2002). Flavonoid antioxidants: Chemistry, metabolism and structure-activity relationships. *The Journal of Nutritional Biochemistry, 13*(10), 572–584. https://doi.org/10.1016/S0955-2863(02)00208-5

Ignat, I., Volf, I., & Popa, V. I. (2011). A critical review of methods for characterisation of polyphenolic compounds in fruits and vegetables. *Food Chemistry, 126*(4), 1821–1835. https://doi.org/10.1016/J.FOODCHEM.2010.12.026

Ishangulyyev, R., Kim, S., & Lee, S. H. (2019). Understanding food loss and waste—Why are we losing and wasting food? *Foods, 8*(8), 297. https://doi.org/10.3390/FOODS8080297

Ismail, H. I., Chan, K. W., Mariod, A. A., & Ismail, M. (2010). Phenolic content and antioxidant activity of cantaloupe (cucumis melo) methanolic extracts. *Food Chemistry, 119*(2), 643–647. https://doi.org/10.1016/J.FOODCHEM.2009.07.023

Jahurul, M. H. A., Zaidul, I. S. M., Ghafoor, K., Al-Juhaimi, F. Y., Nyam, K. L., Norulaini, N. A. N., Sahena, F., & Mohd Omar, A. K. (2015). Mango (Mangifera indica L.) by-products and their valuable components: A review. *Food Chemistry, 183*, 173–180. https://doi.org/10.1016/J.FOODCHEM.2015.03.046

Jamrath, T., Lindner, C., Popović, M. K., & Bajpai, R. (2012). Production of amylases and proteases by bacillus caldolyticus from food industry wastes. *Food Technology and Biotechnology, 50*(3), 355–361.

Jara-Palacios, M. J., Hernanz, D., Cifuentes-Gomez, T., Escudero-Gilete, M. L., Heredia, F. J., & Spencer, J. P. E. (2015). Assessment of white grape pomace from winemaking as source of bioactive compounds, and its antiproliferative activity. *Food Chemistry, 183*, 78–82. https://doi.org/10.1016/J.FOODCHEM.2015.03.022

Javier, F., Armenta, V., Thalía Bernal-Mercado, A., González Aguilar, G. A., & Fernando Ayala-Zavala, J. (2018). *Winery and grape juice extraction by-products development of wound dressings-based on biopolymers view project Aprovechamiento de polifenoles de sub-productos vegetales para el desarrollo de alimentos funcionales para acuacultura view project.* https://doi.org/10.1201/b22352-6

Jentzer, J. B., Alignan, M., Vaca-Garcia, C., Rigal, L., & Vilarem, G. (2015). Response surface methodology to optimise accelerated solvent extraction of steviol glycosides from stevia rebaudiana bertoni leaves. *Food Chemistry, 166*, 561–567. https://doi.org/10.1016/J.FOODCHEM.2014.06.078

Jesus, S. P., & Meireles, M. A. A. (2014). *Supercritical fluid extraction: A global perspective of the fundamental concepts of this eco-friendly extraction technique* (pp. 39–72). https://doi.org/10.1007/978-3-662-43628-8_3

Jurgoński, A., Juśkiewicz, J., Zduńczyk, Z., Matusevicius, P., & Kołodziejczyk, K. (2014). Polyphenol-rich extract from blackcurrant pomace attenuates the intestinal tract and serum lipid changes induced by a high-fat diet in rabbits. *European Journal of Nutrition, 53*(8), 1603–1613. https://doi.org/10.1007/S00394-014-0665-4/TABLES/7

Käferböck, A., Smetana, S., de Vos, R., Schwarz, C., Toepfl, S., & Parniakov, O. (2020). Sustainable extraction of valuable components from Spirulina assisted by pulsed electric fields technology. *Algal Research, 48*, 101914. https://doi.org/10.1016/J.ALGAL.2020.101914

Kandemir, K., Piskin, E., Xiao, J., Tomas, M., & Capanoglu, E. (2022). Fruit juice industry wastes as a source of bioactives. *Cite This: J. Agric. Food Chem., 2022*, 6805–6832. https://doi.org/10.1021/acs.jafc.2c00756

Kasaai, M. R., & Moosavi, A. (2017). Treatment of kraft paper with citrus wastes for food packaging applications: Water and oxygen barrier properties improvement. *Food Packaging and Shelf Life, 12*, 59–65. https://doi.org/10.1016/J.FPSL.2017.02.006

Kasapoğlu, E. D., Kahraman, S., & Tornuk, F. (2021). Optimization of ultrasound assisted antioxidant extraction from apricot pomace using response surface methodology. *Journal of Food Measurement and Characterization, 15*(6), 5277–5287. https://doi.org/10.1007/S11694-021-01089-0/TABLES/6

Katajajuuri, J. M., Silvennoinen, K., Hartikainen, H., Heikkilä, L., & Reinikainen, A. (2014). Food waste in the finnish food chain. *Journal of Cleaner Production, 73*, 322–329. https://doi.org/10.1016/J.JCLEPRO.2013.12.057

Kaur, S., Panesar, P. S., & Chopra, H. K. (2021). Standardization of ultrasound-assisted extraction of bioactive compounds from kinnow mandarin peel. *Biomass Conversion and Biorefinery, 13*(10), 8853–8863. https://doi.org/10.1007/S13399-021-01674-9/FIGURES/3

Kavas, N., & Kavas, G. (2016). Physical-chemical and antimicrobial properties of Egg White Protein Powder films incorporated with orange essential oil on Kashar Cheese. *Food Science and Technology, 36*(4), 672–678. https://doi.org/10.1590/1678-457X.12516

Kheirkhah, H., Baroutian, S., & Quek, S. Y. (2019). Evaluation of bioactive compounds extracted from Hayward kiwifruit pomace by subcritical water extraction. *Food and Bioproducts Processing, 115*, 143–153. https://doi.org/10.1016/J.FBP.2019.03.007

Khoddami, A., Wilkes, M. A., & Roberts, T. H. (2013). Techniques for analysis of plant phenolic compounds. *Molecules, 18*(2), 2328–2375. https://doi.org/10.3390/MOLECULES18022328

Kittiphoom, S. (2012). Utilization of mango seed. *International Food Research Journal, 19*(4), 1325–1335.

Kolawole, A. T., Dapper, V. D., & Eziuzo, C. I. (2017). Effects of the methanolic extract of the rind of Citrullus lanatus (watermelon) on some erythrocyte parameters and indices of oxidative status in phenylhydrazine-treated male Wistar rats. *Journal of African Association of Physiological Sciences, 5*(1), 22–28. https://doi.org/10.4314/JAAPS.V5I1

Koubaa, M., Barba, F. J., Grimi, N., Mhemdi, H., Koubaa, W., Boussetta, N., & Vorobiev, E. (2016). Recovery of colorants from red prickly pear peels and pulps enhanced by pulsed electric field and ultrasound. *Innovative Food Science & Emerging Technologies, 37*, 336–344. https://doi.org/10.1016/J.IFSET.2016.04.015

Koutsoukos, S., Tsiaka, T., Tzani, A., Zoumpoulakis, P., & Detsi, A. (2019). Choline chloride and tartaric acid, a Natural Deep Eutectic Solvent for the efficient extraction of phenolic and carotenoid compounds. *Journal of Cleaner Production, 241*, 118384. https://doi.org/10.1016/J.JCLEPRO.2019.118384

Krivokapić, S., Vlaović, M., Vratnica, B. D., Perović, A., & Perovic, S. (2021). Biowaste as a potential source of bioactive compound-a case study of raspberry fruit pomace. *Foods, 10*(4). https://doi.org/10.3390/foods10040706

Kumar, A. (2020). Aspergillus nidulans: A potential resource of the production of the native and heterologous enzymes for industrial applications. *International Journal of Microbiology, 2020*. https://doi.org/10.1155/2020/8894215

Kumar, A. (2021). Utilization of bioactive components present in pineapple waste: A review. *The Pharma Innovation Journal, 10*(5), 954–961. www.thepharmajournal.com

Kumar, R. S., & Manimegalai, G. (2004). Fruit and vegetable processing industries and environment. In A. Kumar (Ed.), *Industrial pollution & management* (Vol. 97, pp. 97–117). APH.

Kupnik, K., Primožič, M., Vasić, K., Knez, Ž., & Leitgeb, M. (2021). A comprehensive study of the antibacterial activity of bioactive juice and extracts from pomegranate (Punica granatum l.) peels and seeds. *Plants, 10*(8). https://doi.org/10.3390/plants10081554

Larrauri, J. A., Rupérez, P., & Saura Calixto, F. (1997). Pineapple shell as a source of dietary fiber with associated polyphenols. *Journal of Agricultural and Food Chemistry, 45*(10), 4028–4031. https://doi.org/10.1021/JF970450J

Larrea, M. A., Chang, Y. K., & Martinez-Bustos, F. (2005). Some functional properties of extruded orange pulp and its effect on the quality of cookies. *LWT—Food Science and Technology, 38*(3), 213–220. https://doi.org/10.1016/J.LWT.2004.05.014

Ledesma-Escobar, C. A., & de Castro, M. D. L. (2014). Towards a comprehensive exploitation of citrus. *Trends in Food Science & Technology, 39*(1), 63–75.

Leontowicz, H., Leontowicz, M., Gorinstein, S., Martin-belloso, O., & Trakhtenberg, S. (2007). Apple peels and pulp as a source of bioactive compounds and their influence on digestibility and lipid profile in normal and atherogenic rats. *Medycyna Weterynaryjna, 63*(11).

Li, Y., Fabiano-Tixier, A. S., Vian, M. A., & Chemat, F. (2013). Solvent-free microwave extraction of bioactive compounds provides a tool for green analytical chemistry. *TrAC Trends in Analytical Chemistry, 47*, 1–11. https://doi.org/10.1016/J.TRAC.2013.02.007

Liu, Y., Friesen, J. B., McAlpine, J. B., Lankin, D. C., Chen, S. N., & Pauli, G. F. (2018). Natural deep eutectic solvents: Properties, applications, and perspectives. *Journal of Natural Products, 81*(3), 679–690. https://doi.org/10.1021/ACS.JNATPROD.7B00945/SUPPL_FILE/NP7B00945_SI_001.PDF

Llobera, A., & Cañellas, J. (2007). Dietary fibre content and antioxidant activity of Manto Negro red grape (Vitis vinifera): Pomace and stem. *Food Chemistry, 101*(2), 659–666. https://doi.org/10.1016/J.FOODCHEM.2006.02.025

Llobera, A., & Cañellas, J. (2008). Antioxidant activity and dietary fibre of Prensal Blanc white grape (Vitis vinifera) by-products. *International Journal of Food Science & Technology, 43*(11), 1953–1959. https://doi.org/10.1111/J.1365-2621.2008.01798.X

Lohani, U. C., & Muthukumarappan, K. (2017). Process optimization for antioxidant enriched sorghum flour and apple pomace based extrudates using liquid CO_2 assisted extrusion. *LWT, 86*, 544–554. https://doi.org/10.1016/J.LWT.2017.08.034

Loizzo, M. R., Pacetti, D., Lucci, P., Núñez, O., Menichini, F., Frega, N. G., & Tundis, R. (2015). Prunus persica var. platycarpa (Tabacchiera Peach): Bioactive compounds and antioxidant activity of pulp, peel and seed ethanolic extracts. *Plant Foods for Human Nutrition, 70*(3), 331–337. https://doi.org/10.1007/S11130-015-0498-1/TABLES/4

Lončarić, A., Matanović, K., Ferrer, P., Kovač, T., Šarkanj, B., Babojelić, M. S., & Lores, M. (2020). Peel of traditional apple varieties as a great source of bioactive compounds: Extraction by micro-matrix solid-phase dispersion. *Foods, 9*(1), 80. https://doi.org/10.3390/FOODS9010080

Lu, Y., & Foo, L. Y. (1997). Identification and quantification of major polyphenols in apple pomace. *Food Chemistry, 59*(2), 187–194. https://doi.org/10.1016/S0308-8146(96)00287-7

Lu, Y., & Yeap Foo, L. (2000). Antioxidant and radical scavenging activities of polyphenols from apple pomace. *Food Chemistry, 68*(1), 81–85. https://doi.org/10.1016/S0308-8146(99)00167-3

Luchese, C. L., Uranga, J., Spada, J. C., Tessaro, I. C., & de la Caba, K. (2018). Valorisation of blueberry waste and use of compression to manufacture sustainable starch films with enhanced properties. *International Journal of Biological Macromolecules, 115*, 955–960. https://doi.org/10.1016/J.IJBIOMAC.2018.04.162

Luo, F., Fu, Y., Xiang, Y., Yan, S., Hu, G., Huang, X., Huang, G., Sun, C., Li, X., & Chen, K. (2014). Identification and quantification of gallotannins in mango (Mangifera indica L.) kernel and peel and their antiproliferative activities. *Journal of Functional Foods*, *8*(1), 282–291. https://doi.org/10.1016/J.JFF.2014.03.030

Lyu, F., Luiz, S. F., Azeredo, D. R. P., Cruz, A. G., Ajlouni, S., & Ranadheera, C. S. (2020). Apple pomace as a functional and healthy ingredient in food products: A review. *Processes*, *8*(3), 319. https://doi.org/10.3390/PR8030319

Ma, H., Johnson, S. L., Liu, W., Dasilva, N. A., Meschwitz, S., Dain, J. A., & Seeram, N. P. (2018). Evaluation of polyphenol anthocyanin-enriched extracts of blackberry, black raspberry, blueberry, cranberry, red raspberry, and strawberry for free radical scavenging, reactive carbonyl species trapping, anti-glycation, anti-β-amyloid aggregation, and microglial neuroprotective effects. *International Journal of Molecular Sciences*, *19*(2), 461. https://doi.org/10.3390/IJMS19020461

Machado, A. P. D. F., Pasquel-Reátegui, J. L., Barbero, G. F., & Martínez, J. (2015). Pressurized liquid extraction of bioactive compounds from blackberry (Rubus fruticosus L.) residues: A comparison with conventional methods. *Food Research International*, *77*, 675–683. https://doi.org/10.1016/J.FOODRES.2014.12.042

Mahato, N., Sinha, M., Sharma, K., Koteswararao, R., & Cho, M. H. (2019). Modern extraction and purification techniques for obtaining high purity food-grade bioactive compounds and value-added co-products from citrus wastes. *Foods*, *8*(11), 523. https://doi.org/10.3390/FOODS8110523

Majerska, J., Michalska, A., & Figiel, A. (2019). A review of new directions in managing fruit and vegetable processing by-products. *Trends in Food Science & Technology*, *88*, 207–219. https://doi.org/10.1016/J.TIFS.2019.03.021

Makris, D. P., & Lalas, S. (2020). Glycerol and glycerol-based deep eutectic mixtures as emerging green solvents for polyphenol extraction: The evidence so far. *Molecules*, *25*(24), 5842. https://doi.org/10.3390/MOLECULES25245842

Mallek-Ayadi, S., Bahloul, N., & Kechaou, N. (2018). Chemical composition and bioactive compounds of Cucumis melo L. seeds: Potential source for new trends of plant oils. *Process Safety and Environmental Protection*, *113*, 68–77. https://doi.org/10.1016/J.PSEP.2017.09.016

Maragò, E., Iacopini, P., Camangi, F., Scattino, C., Ranieri, A., Stefani, A., & Sebastiani, L. (2015). Phenolic profile and antioxidant activity in apple juice and pomace: Effects of different storage conditions. *Fruits*, *70*(4), 213–223. https://doi.org/10.1051/FRUITS/2015015

Marín, F. R., Soler-Rivas, C., Benavente-García, O., Castillo, J., & Pérez-Alvarez, J. A. (2007). By-products from different citrus processes as a source of customized functional fibres. *Food Chemistry*, *100*(2), 736–741. https://doi.org/10.1016/J.FOODCHEM.2005.04.040

Marinelli, V., Padalino, L., Nardiello, D., Del Nobile, M. A., & Conte, A. (2015). New approach to enrich pasta with polyphenols from Grape Marc. *Journal of Chemistry*, *2015*. https://doi.org/10.1155/2015/734578

Mateos-Aparicio, I., & Matias, A. (2019). Food industry processing by-products in foods. In *The role of alternative and innovative food ingredients and products in consumer wellness* (pp. 239–281). https://doi.org/10.1016/B978-0-12-816453-2.00009-7

Mattos, G. N., Tonon, R. V., Furtado, A. A. L., & Cabral, L. M. C. (2017). Grape by-product extracts against microbial proliferation and lipid oxidation: A review. *Journal of the Science of Food and Agriculture*, *97*(4), 1055–1064. https://doi.org/10.1002/JSFA.8062

Maurya, A. K., Pandey, K. R., Rai, D., Porwal, P., & Rai, D. C. (2015). Waste product of fruits and vegetables processing as a source of dietaryFibre: A review. *Trends in Biosciences*, *8*(19), 5129–5140. www.researchgate.net/publication/319350706

McDaniel, J. G., & Yethiraj, A. (2019). Understanding the properties of ionic liquids: Electrostatics, structure factors, and their sum rules. *Journal of Physical Chemistry B*, *123*(16), 3499–3512. https://doi.org/10.1021/ACS.JPCB.9B00963/SUPPL_FILE/JP9B00963_SI_002.PDF

Melgar, B., Dias, M. I., Barros, L., Ferreira, I. C. F. R., Rodriguez-Lopez, A. D., & Garcia-Castello, E. M. (2019). Ultrasound and microwave assisted extraction of opuntia fruit peels bio-compounds: Optimization and comparison using RSM-CCD. *Molecules*, *24*(19), 3618. https://doi.org/10.3390/MOLECULES24193618

Mena-García, A., Ruiz-Matute, A. I., Soria, A. C., & Sanz, M. L. (2019). Green techniques for extraction of bioactive carbohydrates. *TrAC Trends in Analytical Chemistry*, *119*, 115612. https://doi.org/10.1016/J.TRAC.2019.07.023

Menichini, F., Tundis, R., Bonesi, M., De Cindio, B., Loizzo, M. R., Conforti, F., Statti, G. A., Menabeni, R., Bettini, R., & Menichini, F. (2011). Chemical composition and bioactivity of Citrus medica L. cv. Diamante essential oil obtained by hydrodistillation, cold-pressing and supercritical carbon dioxide extraction. *Natural Product Research*, *25*(8), 789–799. https://doi.org/10.1080/14786410902900085

Meyer, A. S., Jepsen, S. M., & Sørensen, N. S. (1998). Enzymatic release of antioxidants for human low-density lipoprotein from grape pomace. *Journal of Agricultural and Food Chemistry*, *46*(7), 2439–2446. https://doi.org/10.1021/JF971012F

Milala, J., Kosmala, M., Sójka, M., Kołodziejczyk, K., Zbrzeåniak, M., & Markowski, J. (2013). Plum pomaces as a potential source of dietary fibre: Composition and antioxidant properties. *Journal of Food Science and Technology*, *50*(5), 1012–1017. https://doi.org/10.1007/S13197-011-0601-Z/TABLES/3

Mildner-Szkudlarz, S., Bajerska, J., Zawirska-Wojtasiak, R., & Górecka, D. (2013). White grape pomace as a source of dietary fibre and polyphenols and its effect on physical and nutraceutical characteristics of wheat biscuits. *Journal of the Science of Food and Agriculture*, *93*(2), 389–395. https://doi.org/10.1002/JSFA.5774

Milea, A. S., Vasile, A. M., Cîrciumaru, A., Dumitrascu, L., Barbu, V., Râpeanu, G., Bahrim, G. E., & Stanciuc, N. (2019). Valorizations of sweet cherries skins phytochemicals by extraction, microencapsulation and development of value-added food products. *Foods*, *8*(6), 188. https://doi.org/10.3390/FOODS8060188

Mir, S. A., Bosco, S. J. D., Shah, M. A., Santhalakshmy, S., & Mir, M. M. (2017). Effect of apple pomace on quality characteristics of brown rice based cracker. *Journal of the Saudi Society of Agricultural Sciences*, *16*(1), 25–32. https://doi.org/10.1016/J.JSSAS.2015.01.001

Misran, A., Padmanabhan, P., Sullivan, J. A., Khanizadeh, S., & Paliyath, G. (2015). Nature des composé s phénoliques et volatils chez la fraise et conséquences d'un traitement à l'hexanal avant la cueillette sur leur profil. *Canadian Journal of Plant Science*, *95*(1), 115–126. https://doi.org/10.4141/CJPS-2014-245/ASSET/IMAGES/CJPS-2014-245TAB4.GIF

Mohapatra, D., & Kate, A. (2019). Extraction techniques of color pigments from fruits and vegetables. In K. A. Khan, M. R. Goyal, & A. A. Kalne (Eds.), *Processing of fruits and-vegetables from farm to fork* (pp. 175–200). Apple Academic Press, CRC Press, Taylor and Francis.

Molet-Rodríguez, A., Salvia-Trujillo, L., & Martín-Belloso, O. (2018). Beverage emulsions: Key aspects of their formulation and physicochemical stability. *Beverages*, *4*(3), 70. https://doi.org/10.3390/BEVERAGES4030070

Monagas, M., Gómez-Cordovés, C., Bartolomé, B., Laureano, O., & Ricardo da Silva, J. M. (2003). Monomeric, oligomeric, and polymeric flavan-3-ol composition of wines and grapes from Vitis vinifera L. Cv. Graciano, Tempranillo, and Cabernet Sauvignon. *J. Agric. Food Chem*. https://doi.org/10.1021/jf030325

Montero-Calderon, A., Cortes, C., Zulueta, A., Frigola, A., & Esteve, M. J. (2019). Green solvents and ultrasound-assisted extraction of bioactive orange (Citrus sinensis) peel compounds. *Scientific Reports*, *9*(1), 1–8. https://doi.org/10.1038/s41598-019-52717-1

Montevecchi, G., Vasile Simone, G., Masino, F., Bignami, C., & Antonelli, A. (2012). Physical and chemical characterization of Pescabivona, a Sicilian white flesh peach cultivar [Prunus persica (L.) Batsch]. *Food Research International*, *45*(1), 123–131. https://doi.org/10.1016/J.FOODRES.2011.10.019

Moorthy, I. G., Maran, J. P., Surya, S. M., Naganyashree, S., & Shivamathi, C. S. (2015). Response surface optimization of ultrasound assisted extraction of pectin from pomegranate peel. *International Journal of Biological Macromolecules, 72,* 1323–1328. https://doi.org/10.1016/J.IJBIOMAC.2014.10.037

Mrudula, S., & Anitharaj, R. (2011). *Pectinase production in solid state fermentation by Aspergillus niger using orange peel as substrate.* www.researchgate.net/publication/289401264

Muradoğlu, F., & Küçük, O. (2018). Determination of bioactive composition of some peach cultivars. *Journal of Animal and Plant Sciences, 28*(2), 533–538.

Murador, D. C., Braga, A. R. C., Martins, P. L. G., Mercadante, A. Z., & de Rosso, V. V. (2019). Ionic liquid associated with ultrasonic-assisted extraction: A new approach to obtain carotenoids from orange peel. *Food Research International, 126,* 108653. https://doi.org/10.1016/J.FOODRES.2019.108653

Mushtaq, M., Gani, A., Gani, A., Punoo, H. A., & Masoodi, F. A. (2018). Use of pomegranate peel extract incorporated zein film with improved properties for prolonged shelf life of fresh Himalayan cheese (Kalari/kradi). *Innovative Food Science & Emerging Technologies, 48,* 25–32. https://doi.org/10.1016/J.IFSET.2018.04.020

Naczk, M., & Shahidi, F. (2004). Extraction and analysis of phenolics in food. *Journal of Chromatography A, 1054*(1–2), 95–111. https://doi.org/10.1016/J.CHROMA.2004.08.059

Nawirska, A., & Kwaśniewska, M. (2005). Dietary fibre fractions from fruit and vegetable processing waste. *Food Chemistry, 91*(2), 221–225. https://doi.org/10.1016/J.FOODCHEM.2003.10.005

Ndayishimiye, J., & Chun, B. S. (2017). Optimization of carotenoids and antioxidant activity of oils obtained from a co-extraction of citrus (Yuzu ichandrin) by-products using supercritical carbon dioxide. *Biomass and Bioenergy, 106,* 1–7. https://doi.org/10.1016/j.biombioe.2017.08.014

Nemetz, N. J., Schieber, A., & Weber, F. (2021). Application of crude pomace powder of chokeberry, bilberry, and elderberry as a coloring foodstuff. *Molecules, 26*(9), 2689. https://doi.org/10.3390/MOLECULES26092689

Noor Aziah, A. A., Lee Min, W., & Bhat, R. (2011). Nutritional and sensory quality evaluation of sponge cake prepared by incorporation of high dietary fiber containing mango (Mangifera indica var. Chokanan) pulp and peel flours. *International Journal of Food Sciences and Nutrition, 62*(6), 559–567. https://doi.org/10.3109/09637486.2011.562883

Nordin, N. H., Kaida, N., Othman, N. A., Akhir, F. N. M., & Hara, H. (2020). Reducing food waste: Strategies for household waste management to minimize the impact of climate change and contribute to Malaysia's sustainable development. *IOP Conference Series: Earth and Environmental Science, 479*(1), 012035. https://doi.org/10.1088/1755-1315/479/1/012035

Ochoa-Velasco, C. E., & Guerrero-Beltrán, J. Á. (2016). The effects of modified atmospheres on prickly pear (Opuntia albicarpa) stored at different temperatures. *Postharvest Biology and Technology, 111,* 314–321. https://doi.org/10.1016/J.POSTHARVBIO.2015.09.028

Ordóñez-Santos, L. E., Esparza-Estrada, J., & Vanegas-Mahecha, P. (2021). Ultrasound-assisted extraction of total carotenoids from mandarin epicarp and application as natural colorant in bakery products. *LWT, 139,* 110598. https://doi.org/10.1016/J.LWT.2020.110598

Otoni, C. G., Avena-Bustillos, R. J., Azeredo, H. M. C., Lorevice, M. V., Moura, M. R., Mattoso, L. H. C., & McHugh, T. H. (2017). Recent advances on edible films based on fruits and vegetables—A review. *Comprehensive Reviews in Food Science and Food Safety, 16*(5), 1151–1169. https://doi.org/10.1111/1541-4337.12281

Özvural, E. B., & Vural, H. (2011). Grape seed flour is a viable ingredient to improve the nutritional profile and reduce lipid oxidation of frankfurters. *Meat Science, 88*(1), 179–183. https://doi.org/10.1016/J.MEATSCI.2010.12.022

Pal, C. B. T., & Jadeja, G. C. (2020). Microwave-assisted extraction for recovery of polyphenolic antioxidants from ripe mango (Mangifera indica L.) peel using lactic acid/sodium acetate deep eutectic mixtures. *Food Science and Technology International*, *26*(1), 78–92. https://doi.org/10.1177/1082013219870010/ASSET/IMAGES/LARGE/10.1177_1082013219870010-FIG4.JPEG

Panzella, L., Moccia, F., Nasti, R., Marzorati, S., Verotta, L., & Napolitano, A. (2020). Bioactive phenolic compounds from agri-food wastes: An update on green and sustainable extraction methodologies. *Frontiers in Nutrition*, 7, 538169. https://doi.org/10.3389/FNUT.2020.00060/BIBTEX

Pap, N., Beszédes, S., Pongrácz, E., Myllykoski, L., Gábor, M., Gyimes, E., Hodúr, C., & Keiski, R. L. (2013). Microwave-assisted extraction of anthocyanins from black currant marc. *Food and Bioprocess Technology*, *6*(10), 2666–2674. https://doi.org/10.1007/S11947-012-0964-9/TABLES/3

Pazos, M., Gallardo, J. M., Torres, J. L., & Medina, I. (2005). Activity of grape polyphenols as inhibitors of the oxidation of fish lipids and frozen fish muscle. *Food Chemistry*, *92*(3), 547–557. https://doi.org/10.1016/J.FOODCHEM.2004.07.036

Peschel, W., Sánchez-Rabaneda, F., Diekmann, W., Plescher, A., Gartzía, I., Jiménez, D., Lamuela-Raventós, R., Buxaderas, S., & Codina, C. (2006). An industrial approach in the search of natural antioxidants from vegetable and fruit wastes. *Food Chemistry*, *97*(1), 137–150. https://doi.org/10.1016/J.FOODCHEM.2005.03.033

Picot-Allain, C., Mahomoodally, M. F., Ak, G., & Zengin, G. (2021). Conventional versus green extraction techniques—A comparative perspective. *Current Opinion in Food Science*, *40*, 144–156. https://doi.org/10.1016/J.COFS.2021.02.009

Plainfossé, H., Trinel, M., Verger-Dubois, G., Azoulay, S., Burger, P., & Fernandez, X. (2020). Valorisation of ribes nigrum L. Pomace, an agri-food by-product to design a new cosmetic active. *Cosmetics*, *7*(3), 56. https://doi.org/10.3390/COSMETICS7030056

Plazzotta, S., Ibarz, R., Manzocco, L., & Martín-Belloso, O. (2020). Optimizing the antioxidant biocompound recovery from peach waste extraction assisted by ultrasounds or microwaves. *Ultrasonics Sonochemistry*, *63*, 104954. https://doi.org/10.1016/J.ULTSONCH.2019.104954

Plazzotta, S., Ibarz, R., Manzocco, L., & Martín-Belloso, O. (2021). Modelling the recovery of biocompounds from peach waste assisted by pulsed electric fields or thermal treatment. *Journal of Food Engineering*, *290*. https://doi.org/10.1016/j.jfoodeng.2020.110196

Pourzaki, A., Mirzaee, H., & Hemmati Kakhki, A. (2013). Using pulsed electric field for improvement of components extraction of saffron (Crocus sativus) stigma and its pomace. *Journal of Food Processing and Preservation*, *37*(5), 1008–1013. https://doi.org/10.1111/J.1745-4549.2012.00749.X

Prakash Maran, J., Sivakumar, V., Thirugnanasambandham, K., & Sridhar, R. (2013). Optimization of microwave assisted extraction of pectin from orange peel. *Carbohydrate Polymers*, *97*(2), 703–709. https://doi.org/10.1016/J.CARBPOL.2013.05.052

Priefert, H., Rabenhorst, J., & Steinbüchel, A. (2001). Biotechnological production of vanillin. *Applied Microbiology and Biotechnology*, *56*(3–4), 296–314. https://doi.org/10.1007/S002530100687/METRICS

Priyadarsani, S., Patel, A. S., Kar, A., & Dash, S. (2021). Process optimization for the supercritical carbondioxide extraction of lycopene from ripe grapefruit (Citrus paradisi) endocarp. *Scientific Reports*, *11*(1), 1–8. https://doi.org/10.1038/s41598-021-89772-6

Priyadarshi, R., Sauraj Kumar, B., Deeba, F., Kulshreshtha, A., & Negi, Y. S. (2018). Chitosan films incorporated with Apricot (Prunus armeniaca) kernel essential oil as active food packaging material. *Food Hydrocolloids*, *85*, 158–166. https://doi.org/10.1016/J.FOODHYD.2018.07.003

Puértolas, E., & Barba, F. J. (2016). Electrotechnologies applied to valorization of by-products from food industry: Main findings, energy and economic cost of their industrialization. *Food and Bioproducts Processing*, *100*, 172–184. https://doi.org/10.1016/J.FBP.2016.06.020

Puri, M., Sharma, D., & Barrow, C. J. (2012). Enzyme-assisted extraction of bioactives from plants. *Trends in Biotechnology*, *30*(1), 37–44. https://doi.org/10.1016/j.tibtech.2011.06.014

Quiles, A., Llorca, E., Schmidt, C., Reißner, A. M., Struck, S., Rohm, H., & Hernando, I. (2018). Use of berry pomace to replace flour, fat or sugar in cakes. *International Journal of Food Science & Technology*, *53*(6), 1579–1587. https://doi.org/10.1111/IJFS.13765

Radošević, K., Ćurko, N., Gaurina Srček, V., Cvjetko Bubalo, M., Tomašević, M., Kovačević Ganić, K., & Radojčić Redovniković, I. (2016). Natural deep eutectic solvents as beneficial extractants for enhancement of plant extracts bioactivity. *LWT*, *73*, 45–51. https://doi.org/10.1016/J.LWT.2016.05.037

Raeissi, S., Diaz, S., Espinosa, S., Peters, C. J., & Brignole, E. A. (2008). Ethane as an alternative solvent for supercritical extraction of orange peel oils. *The Journal of Supercritical Fluids*, *45*(3), 306–313. https://doi.org/10.1016/J.SUPFLU.2008.01.008

Rainero, G., Bianchi, F., Rizzi, C., Cervini, M., Giuberti, G., & Simonato, B. (2022). Breadstick fortification with red grape pomace: Effect on nutritional, technological and sensory properties. *Journal of the Science of Food and Agriculture*, *102*(6), 2545–2552. https://doi.org/10.1002/JSFA.11596

Rajha, H. N., Boussetta, N., Louka, N., Maroun, R. G., & Vorobiev, E. (2015). Effect of alternative physical pretreatments (pulsed electric field, high voltage electrical discharges and ultrasound) on the dead-end ultrafiltration of vine-shoot extracts. *Separation and Purification Technology*, *146*, 243–251. https://doi.org/10.1016/J.SEPPUR.2015.03.058

Ran, X., Zhang, M., Wang, Y., & Adhikari, B. (2018). Novel technologies applied for recovery and value addition of high value compounds from plant byproducts: A review. *Critical Reviews in Food Science and Nutrition*, *59*(3), 450–461. https://doi.org/10.1080/10408398.2017.1377149

Rana, S., Gupta, S., Rana, A., & Bhushan, S. (2015). Functional properties, phenolic constituents and antioxidant potential of industrial apple pomace for utilization as active food ingredient. *Food Science and Human Wellness*, *4*(4), 180–187. https://doi.org/10.1016/J.FSHW.2015.10.001

Rao, P., & Rathod, V. (2019). Valorization of food and agricultural waste: A step towards greener guture. *The Chemical Record*, *19*(9), 1858–1871. https://doi.org/10.1002/TCR.201800094

Rashid, F., Bao, Y., Ahmed, Z., & Huang, J. Y. (2020). Effect of high voltage atmospheric cold plasma on extraction of fenugreek galactomannan and its physicochemical properties. *Food Research International*, *138*, 109776. https://doi.org/10.1016/J.FOODRES.2020.109776

Redondo, D., Venturini, M. E., Luengo, E., Raso, J., & Arias, E. (2018). Pulsed electric fields as a green technology for the extraction of bioactive compounds from thinned peach by-products. *Innovative Food Science & Emerging Technologies*, *45*, 335–343. https://doi.org/10.1016/J.IFSET.2017.12.004

Reißner, A. M., Al-Hamimi, S., Quiles, A., Schmidt, C., Struck, S., Hernando, I., Turner, C., & Rohm, H. (2019). Composition and physicochemical properties of dried berry pomace. *Journal of the Science of Food and Agriculture*, *99*(3), 1284–1293. https://doi.org/10.1002/JSFA.9302

Rezzadori, K., Benedetti, S., & Amante, E. R. (2012). Proposals for the residues recovery: Orange waste as raw material for new products. *Food and Bioproducts Processing*, *90*(4), 606–614. https://doi.org/10.1016/J.FBP.2012.06.002

Ribeiro, S. M. R., Barbosa, L. C. A., Queiroz, J. H., Knödler, M., & Schieber, A. (2008). Phenolic compounds and antioxidant capacity of Brazilian mango (Mangifera indica L.) varieties. *Food Chemistry*, *110*(3), 620–626. https://doi.org/10.1016/J.FOODCHEM.2008.02.067

Rico, X., Gullón, B., Alonso, J. L., & Yáñez, R. (2020). Recovery of high value-added compounds from pineapple, melon, watermelon and pumpkin processing by-products: An overview. *Food Research International*, *132*, 109086. https://doi.org/10.1016/J.FOODRES.2020.109086

Rimando, A. M., & Perkins-Veazie, P. M. (2005). Determination of citrulline in watermelon rind. *Journal of Chromatography A*, *1078*(1–2), 196–200. https://doi.org/10.1016/J.CHROMA.2005.05.009

Rivera, S. M., & Canela-Garayoa, R. (2012). Analytical tools for the analysis of carotenoids in diverse materials. *Journal of Chromatography A*, *1224*, 1–10. https://doi.org/10.1016/J.CHROMA.2011.12.025

Robbins, R. J. (2003). Phenolic acids in foods: An overview of analytical methodology. *Journal of Agricultural and Food Chemistry*, *51*(10), 2866–2887. https://doi.org/10.1021/JF026182T

Rodriguez, E. B., Vidallon, M. L. P., Mendoza, D. J. R., & Reyes, C. T. (2016). Health-promoting bioactivities of betalains from red dragon fruit (Hylocereus polyrhizus (Weber) Britton and Rose) peels as affected by carbohydrate encapsulation. *Journal of the Science of Food and Agriculture*, *96*(14), 4679–4689. https://doi.org/10.1002/JSFA.7681

Rodriguez-Amaya, D. B. (2015). Nomenclature, structures, and physical and chemical properties. In *Food carotenoids: Chemistry, biology and technology* (pp. 1–23). IFT Press-Wiley.

Rodríguez García, S. L., & Raghavan, V. (2021). Green extraction techniques from fruit and vegetable waste to obtain bioactive compounds—A review. *Critical Reviews in Food Science and Nutrition*, *62*(23), 6446–6466. https://doi.org/10.1080/10408398.2021.1901651

Rohilla, S., & Mahanta, C. L. (2021). Optimization of extraction conditions for ultrasound-assisted extraction of phenolic compounds from tamarillo fruit (Solanum betaceum) using response surface methodology. *Journal of Food Measurement and Characterization*, *15*(2), 1763–1773. https://doi.org/10.1007/S11694-020-00751-3/FIGURES/4

Rohm, H., Brennan, C., Turner, C., Günther, E., Campbell, G., Hernando, I., Struck, S., & Kontogiorgos, V. (2015). Adding value to fruit processing waste: Innovative ways to incorporate fibers from berry pomace in baked and extruded cereal-based foods—A SUSFOOD project. *Foods*, *4*(4), 690–697. https://doi.org/10.3390/FOODS4040690

Roy, B. C., Hoshino, M., Ueno, H., Sasaki, M., & Goto, M. (2011). Supercritical carbon dioxide extraction of the volatiles from the peel of Japanese citrus fruits. *Journal of Essential Oil Research*, *19*(1), 78–84. https://doi.org/10.1080/10412905.2007.9699234

Ruiz, D., Egea, J., Tomás-Barberán, F. A., & Gil, M. I. (2005). Carotenoids from new apricot (Prunus armeniaca L.) varieties and their relationship with flesh and skin color. *Journal of Agricultural and Food Chemistry*, *53*(16), 6368–6374. https://doi.org/10.1021/JF0480703

Rupasinghe, H. P. V., & Kean, C. (2011). Polyphenol concentrations in apple processing by-products determined using electrospray ionization mass spectrometry. *Canadian Journal of Plant Science*, *88*(4), 759–762. https://doi.org/10.4141/CJPS07146

Russo, C., Maugeri, A., Lombardo, G. E., Musumeci, L., Barreca, D., Rapisarda, A., Cirmi, S., & Navarra, M. (2021). The second life of citrus fruit waste: A valuable source of bioactive compounds. *Molecules*, *26*(19), 5991. https://doi.org/10.3390/MOLECULES26195991

Saad, A. M., El-Saadony, M. T., El-Tahan, A. M., Sayed, S., Moustafa, M. A. M., Taha, A. E., Taha, T. F., & Ramadan, M. M. (2021). Polyphenolic extracts from pomegranate and watermelon wastes as substrate to fabricate sustainable silver nanoparticles with larvicidal effect against Spodoptera littoralis. *Saudi Journal of Biological Sciences*, *28*(10), 5674–5683. https://doi.org/10.1016/J.SJBS.2021.06.011

Šafranko, S., Ćorković, I., Jerković, I., Jakovljević, M., Aladić, K., Šubarić, D., & Jokić, S. (2021). Green extraction techniques for obtaining bioactive compounds from mandarin peel (Citrus unshiu var. Kuno): Phytochemical analysis and process optimization. *Foods*, *10*(5), 1043. https://doi.org/10.3390/FOODS10051043/S1

Sagar, N. A., Pareek, S., Sharma, S., Yahia, E. M., & Lobo, M. G. (2018). Fruit and vegetable waste: Bioactive compounds, their extraction, and possible utilization. *Comprehensive Reviews in Food Science and Food Safety*, *17*(3), 512–531. https://doi.org/10.1111/1541-4337.12330

Sahraoui, N., Vian, M. A., El Maataoui, M., Boutekedjiret, C., & Chemat, F. (2011). Valorization of citrus by-products using Microwave Steam Distillation (MSD). *Innovative Food Science & Emerging Technologies*, *12*(2), 163–170. https://doi.org/10.1016/J.IFSET.2011.02.002

Saini, A., Panesar, P. S., & Bera, M. B. (2021). Valuation of citrus reticulata (kinnow) peel for the extraction of lutein using ultrasonication technique. *Biomass Conversion and Biorefinery*, *11*(5), 2157–2165. https://doi.org/10.1007/S13399-020-00605-4/FIGURES/7

Saldaña, M. D. A., Martinez, E. R., Sekhon, J. K., & Vo, H. (2021). The effect of different pressurized fluids on the extraction of anthocyanins and total phenolics from cranberry pomace. *The Journal of Supercritical Fluids*, *175*, 105279. https://doi.org/10.1016/J.SUPFLU.2021.105279

Samsuri, S., Li, T. H., Ruslan, M. S. H., & Amran, N. A. (2020). Antioxidant recovery from pomegranate peel waste by integrating maceration and freeze concentration technology. *International Journal of Food Engineering*, *16*(10). https://doi.org/10.1515/IJFE-2019-0232/MACHINEREADABLECITATION/RIS

Sandhu, H. K., Sinha, P., Emanuel, N., Kumar, N., Sami, R., Khojah, E., & Al-Mushhin, A. A. M. (2021). Effect of ultrasound-assisted pretreatment on extraction efficiency of essential oil and bioactive compounds from citrus waste by-products. *Separations*, *8*(12), 244. https://doi.org/10.3390/SEPARATIONS8120244

Santos, E., Andrade, R., & Gouveia, E. (2017). Utilization of the pectin and pulp of the passion fruit from Caatinga as probiotic food carriers. *Food Bioscience*, *20*, 56–61. https://doi.org/10.1016/J.FBIO.2017.08.005

Sarris, D., & Papanikolaou, S. (2016). Biotechnological production of ethanol: Biochemistry, processes and technologies. *Engineering in Life Sciences*, *16*(4), 307–329. https://doi.org/10.1002/ELSC.201400199

Savić, I. M., Nikolić, V. D., Savić-Gajić, I. M., Kundaković, T. D., Stanojković, T. P., & Najman, S. J. (2016). Chemical composition and biological activity of the plum seed extract. *Advanced Technologies*, *5*(2), 38–45. https://doi.org/10.5937/SAVTEH1602038S

Schanes, K., Dobernig, K., & Gözet, B. (2018). Food waste matters—A systematic review of household food waste practices and their policy implications. *Journal of Cleaner Production*, *182*, 978–991. https://doi.org/10.1016/J.JCLEPRO.2018.02.030

Schieber, A., Stintzing, F. C., & Carle, R. (2001). By-products of plant food processing as a source of functional compounds—recent developments. *Trends in Food Science & Technology*, *12*(11), 401–413. https://doi.org/10.1016/S0924-2244(02)00012-2

Segat, A., Misra, N. N., Cullen, P. J., & Innocente, N. (2016). Effect of atmospheric pressure cold plasma (ACP) on activity and structure of alkaline phosphatase. *Food and Bioproducts Processing*, *98*, 181–188. https://doi.org/10.1016/J.FBP.2016.01.010

Selvam, K., Ameen, F., Amirul Islam, M., Sudhakar, C., & Selvankumar, T. (2022). Laccase production from Bacillus aestuarii KSK using Borassus flabellifer empty fruit bunch waste as a substrate and assessing their malachite green dye degradation. *Journal of Applied Microbiology*, *133*(6), 3288–3295. https://doi.org/10.1111/JAM.15670

Senica, M., Stampar, F., Veberic, R., & Mikulic-Petkovsek, M. (2017). Fruit seeds of the rosaceae family: A waste, new life, or a danger to human health? *Journal of Agricultural and Food Chemistry*, *65*(48), 10621–10629. https://doi.org/10.1021/ACS.JAFC.7B03408/ASSET/IMAGES/MEDIUM/JF-2017-03408Z_0003.GIF

Senica, M., Stampar, F., Veberic, R., & Mikulic-Petkovsek, M. (2019). Cyanogenic glycosides and phenolics in apple seeds and their changes during long term storage. *Scientia Horticulturae*, *255*, 30–36. https://doi.org/10.1016/J.SCIENTA.2019.05.022

Sharma, K., Mahato, N., Cho, M. H., & Lee, Y. R. (2017). Converting citrus wastes into value-added products: Economic and environmently friendly approaches. *Nutrition*, *34*, 29–46. https://doi.org/10.1016/J.NUT.2016.09.006

Sharma, M., Usmani, Z., Gupta, V. K., & Bhat, R. (2021). Valorization of fruits and vegetable wastes and by-products to produce natural pigments. *Critical Reviews in Biotechnology*, *41*(4), 535–563. https://doi.org/10.1080/07388551.2021.1873240

Shen, L., Xiong, X., Zhang, D., Zekrumah, M., Hu, Y., Gu, X., Wang, C., & Zou, X. (2019). Optimization of betacyanins from agricultural by-products using pressurized hot water extraction for antioxidant and in vitro oleic acid-induced steatohepatitis inhibitory activity. *Journal of Food Biochemistry, 43*(12), e13044. https://doi.org/10.1111/JFBC.13044

Shinwari, K. J., & Rao, P. S. (2018). Thermal-assisted high hydrostatic pressure extraction of nutraceuticals from saffron (Crocus sativus): Process optimization and cytotoxicity evaluation against cancer cells. *Innovative Food Science & Emerging Technologies, 48*, 296–303. https://doi.org/10.1016/J.IFSET.2018.07.003

Sicari, V., & Poiana, M. (2017). Comparison of the volatile component of the essential oil of kumquat (Fortunella margarita swingle) extracted by supercritical carbon dioxide, hydrodistillation and conventional solvent extraction. *Journal of Essential Oil Bearing Plants, 20*(1), 87–94. https://doi.org/10.1080/0972060X.2017.1282841

Sikdar, D. C., & Baruah, R. (2017). Comparative study on solvent extraction of oil from citrus fruit peels by steam distillation and its characterizations. *International Journal of Technical Research and Applications, 5*(5), 31–37. www.ijtra.com

Singh, H., Lily, M. K., & Dangwal, K. (2016). Viburnum mullaha D. DON fruit (Indian cranberry): A potential source of polyphenol with rich antioxidant, anti-elastase, anticollagenase, and anti-tyrosinase activities. *International Journal of Food Properties, 20*(8), 1729–1739. https://doi.org/10.1080/10942912.2016.1217878

Sogi, D. S., Siddiq, M., Greiby, I., & Dolan, K. D. (2013). Total phenolics, antioxidant activity, and functional properties of 'Tommy Atkins' mango peel and kernel as affected by drying methods. *Food Chemistry, 141*(3), 2649–2655. https://doi.org/10.1016/J.FOODCHEM.2013.05.053

Sogut, E., & Seydim, A. C. (2018). The effects of Chitosan and grape seed extract-based edible films on the quality of vacuum packaged chicken breast fillets. *Food Packaging and Shelf Life, 18*, 13–20. https://doi.org/10.1016/J.FPSL.2018.07.006

Sójka, M., Kołodziejczyk, K., Milala, J., Abadias, M., Viñas, I., Guyot, S., & Baron, A. (2015). Composition and properties of the polyphenolic extracts obtained from industrial plum pomaces. *Journal of Functional Foods, 12*, 168–178. https://doi.org/10.1016/J.JFF.2014.11.015

Sójka, M., & Król, B. (2009). Composition of industrial seedless black currant pomace. *European Food Research and Technology, 228*(4), 597–605. https://doi.org/10.1007/S00217-008-0968-X/FIGURES/2

Soong, Y. Y., & Barlow, P. J. (2004). Antioxidant activity and phenolic content of selected fruit seeds. *Food Chemistry, 88*(3), 411–417. https://doi.org/10.1016/J.FOODCHEM.2004.02.003

Soquetta, M. B., Terra, L. de M., & Bastos, C. P. (2018). Green technologies for the extraction of bioactive compounds in fruits and vegetables. *CyTA: Journal of Food, 16*(1), 400–412. Http://Mc.Manuscriptcentral.Com/Tcyt. https://doi.org/10.1080/19476337.2017.1411978

Stintzing, F. C., & Carle, R. (2007). Betalains—Emerging prospects for food scientists. *Trends in Food Science & Technology, 18*(10), 514–525. https://doi.org/10.1016/J.TIFS.2007.04.012

Suchan, P., Pulkrabová, J., Hajšlová, J., & Kocourek, V. (2004). Pressurized liquid extraction in determination of polychlorinated biphenyls and organochlorine pesticides in fish samples. *Analytica Chimica Acta, 520*(1–2), 193–200. https://doi.org/10.1016/J.ACA.2004.02.061

Suri, S., Singh, A., & Nema, P. K. (2021). Recent advances in valorization of citrus fruits processing waste: A way forward towards environmental sustainability. *Food Science and Biotechnology, 30*(13), 1601–1626.

Syed, Q. A., Ishaq, A., Rahman, U. U., Aslam, S., & Shukat, R. (2017). Pulsed electric field technology in food preservation: A review. *Journal of Nutritional Health & Food Engineering, 6*(6), 168–172. https://doi.org/10.15406/jnhfe.2017.06.00219

Tańska, M., Roszkowska, B., Czaplicki, S., Borowska, E. J., Bojarska, J., & Dąbrowska, A. (2016). Effect of fruit pomace addition on shortbread cookies to improve their physical and nutritional values. *Plant Foods for Human Nutrition, 71*(3), 307–313. https://doi.org/10.1007/S11130-016-0561-6/FIGURES/2

Tejada-Ortigoza, V., Garcia-Amezquita, L. E., Serment-Moreno, V., Torres, J. A., & Welti-Chanes, J. (2017). Moisture sorption isotherms of high pressure treated fruit peels used as dietary fiber sources. *Innovative Food Science & Emerging Technologies, 43*, 45–53. https://doi.org/10.1016/J.IFSET.2017.07.023

Tengku Norsalwani, T. L., & Nik Norulaini, N. A. (2012). Utilization of lignocellulosic wastes as a carbon source for the production of bacterial cellulases under solid state fermentation. *International Journal of Environmental Science and Development, 3*(2), 136–140.

Teuber, R., & Jensen, J. D. (2020). Definitions, measurement, and drivers of food loss and waste. *Food Industry Wastes: Assessment and Recuperation of Commodities*, 3–18. https://doi.org/10.1016/B978-0-12-817121-9.00001-2

Thirugnanasambandham, K., & Sivakumar, V. (2017). Microwave assisted extraction process of betalain from dragon fruit and its antioxidant activities. *Journal of the Saudi Society of Agricultural Sciences, 16*(1), 41–48. https://doi.org/10.1016/J.JSSAS.2015.02.001

Tilay, A., Bule, M., Kishenkumar, J., & Annapure, U. (2008). Preparation of ferulic acid from agricultural wastes: Its improved extraction and purification. *Journal of Agricultural and Food Chemistry, 56*(17), 7644–7648. https://doi.org/10.1021/JF801536T

Tiwari, B. K. (2015). Ultrasound: A clean, green extraction technology. *TrAC Trends in Analytical Chemistry, 71*, 100–109. https://doi.org/10.1016/J.TRAC.2015.04.013

Trigo, J. P., Alexandre, E. M. C., Silva, S., Costa, E., Saraiva, J. A., & Pintado, M. (2020). Study of viability of high pressure extract from pomegranate peel to improve carrot juice characteristics. *Food & Function, 11*(4), 3410–3419. https://doi.org/10.1039/C9FO02922B

Trusheva, B., Petkov, H., Popova, M., Dimitrova, L., Zaharieva, M., Tsvetkova, I., Najdenski, H., & Bankova, V. (2019). 'Green' approach to propolis extraction: Natural deep eutectic solvents. *Comptes Rendus de L'Academie Bulgare Des Sciences, 72*(7), 897–905. https://doi.org/10.7546/CRABS.2019.07.06

Tseng, A., & Zhao, Y. (2013). Wine grape pomace as antioxidant dietary fibre for enhancing nutritional value and improving storability of yogurt and salad dressing. *Food Chemistry, 138*(1), 356–365. https://doi.org/10.1016/J.FOODCHEM.2012.09.148

Tumbas Šaponjac, V., Ćetković, G., Čanadanović-Brunet, J., Pajin, B., Djilas, S., Petrović, J., Lončarević, I., Stajčić, S., & Vulić, J. (2016). Sour cherry pomace extract encapsulated in whey and soy proteins: Incorporation in cookies. *Food Chemistry, 207*, 27–33. https://doi.org/10.1016/J.FOODCHEM.2016.03.082

Uma, G., Gomathi, D., Muthulakshmi, C., & Gopalakrishnan, V. K. (2010). *Production, purification and characterization of invertase by Aspergillus flavus using fruit peel waste as substrate.* www.researchgate.net/publication/284079143

Umesh Hebbar, H., Sumana, B., & Raghavarao, K. S. M. S. (2008). Use of reverse micellar systems for the extraction and purification of bromelain from pineapple wastes. *Bioresource Technology, 99*(11), 4896–4902. https://doi.org/10.1016/J.BIORTECH.2007.09.038

U.S. Department of Agriculture, Foreign Agricultural Service (USDA/FAS), (2014). *Fresh Deciduous Fruit (Apples, Grapes, & Pears): World Markets and Trade*, October 31, Available at: https://apps.fas.usda.gov/newgainapi/api/Report/DownloadReportByFile Name?fileName=Fresh+Deciduous+Fruit+Annual_Buenos+Aires_Argentina_10-31-2014. pdf (Accessed May 2023).

Uwineza, P. A., & Waśkiewicz, A. (2020). Recent advances in supercritical fluid extraction of natural bioactive compounds from natural plant materials. *Molecules, 25*(17), 3847. https://doi.org/10.3390/MOLECULES25173847

Van Heerden, I., Cronjé, C., Swart, S. H., & Kotzé, J. M. (2002). Microbial, chemical and physical aspects of citrus waste composting. *Bioresource Technology, 81*(1), 71–76. https://doi.org/10.1016/S0960-8524(01)00058-X

Venkatesh, M., Pushpalatha, P. B., Sheela, K. B., & Girija, D. (2009). Microbial pectinase from tropical fruit wastes. *Journal of Tropical Agriculture, 47*(1), 67–69. www.jtropag.kau.in/index.php/ojs2/article/view/207

Vernes, L., Vian, M., & Chemat, F. (2019). Ultrasound and microwaveas green tools for solid-liquid extraction. In C. F. Poole (Ed.), *InLiquid-phase extraction* (pp. 355–374). Elsevier.

Viladomiu, M., Hontecillas, R., Lu, P., & Bassaganya-Riera, J. (2013). Preventive and prophylactic mechanisms of action of pomegranate bioactive constituents. *Evidence-Based Complementary and Alternative Medicine, 2013.* https://doi.org/10.1155/2013/789764

Vilariño, M. V., Franco, C., & Quarrington, C. (2017). Food loss and waste reduction as an integral part of a circular economy. *Frontiers in Environmental Science, 5*(May), 257971. https://doi.org/10.3389/FENVS.2017.00021/BIBTEX

Vorobyova, V., Skiba, M., Miliar, Y., & Frolenkova, S. (2021). Enhanced phenolic compounds extraction from apricot pomace by natural deep eutectic solvent combined with ultrasonic-assisted extraction. *Journal of Chemical Technology and Metallurgy, 56*(5), 919–931.

Vu, H. T., Scarlett, C. J., & Vuong, Q. V. (2018). Phenolic compounds within banana peel and their potential uses: A review. *Journal of Functional Foods, 40*, 238–248. https://doi.org/10.1016/J.JFF.2017.11.006

Vu, H. T., Scarlett, C. J., & Vuong, Q. V. (2019). Maximising recovery of phenolic compounds and antioxidant properties from banana peel using microwave assisted extraction and water. *Journal of Food Science and Technology, 56*(3), 1360–1370. https://doi.org/10.1007/S13197-019-03610-2/FIGURES/3

Wadhwa, M., & Bakshi, M. P. S. (2013). *Utilization of fruit and vegetable wastes as livestock feed and as substrates for generation of other value-added products.* www.fao.org/

Walker, R. P., Battistelli, A., Bonghi, C., Drincovich, M. F., Falchi, R., Lara, M. V., Moscatello, S., Vizzotto, G., & Famiani, F. (2020). Non-structural carbohydrate metabolism in the flesh of stone fruits of the genus prunus (Rosaceae)—A review. *Frontiers in Plant Science, 11*, 549921. https://doi.org/10.3389/FPLS.2020.549921/BIBTEX

Walker, R. P., Tseng, A., Cavender, G., Ross, A., & Zhao, Y. (2014). Physicochemical, nutritional, and sensory qualities of wine grape pomace fortified baked goods. *Journal of Food Science, 79*(9), S1811–S1822. https://doi.org/10.1111/1750-3841.12554

Wang, L., Boussetta, N., Lebovka, N., & Vorobiev, E. (2020). Cell disintegration of apple peels induced by pulsed electric field and efficiency of bio-compound extraction. *Food and Bioproducts Processing, 122*, 13–21. https://doi.org/10.1016/J.FBP.2020.03.004

Wang, W., Ma, X., Xu, Y., Cao, Y., Jiang, Z., Ding, T., Ye, X., & Liu, D. (2015). Ultrasound-assisted heating extraction of pectin from grapefruit peel: Optimization and comparison with the conventional method. *Food Chemistry, 178*, 106–114. https://doi.org/10.1016/J.FOODCHEM.2015.01.080

Wang, X., Kristo, E., & LaPointe, G. (2019). The effect of apple pomace on the texture, rheology and microstructure of set type yogurt. *Food Hydrocolloids, 91*, 83–91. https://doi.org/10.1016/J.FOODHYD.2019.01.004

Wikiera, A., Mika, M., & Grabacka, M. (2015). Multicatalytic enzyme preparations as effective alternative to acid in pectin extraction. *Food Hydrocolloids, 44*, 156–161. https://doi.org/10.1016/J.FOODHYD.2014.09.018

Wilts, H., Schinkel, J., & Koop, C. (2020). Effectiveness and efficiency of food-waste prevention policies, circular economy, and food industry. In M. R. Kosseva & C. Webb (Eds.), *In food industry wastes* (pp. 19–35). Academic Press.

Wu, M., Ma, H., Ma, Z., Jin, Y., Chen, C., Guo, X., Qiao, Y., Pedersen, C. M., Hou, X., & Wang, Y. (2018). Deep eutectic solvents: Green solvents and catalysts for the preparation of pyrazine derivatives by self-condensation of d-glucosamine. *ACS Sustainable Chemistry and Engineering, 6*(7), 9434–9441. https://doi.org/10.1021/ACSSUSCHEMENG.8B01788/SUPPL_FILE/SC8B01788_SI_001.PDF

Xi, J., He, L., & Yan, L. gong. (2017). Continuous extraction of phenolic compounds from pomegranate peel using high voltage electrical discharge. *Food Chemistry, 230*, 354–361. https://doi.org/10.1016/J.FOODCHEM.2017.03.072

Xu, Y., Fan, M., Ran, J., Zhang, T., Sun, H., Dong, M., Zhang, Z., & Zheng, H. (2016). Variation in phenolic compounds and antioxidant activity in apple seeds of seven cultivars. *Saudi Journal of Biological Sciences, 23*(3), 379–388. https://doi.org/10.1016/J.SJBS.2015.04.002

Xue, L., Liu, G., Parfitt, J., Liu, X., Van Herpen, E., Stenmarck, Å., O'Connor, C., Östergren, K., & Cheng, S. (2017). Missing food, missing data? A critical review of global food losses and food waste data. *Environmental Science and Technology*, *51*(12), 6618–6633. https://doi.org/10.1021/ACS.EST.7B00401/SUPPL_FILE/ES7B00401_SI_002.XLSX

Yadav, P. K., Singh, V. K., Yadav, S., Yadav, K. D. S., & Yadav, D. (2009). In silico analysis of pectin lyase and pectinase sequences. *Biochemistry (Moscow)*, *74*(9), 1049–1055. https://doi.org/10.1134/S0006297909090144/METRICS

Yan, H., & Kerr, W. L. (2013). Total phenolics content, anthocyanins, and dietary fiber content of apple pomace powders produced by vacuum-belt drying. *Journal of the Science of Food and Agriculture*, *93*(6), 1499–1504. https://doi.org/10.1002/JSFA.5925

Yiğit, D., Yiğit, N., & Mavi, A. (2009). Antioxidant and antimicrobial activities of bitter and sweet apricot (Prunus armeniaca L.) kernels. *Brazilian Journal of Medical and Biological Research*, *42*(4), 346–352. https://doi.org/10.1590/S0100-879X2009000400006

Yılmaz, F. M., Görgüç, A., Karaaslan, M., Vardin, H., Ersus Bilek, S., Uygun, Ö., & Bircan, C. (2018). Sour cherry by-products: Compositions, functional properties and recovery potentials—A review. *Critical Reviews in Food Science and Nutrition*, *59*(22), 3549–3563. https://doi.org/10.1080/10408398.2018.1496901

Yılmaz, F. M., Karaaslan, M., & Vardin, H. (2015). Optimization of extraction parameters on the isolation of phenolic compounds from sour cherry (Prunus cerasus L.) pomace. *Journal of Food Science and Technology*, *52*(5), 2851–2859. https://doi.org/10.1007/S13197-014-1345-3/TABLES/3

Yu, J., & Smith, I. N. (2015). Nutritional and sensory quality of bread containing different quantities of grape pomace from different grape cultivars. *EC Nutrition*, *2*(1), 291–301. www.researchgate.net/publication/282849763

Yu, X., Van De Voort, F. R., Li, Z., & Yue, T. (2007). Proximate composition of the apple seed and characterization of its oil. *International Journal of Food Engineering*, *3*(5). https://doi.org/10.2202/1556-3758.1283/MACHINEREADABLECITATION/RIS

Yuan, G., Lv, H., Tang, W., Zhang, X., & Sun, H. (2016). Effect of chitosan coating combined with pomegranate peel extract on the quality of Pacific white shrimp during iced storage. *Food Control*, *59*, 818–823. https://doi.org/10.1016/J.FOODCONT.2015.07.011

Yuliarti, O., Goh, K. K. T., Matia-Merino, L., Mawson, J., & Brennan, C. (2015). Extraction and characterisation of pomace pectin from gold kiwifruit (Actinidia chinensis). *Food Chemistry*, *187*, 290–296. https://doi.org/10.1016/J.FOODCHEM.2015.03.148

Zafra-Rojas, Q. Y., González-Martínez, B. E., Cruz-Cansino, N. del S., López-Cabanillas, M., Suárez-Jacobo, Á., Cervantes-Elizarrarás, A., & Ramírez-Moreno, E. (2020). Effect of ultrasound on in vitro bioaccessibility of phenolic compounds and antioxidant capacity of blackberry (Rubus fruticosus) residues cv. tupy. *Plant Foods for Human Nutrition*, *75*(4), 608–613. https://doi.org/10.1007/S11130-020-00855-7/TABLES/2

Zhang, L., Fan, G., Khan, M. A., Yan, Z., & Beta, T. (2020). Ultrasonic-assisted enzymatic extraction and identification of anthocyanin components from mulberry wine residues. *Food Chemistry*, *323*, 126714. https://doi.org/10.1016/J.FOODCHEM.2020.126714

Zhao, X., Zhang, S. S., Zhang, X. K., He, F., & Duan, C. Q. (2020). An effective method for the semi-preparative isolation of high-purity anthocyanin monomers from grape pomace. *Food Chemistry*, *310*, 125830. https://doi.org/10.1016/J.FOODCHEM.2019.125830

Zhu, M. T., Huang, Y. S., Wang, Y. L., Shi, T., Zhang, L. L., Chen, Y., & Xie, M. Y. (2019). Comparison of (poly)phenolic compounds and antioxidant properties of pomace extracts from kiwi and grape juice. *Food Chemistry*, *271*, 425–432. https://doi.org/10.1016/J.FOODCHEM.2018.07.151

Žlabur, J. Š., Voća, S., Brnč, M., & Rimac-Brnč, S. (2018). New trends in food technology for green recovery of bioactive compounds from plant materials. In A. M. Grumezescu & A. M. Holban (Eds.), *Role of materials science in food bioengineering* (Vol. 19, pp. 1–36). Academic Press.

6

Value Addition of Vegetable Processing Waste

Anil S. Nandane

6.1 Introduction

Food waste can be considered as a non-edible portion of the food supply chain. Food waste occurs at different stages, such as the farm, processing line, storage, transportation, and utilization. All over the world food losses are to the tune of nearly 1.3 billion tons. The most food waste–producing countries globally include India, China, the United States, and Brazil. Food waste will increase environmental, health, and also profitable issues.

There are four types of vegetable waste subtypes: grains (such as brewing grains, wheat bran, and rice external layer), roots and stolons (e.g. potato skins, sugar beet), legumes and oilseeds (which include sunflower seeds, soybean seeds, as well as pressed olive residue), and vegetables and fruits (e.g. orange rind, grape pomace, apple skin, tomato pomace). To effectively utilize this food waste into value-added products, there is a necessity to introduce methods and techniques for it. These food waste sources are rich in various bioactive compounds, including biofuel, bioenergy, antibiotics, and enzymes.

Vegetable by-products are the secondary products that are often discarded or wasted during manufacturing or other stages of food processing. Up to one-third of vegetables could be wasted in the preparation process. Interestingly, certain parts of the vegetables are knowingly wasted due to their unfavorable taste or texture. For example, vegetable hulls, bagasse, and seeds are mostly discarded on the production line. For certain types of vegetables, such as broccoli, cauliflower, and pumpkin, the stem and leaves are not consumed and are discarded. The seed coat, which is the outer layer that covers the seeds or beans, functions to protect them from external damage and is a type of vegetable by-product. Seed coats are usually not consumed due to their texture and taste, except for some thin seed coats, such as that of peanuts. The wastage of oilseeds and pulses are highest in the North Africa, West Asia,

DOI: 10.1201/9781003269199-6

147

and Central Asia regions, and these by-products are mostly lost during the agriculture stage. Unlike the seed coat, the hull is the hard-protecting cover of the seeds or grains that protects them during the growing period. Due to their hard texture, hulls are removed before cooking or manufacturing. Globally, hulls contribute to a large amount of food waste, especially rice hulls, due to the high consumption of rice among the Asian countries. It is observed that 30% of whole cereals are lost, and most of the wastage is due to human consumption and postharvest practices. To overcome this problem, hulls are used as a building material, fuel, and fertilizer as a part of different strategies to reduce wastage. Owing to its high content of fiber and protein, the hull can be an alternative source of functional ingredient. Not all vegetables have peels, unlike fruits. Tubers, gourds, and allium families contain peels, yet they are not normally consumed or used in food preparation. Therefore, vegetable peels are often discarded as kitchen or production waste during manufacturing.

In India about 85 million tons of fruits and 170 million tons of vegetables are produced annually. However, about 35–45% of the harvested fruits and vegetables are lost during handling, storage, transportation, etc., leading to the loss of approximately 40,000 crore rupees per year. The estimated losses in vegetables are: onion, 25–40%; garlic 8–22%; potato, 30–40%; tomato, 5–47%; cabbage and cauliflower, 7.08–25%; chili pepper, 4–35%; radish, 3–5%; and carrot, 5–9%. Vegetable loss is a serious matter of concern for India's agricultural sector. Due to their high perishability, vegetable commodities are lost after harvest due to factors like insufficient methods of harvest, decay, over-ripening, mechanical injury, weight loss, trimming, and sprouting (Bala et al. 2020).

India is the second largest producer of fruits and vegetables, but it lacks proper postharvest handling of these products. However, during processing it produces numerous by-products, which have the potential to for various applications. The bioactive components extracted from fruit and vegetable by-products are now becoming popular and also help to reduce effluent treatment loads. Thus, waste utilization and value addition for large-scale application has to be strengthened and also the regulations have to be enforced for the same. This also leads to applications in new product development with enriched nutrients, structural modification of food, and minimized waste from the fruits and vegetables industry.

The food processing industry, including fruit and vegetable processing, is the second largest generator of waste into the environment, second only to household waste. A number of value-added products, such as essential oils, starch, pectin, dietary fibers, acids, wine, ethanol, vinegar, microbial pigments, flavors and gums, enzymes, single-cell proteins, amino acids, vitamins, organic compounds, colors, and animal feed can be made out of the waste from processing industries (Joshi et al. 2020).

6.2 Methods of Waste Conversion

The vegetable wastes and other by-products produced from food industries can be used in the production of different value-added products, mainly through three different conversion processes, i.e., thermal, chemical, and biological conversion (Singh et al. 2019). The most suitable conversion method depends on the composition of waste and by-products and the purpose of the recovery process.

6.2.1 Thermal Conversion Methods

Thermal conversion methods include incineration, hydrothermal carbonization, pyrolysis, and gasification. Incineration involves the burning of waste constituents under extreme heat. It also results in a decrease of the volume of solid waste up to 85%. This technique of combusting solid waste is very old, and food waste seems to be unsuitable for incineration due to the high moisture content of vegetable waste. However, this technique may be beneficially utilized after drying the waste. The heat generated from the combustion process is generally consumed by steam turbines for producing energy or for exchanging heat (Pham et al. 2015). Thus, the thermal treatment of waste is applied with the primary purpose of generating power (Singh et al. 2019).

Hydrothermal carbonization is an aqueous carbonization process performed at a relatively lower temperature (180–350°C) and atmospheric pressure. This process is suitable for wet or high-moisture wastes, which transforms the waste into an energy-rich valuable resource (Pham et al. 2015). This process has various advantages, including being faster than biological processes, removing many organic impurities and pathogens, and reducing waste volume. This process results in the production of a highly carbonized and energy-containing material known as hydrochar, which is equivalent to lignite coal. The hydrochars can be used in removing dyes from polluted water (Singh et al. 2019; Pham et al. 2015).

Gasification and pyrolysis are also thermal processes that can be effectively used for the conversion of carbon-rich wastes. In this process waste is converted into a mixture of combustible gas by partial oxidation at temperatures of 800–900°C. Similarly, pyrolysis is a process that converts waste into bio-oil, solid biochar, and syngas. The produced combustible gas can be burned directly or can be used as a raw material for methanol production (Pham et al. 2015).

6.2.2 Chemical Conversion Methods

This process includes techniques like hydrolysis, oxidation for producing value-added products from food waste (Singh et al. 2019). The chemicals (acid or alkali) are used to rupture the cell and extract the compounds. In this technique solvents such as propane, butane, and dimethyl ether can be used for the extraction of natural products like oils, antioxidants, and aromas.

6.2.3 Biological Conversion Methods

Biological conversion methods are becoming more popular day by day. Energy, bioactive compounds, and other value-added products can be obtained from organic wastes by biological conversion through anaerobic digestion and fermentation. Anaerobic digestion is the process of microbial catabolism in which organic wastes decompose and produce biogas, mainly methane, and traces of nitrogen, CO_2, and hydrogen sulfide in absence of oxygen. The fermentation process includes either solid-state fermentation or submerged-state fermentation. Different factors like type of pretreatment, types and quality of substrates, and microbes employed play an important role. Biological conversion is eco-friendly and can reduce carbon dioxide, a methane-like gas emission (Singh et al. 2019; Awasthi et al. 2019). Vegetable wastes such as tomatoes, fennel, carrots, and others can be used as eco-friendly and low-cost substrates in culture media for the manufacture of biomolecules such as enzymes and biopolymers using some specific microorganisms (Di Donato et al. 2011).

6.2.4 Other Conversion Methods

Valuable by-products and materials can be extracted by alternative extraction techniques, such as high-voltage electrical discharge and pulsed electric field technology. In both technologies, electroporation occurs, resulting in increased cell permeability, which enhances the extraction of intracellular compounds (Chemat et al. 2020; Sarkis et al. 2015).

6.3 Bioconversion of Vegetable Wastes into Different Value-Added Products

Different biomasses in the form of organic wastes are attracting interest as alternative renewable resources. These organic wastes can be used for a number of commercially important intermediates and products. For example, some furans and organic acids are usually used as chemical precursors to manufacture various products, including polymers, biosurfactants, biolubricants, and nanoparticles (Esteban and Ladero 2018).

The waste management systems begin with a focus on waste reduction. Therefore, to achieve waste minimization in the industry, it is beneficial to use further effective approaches, such as in-house recycling or reuse of waste products. Vegetable wastes are rich in starch, inulin, hemicellulose, pectin, and cellulose like polysaccharides and thus can be used as resources for producing a wide array of products, such as antibiotics, biofuels, vitamins, enzymes, pigments, livestock feed, and more (Sadh et al. 2018).

6.3.1 Bioactive Compounds

Bioactive compounds consist of a number of natural compounds that can be found mainly in different colorful vegetables. These can be used as a source of food additives, nutraceuticals, and functional foods. Natural sources of bioactive compounds are plants, fruits, tea, olives, algae, bacteria, and fungi. Polyphenols and other similar compounds can be found in the environment at relatively higher concentrations. So, to extract these bioactive compounds advanced and newer technologies need to be used. Some examples of advanced extraction techniques include pressurized liquid extraction, solid–liquid, or liquid–liquid extraction, ultrasound- and microwave-assisted extractions, enzyme and instant controlled pressure drop-assisted extractions, and supercritical and subcritical extractions (Gil-Chávez et al. 2013). The vegetable wastes mainly contain sterols, tocopherols, carotenes, terpenes, polyphenols, and dietary fibers like bioactive compounds, which are value-added compounds (Kumar et al. 2017). With an increase in by-products from the food processing industry and losses of fruits and vegetables, there is an increase in the amount of vegetable wastes. Therefore, these alternative sources of bioactive compounds will help make farmers financially stronger and also reduce the problem of waste management.

6.3.2 Phenolic Compounds

Phenolic compounds are a group of structurally diverse compounds categorized as secondary metabolites that are mainly found in plants. The bioactive phenolic compounds extracted from vegetable waste are found to have health benefits, such as cardioprotective, anticarcinogenic, antioxidant, and anti-inflammatory properties (Haminiuk et al. 2012; Balasundram et al. 2006). The phenolic compounds normally contain an aromatic ring with hydroxy substituents. These substances are produced by plants as a part of normal growth metabolism as well as a response to adverse situations like biotic stress and UV radiation (Rispail et al. 2005). Varieties of FVWs produced as a residue of asparagus, onion, potato, carrot, etc. can be worthy sources of phenolic compounds (Kumar et al. 2017). The phenolic compounds act as antioxidants and as a substrate for oxidation reaction. The extraction and recovery of phenolic compounds is a difficult process, as these compounds are highly reactive and are unevenly distributed in various forms. The soluble form of phenolic compounds is mainly located in vacuoles (Rispail et al. 2005). Submerged or solid-state fermentation methods can be employed for the extraction, as well as recovery of phenolic compounds. However, the most commonly used is solid-state fermentation due to high efficiency, more yield, fast, and low cost (Martins et al. 2011).

Phenolic acids include hydroxybenzoic acid and others, such as caffeic, sinapic, and ferulic acids. Flavonoids are the largest group of plant phenolics, having few molecular compounds. Tannins are a third important group of phenolics and have relatively high molecular weight and can be hydrolyzable and condensed (Balasundram et al. 2006).

Flavonols and flavones are commonly found in plants. One or more hydroxyl group is bound to a sugar unit (most commonly glucose) with rhamnose and the disaccharide (Balasundram et al. 2006). Anthocyanins are another common flavonoid and are responsible for the blue, red, and violet colors of some fruits and vegetables, although the red color of oranges and tomatoes is due to carotenoids.

6.3.3 Enzymes

Enzymes are bio-catalysts and catalyze a number of metabolic processes. Enzymes are used in different industries for the production of a wide variety of products. For instance, pectinases and amylases are used in food industries, cellulases in biofuel industries, and tannases in reducing the tannic acid amount in effluents. However, raw materials used for the production of different enzymes account for about 30% of the operation cost (Ravindran and Jaiswal 2016). Plant-related food wastes mainly contain cellulose, hemicellulose, lignin, starch, xylan, pectin, glucan, etc., depending on the nature of waste products. The most commonly applied enzymes are amylases, cellulases, hemicellulases, ligninases, pectinases, tannases, proteases, and lipases. Microorganisms can be utilized for the production of different enzymes, and the rate of enzyme production varies, with the organisms growing on different substrates and different methods of fermentation.

6.3.3.1 Amylases

This group of enzymes is composed of glucoamylase, b-amylase, and a-amylase, which hydrolyze starch, oligosaccharides, and polysaccharides into glucose, fructose, and maltose sugars. Amylases are classified in exo-amylase and endo-amylase based on the hydrolysis of starch (Panda et al. 2016). Many vegetable wastes, such as potato peels, have been used in amylase production (Mushtaq et al. 2017). Microorganisms like *Candida guilliermondii, Aspergillus niger, A. tamarii, A. oryzae, Bacillus licheniformis, B. subtilis, Thermomyces lanuginosus*, and *Rhizopus oryzaeare* are also exploited for amylase production (Unakal et al. 2012; Said et al. 2014; Mushtaq et al. 2017; Metha and Satyanarayana 2016). Amylases are widely used in baking, brewing, industry, and the preparation of sugar, paper, moist cakes, fruit juices, chocolate cakes, starch syrup, and more (Metha and Satyanarayana 2016).

6.3.3.2 Cellulases

This group of enzymes comprises exoglucanase (cellobiohydrolase), endo-b-glucanase, and b-glucosidase, which hydrolyze cellulose into glucose, cellobiose, and other oligosaccharides. The three main cellulases act synergistically to cleave the glycosidic linkage of cellulose and completely hydrolyze the cellulose present in plant wastes. Exoglucanase breaks the long chain from endings (reducing or non-reducing), endoglucanase cut the long ligosaccharides into short oligosaccharide chain, and b-glucosidase further hydrolyze to glucose (Juturu and Wu 2014). Vegetable waste like bottle gourd peel can be used as the substrate for producing cellulase by *Neurospora crassa* and *Trichoderma reesei* (Verma and Kumar 2020), potato peel by *Aspergillus niger* (dos Santos et al. 2012). Cellulases are applied in animal feed, food, and brewery production; textile processing; detergent production; and pulp paper manufacture. Recently, the increasing demand for biofuels and chemicals recovered from renewable resources has led to cellulases being exploited for producing fermentable sugars in cellulose biorefinery (Kuhad and Gupta 2011).

6.3.4 Pigments

Cheaply available vegetable wastes can be used effectively for microbial pigment production and are helpful for making such processes economic and eco-friendly. Microbial biotechnology has created new possibilities for the immense utilization of waste in the production of value-added products via the fermentation process rather than conventional applications, like making compost or feeding cattle as fodder (Panesar et al. 2015).

The various synthetic colorants have carcinogenic properties. Therefore, foods containing synthetic colorants are being avoided and those with natural colorants are widely accepted. Examples of natural pigments include betalains, carotenoids, anthocyanins, and carminic acid. Beetroot, leafy or grainy amaranth, cactus fruits, and Swiss chard contain water-soluble nitrogenous pigments known as betalains, which are composed of yellow betaxanthin and red betacyanin. They act as antioxidants and counteract biological molecule oxidation. Betalains are widely used in desserts, confectioneries, dry mixes, dairy, and meat products (Azeredo 2009).

Carotenoid pigment is found in green leafy vegetables, carrots, spinach, squash/ pumpkins, and other vegetables. Pigments like lycopene, carotene, cryptoxanthin, lutein, zeaxanthin, astaxanthin, and fucoxanthin are examples of carotenoids. They perform different functions and are pro-vitamin A and antioxidants. The vegetable wastes from pea pod powder, taro leaves, okra, soybeans, and green gram are considered prospective sources of mineral, carbon, and nitrogen to produce microbial pigments (Panesar et al. 2015). Microbial pigments are produced from bacteria, yeasts, mold, and algae by fermentation processes. Some pigments can fight against protozoal, bacteria, fungi, and are inflammatory and cytotoxic.

6.3.5 Dietary Fiber

Nowadays, people are being more and more health conscious and are more inclined toward fruits rich in minerals, bioactive compounds, and dietary fibers and low in calories, sodium, and fats. Dietary fibers are a type of indigestible carbohydrate present in plant cell walls and are vital for human health (Palafox-Carlos et al. 2011). Plant carbohydrate polymers and other non-carbohydrate components like pectin, hemicellulose, hemicellulose, waxes, polyphenols, and resistant protein are dietary fibers (Elleuch et al. 2011). Many vegetable wastes are rich in dietary fibers and can be used as food supplements (Kowalska et al. 2017). The intake of dietary fibers reduces obesity, diabetes, cardiovascular diseases, hyperlipidemia, hypercholesterolemia, and hyperglycemia (Mann and Cummings 2009). Dietary fiber also helps in the absorption of antioxidants like carotenoids and phenolic compounds.

Dietary fiber has good water- and oil-holding capacity, emulsifying as well as gel-forming capacity, and swelling capacity so that it helps in lowering cholesterol, modifying the viscosity of intestinal contents, and forming gel with bile in the intestine (Palafox-Carlos et al. 2011; Elleuch et al. 2011; Ayala-Zavala et al. 2011). The addition of dietary fibers helps to modify the stickiness, consistency, shelf-life, and sensory properties of food products like bakery items, dairy products, meats, jams, soups, etc. However, the addition of fiber must be in the appropriate amount, otherwise it may produce adverse changes in the color, flavor, and quality of foods (Elleuch et al. 2011). In bakery products, dietary fibers prolong freshness and retain the water, loaf volume, and flexibility of the products, thereby enhancing digestion. Similarly, in dairy products like ice cream, the addition of fiber develops texture and also prevents crystal formation during storage (Elleuch et al. 2011; Ayala-Zavala et al. 2011).

6.3.6 Bioenergy

The depletion of fossil fuels and the ecological concerns associated with greenhouse gas emissions, combined with rising prices of oil, are forcing us to look for alternative resources for the production of transport fuels, energy, and compounds. In such cases, it is possible to utilize organic wastes, particularly from vegetables, as important resources for the generation of hydrogen, ethanol, and biodiesel forms of bioenergy.

6.3.6.1 Bioethanol

Bioethanol, commonly known as ethyl alcohol, is a colorless liquid and is decomposable, less toxic, and used as fuel in automobiles. Bioethanol can be produced from the fermentable sugars like glucose, sucrose, etc. of plant sources using microorganisms. Bioethanol produced from plant sources is CO_2-neutral because the CO_2 released through the combustion of bioethanol is equal to the CO_2 absorbed by the plant during the growing phase (Chin and H'ng 2013). Bioethanol production from wastes comprises steps such as biomass pretreatment and saccharification followed by the fermentation of sugars. Different experiments are carried out by using different vegetable wastes like potato peel, soybean litter, and soybean molasses for bioethanol production using *Saccharomyces cerevisiae*. Mushimiyimana and Tallapragada also used agro-waste, including peels from carrots, onions, sugar beet, and potatoes to produce bioethanol. In this process, *Penicillium* sp. and *Saccharomyces cerevisiae* are used for hydrolysis and fermentation to produce bioethanol, respectively (Mushimiyimana and Tallapragada 2016).

6.3.6.2 Biohydrogen

Biohydrogen is universally recognized as complementary to fossil fuels due to its non-polluting feature, lower cost, and renewable source. Hydrogen gas includes 2.75 times greater energy yield than hydrocarbon fuels, and it is carbon neutral. This can be considered a clean fuel and energy carrier without CO_2 releases and can be easily used in generating electricity (Kapdan and Kargi 2006). With sustainable energy production, biohydrogen production is considered one of the most important renewable sources, also known as green technology. Hydrogen can be obtained by various techniques like the electrolysis of water, biological processes, and thermocatalytic reformation of hydrogen-rich organic compounds (Kapdan and Kargi 2006).

Biohydrogen is produced as a secondary product during the anaerobic digestion of organic wastes, whereas in photosynthetic processes microorganisms use carbon dioxide and water for hydrogen production (Levin et al. 2004). Different wastes like potato waste, pumpkin waste, fennel waste, olive pomace, leafy vegetables like cabbage, water celery, cauliflower, etc., can be used as a substrate for biohydrogen production (Ghimire et al. 2015; Lee et al. 2010). Some of the microorganisms capable of biohydrogen production include *Clostridium butyricum*, *Bacillus* sp., *Escherichia coli*, *Rhodobacter sphaeroides*, *Rhodopseudomonas palustris*, *R. faecalis*, *Rhodospirillum rubrum*, etc. (Rahman et al. 2016).

6.3.6.3 Biomethane

Biomethane is a cheap form of bioenergy that can be produced from the anaerobic digestion of biogenic wastes by different microbes. The process of utilizing vegetable waste to generate biogas is environmentally friendly and reduces the solid waste disposal problem, air and water pollution, and soil contamination. During anaerobic digestion, the acidogenic microbes produce acetate, carbon dioxide, and hydrogen. This produced hydrogen and acetate are converted into water and methane by methanogens. Biomethane production involves hydrolysis, methanogenesis, and acidogenesis that are completed by a sequence of microbial interactions. However, the end products may differ with the type of bacteria utilized (Singh et al. 2012).

6.3.6.4 Biodiesel

Biodiesel is a renewable and clean-burning liquid biofuel made up of low aliphatic alcohols and esters of alkyl groups having high fatty acids. Biodiesel can be considered carbon neutral because this biofuel produces no net output of carbon dioxide. In addition, biodiesel is an inexhaustible source of energy that reduces much faster (four times) than fossil fuels. It is also considered to have a greasing property, which reduces engine wear, and is safe for storage and handling due to low explosiveness and a high flash point of 100–170°C (Ramirez-Arias et al. 2018). Transesterification is a common procedure for producing biodiesel that requires low temperature and pressure and produces a 98% conversion yield (Muniraj et al. 2015). In the transesterification process, triglyceride reacts with alcohol to form biodiesel and crude glycerol. Vegetable wastes that are rich in lipids or oil, such as rapeseed, palm, soybean, and canola, can be used in biodiesel manufacture (Lee et al. 2010; Muniraj et al. 2015).

6.3.7 Single-Cell Protein (SCP)

SCP is a protein that originated from microorganisms such as algae, bacteria, fungi, and yeast. These microorganisms can utilize various carbon sources for SCP synthesis. For human consumption, SCP is commonly produced from filamentous fungi and yeast. Different vegetable wastes can be used as a cheap carbon source for the growth of microorganisms and SCP production (Najafpour 2007; Mondal et al. 2012). Microorganisms like *Aspergillus oryzae, A. flavus, A. niger, Fusarium semitectum, Rhizopus oligosporus, Saccharomyces cerevisiae, Trichoderma harzianum, T. reesei, Penicillium javanicum, Kluyveromyces marxianus*, etc., are capable of SCP production (Malav et al. 2017). The common steps involved in SCP production are (a) preparation of culture media, (b) cultivation, (c) extraction and intensifying SCP, and (d) final processing. SCP is becoming an important element to meet the protein demands of an increasing population, and it can also be used in livestock feed as a protein source.

Algal SCP can be used as source of omega-3 fatty acids, vitamins, and carotenoids, along with protein; thus, SCP is used as a food supplement. Waste from the fermentation industry, e.g., sauerkraut brine with high BOD and salt content, has successfully been used for the growth of a number of yeasts, even in non-sterile conditions, but *Candida utilis* was preferred, as it gave a higher yield in a shorter amount of time. *Saccharomyces cerevisiae* and *Torulopsis utilis* have been successfully grown on the molasses with protein yield of 42–47% and 36–38%, respectively. With the addition of corn steep liquor, the protein yield can be increased to 60% and 52.6%, respectively.

6.4 Challenges for the Extraction of Natural Food Ingredients from Fruit and Vegetable By-Products

Natural food ingredients are becoming more popular, therefore demand for them is likely to increase despite their high cost. However, the extraction of natural food ingredients from vegetable by-products at the commercial scale is much more complex and challenging than at the laboratory scale. It may contain the steps of physical or chemical pretreatments of by-products, the extraction process, separation, purification, and chemical characterization of the value-added ingredients (Arshadi et al. 2016). The challenges faced during the extraction of value-added ingredients from vegetable by-products on the industrial scale are outlined in the following sections.

6.4.1 Availability of Raw Materials at Different Locations

Vegetable by-products are produced in different food factories, thus they are not concentrated in one place. It makes it difficult and expensive to collect all vegetable by-products from small and distant factories and process them at one place, due to their perishable nature (Jin et al. 2018). This problem can be overcome by short value chains and short-distance transportation. To achieve this goal, it is important to take into account that the local production of the by-products should allow a profitable extraction process. In addition, setting up small, localized biorefinery plants near food factories instead of traditional industrial plants with large production capacities can be suggested (Teigiserova et al. 2019). This solution can reduce not only the transportation cost and spoilage of by-products during transportation but also the environmental pollution. Thus, it could be beneficial to extract value-added ingredients from vegetable by-product for both economic and environmental aspects. In this case, the base vegetable by-products should be selected on the basis of their quantity and quality to be processed in localized bio-refinery plants (Teigiserova et al. 2019).

6.4.2 Low Concentrations of Natural Ingredient Vegetable By-Products

It is very important to take into account that vegetable by-products contain many ingredients, but not all of them are applicable for developing a sustainable and economical biorefinery process. The selection of by-products for value addition depends on factors like the concentration of the ingredients in the vegetable by-product, the market value of the extracted ingredients, extraction yields, and costs. Considering that vegetable wastes have high water content, the concentrations of other ingredients, such as phenolic compounds and pigments, are not considerably high. For example, the lycopene concentration found in 67 samples of tomato puree ranged from 10.0 to 50.5 mg/kg (Choudhary et al. 2009). Moreover, it is important to consider that the degradation of natural ingredients during processing and storage of these by-products could decrease their concentrations even further. Therefore, extraction of these value-added ingredients with low concentration from a matrix could be difficult and expensive. The concentration of some value-added ingredients in vegetable by-products can be increased with the use of a fermentation procedure before extraction.

6.4.3 The Variable Sources and Characteristics of Fruit and Vegetable By-Products

Another challenge for extracting the value-added ingredients from vegetable by-products is related to the complicated nature and various sources of these by-products, which leads to their different characteristics (Arshadi et al. 2016). The concentration of different ingredients in these by-products depends on the variety, environmental parameters, and ripening stage. Laaksonen et al. (2017) reported that phenolic contents in different apple cultivars are significantly different. It emphasizes the importance of appropriate cultivar selection for the extraction of desired value-added ingredients (Laaksonen et al. 2017). The concentration of some ingredients in fruits and vegetables also vary with changing the environmental conditions during their cultivation, e.g., lycopene content in tomatoes varies extensively depending on the environmental parameters of light, temperature, maturity stages (Brandt et al. 2006) and the agricultural parameters of water, minerals, nutrients, etc. during the growing process (Dumas et al. 2003). Considering the effect of the ripening stage on ingredient concentrations, lycopene can be mentioned as one example in which its concentration is higher in fully ripe tomatoes (Gebhardt et al. 2020).

Depending on the applied food processes and technology, vegetable by-products can contain different parts of raw materials, such as seeds, cores, peels, skins, rinds, vines, shells, pomace, pods, etc. with different percentages (Rodríguez et al. 2021). It is worth mentioning that by-product characteristics such as pH, moisture content, texture, density, microbial quality, etc. can vary considerably. Therefore, the extraction parameters of one value-added ingredient and the resulting extraction efficiency, especially in the industrial scale, may not be fixed by changing the by-products and, thus, optimization of process parameters could be necessary when vegetable by-products are changed. These optimizations could increase the chance of using different vegetable by-products more efficiently that can simultaneously lead to more economic and environmental profits.

References

Arshadi, M., Attard, T. M., Lukasik, R. M., Brncic, M., da Costa Lopes, A. M., Finell, M., Geladi, P., Gerschenson, L. N., Gogus, F., Herrero, M., et al. (2016). Pre-treatment and extraction techniques for recovery of added value compounds from wastes throughout the agri-food chain. Green Chemistry, 18(23), 6160–6204. doi: 10.1039/C6GC01389A.

Awasthi, M. K., Chen, H., Awasthi, S. K., Liu, T., Wang, M., Duan, Y., & Li, J. (2019). Biological processing of solid waste and their global warming potential. In Kumar, S., Zhang, Z., Awasthi, M., & Li, R. (Eds.), Biological Processing of Solid Waste (pp. 111–128). Boca Raton, Florida: CRC Press.

Ayala-Zavala, J. F., Vega-Vega, V., Rosas-Domínguez, C., Palafox-Carlos, H., Villa-Rodriguez, J. A., Siddiqui, M. W., et al. (2011). Agro-industrial potential of exotic fruit byproducts as a source of food additives. Food Research International, 44(7), 1866–1874.

Azeredo, H. M. (2009). Betalains: Properties, sources, applications, and stability—A review. International Journal of Food Science Technology, 44(12), 2365–2376.

Bala, S., Gautam, K. K., & Sahu, M. (2020, July). A review of Post Harvest Management and value addition of horticultural crops. International Journal of Creative Research Thoughts (JCRT), 8(7). ISSN: 2320–2882. www.ijcrt.org.

Balasundram, N., Sundram, K., & Samman, S. (2006). Phenolic compounds in plants and agri-industrial by-products: Antioxidant activity, occurrence, and potential uses. Food Chemistry, 99(1), 191–203.

Brandt, S., Pek, Z., Barna, E., Lugasi, A., & Helyes, L. 2006. Lycopene content and colour of ripening tomatoes as affected by environmental. Journal of the Science of Food and Agriculture, 86(4), 568–572. doi: 10.1002/jsfa.2390.

Chemat, F., Vian, M. A., Fabiano-Tixier, A. S., Nutizio, M., Jambrak, A. R., Munekata, P. E. S., et al. (2020). A review of sustainable and intensified techniques for extraction of food and natural products. Green Chemistry, 22(8), 2325–2353.

Chin, K. L., & H'ng, P. S. (2013). A real story of bioethanol from biomass: Malaysia perspective. In M. D. Matovic (Ed.), Biomass Now-Sustainable Growth and Use (pp. 329–346). Rijeka, Croatia: InTech Publishers.

Choudhary, R., T.J. Bowser, P. Weckler, N.O. Maness, W. McGlynn. (2009). Rapid estimation of lycopene concentration in watermelon and tomato puree by fiber optic visible reflectance spectroscopy, *Postharvest Biology and Technology, 52*(1), 103–109, ISSN 0925-5214, https://doi.org/10.1016/j.postharvbio.2008.10.002.

Di Donato, P., Fiorentino, G., Anzelmo, G., Tommonaro, G., Nicolaus, B., & Poli, A. (2011). Re-use of vegetable wastes as cheap substrates for extremophile biomass production. Waste & Biomass Valorization, 2(2), 103–111.

dos Santos, T. C., Gomes, D. P. P., Bonomo, R. C. F., & Franco, M. (2012). Optimisation of solid-state fermentation of potato peel for the production of cellulolytic enzymes. Food Chemistry, 133(4), 1299–1304.

Dominika Alexa Teigiserova, Lorie Hamelin, Marianne Thomsen. (2020). Towards transparent valorization of food surplus, waste and loss: Clarifying definitions, food waste hierarchy, and role in the circular economy, Science of The Total Environment, Volume 706, 136033, ISSN 0048-9697, https://doi.org/10.1016/j.scitotenv.2019.136033.

Dumas, Y., Dadomo, M., Lucca, G., & Di Grolier, P. (2003). Effects of environmental factors and agricultural techniques on antioxidant content of tomatoes. Journal of the Science of Food and Agriculture, 83(5), 369–382. doi: 10.1002/jsfa.1370.

Elleuch, M., Bedigian, D., Besbes, S., Roiseux, O., Blecker, C., & Attia, H. (2011). Dietary fiber and fibre-rich by-products of food processing: Characterisation, technological functionality and commercial application: A review. Food Chemistry, 124(2), 411–421.

Esteban, J., & Ladero, M. (2018). Food waste as a source of value-added chemicals and materials: A biorefinery perspective. International Journal of Food Science & Technology, 53(5), 1095–1108.

Gebhardt, B., Sperl, R., Carle, R., & Müller-Maatsch, J. (2020). Assessing the sustainability of natural and artificial food colorants. Journal of Cleaner Production, 260, 120884. doi: 10.1016/j.jclepro.2020.120884.

Ghasem D. Najafpour. (2007). CHAPTER 7 - Downstream Processing, Editor(s): Ghasem D. Najafpour, *Biochemical Engineering and Biotechnology,* Elsevier, 170–198, ISBN 9780444528452, https://doi.org/10.1016/B978-044452845-2/50007-0.

Ghimire, A., Frunzo, L., Pontoni, L., d'Antonio, G., Lens, P. N. L., Esposito, G., et al. (2015). Dark fermentation of complex waste biomass for biohydrogen production by pretreated thermophilic anaerobic digestate. Journal of Environmental Management, 152(2015), 43–48.

Haminiuk, C. H. I., Maciel, G. M., Plata-Oviedo, M. S. V., & Peralta, R. M. (2012). Phenolic compounds in fruits—An overview. International Journal of Food Science & Technology, 47(10), 1–22.

Jin, Q., Yang, L., Poe, N., & Huang, H. (2018). Integrated processing of plant-derived waste to produce value-added products based on the biorefinery concept. Trends in Food Science & Technology, 74, 119–131. doi: 10.1016/j.tifs.2018.02.014.

Joana Gil-Chávez, G., Villa, J.A., Fernando Ayala-Zavala, J., Basilio Heredia, J., Sepulveda, D., Yahia, E.M. and González-Aguilar, G.A. (2013), Technologies for Extraction and Production of Bioactive Compounds to be Used as Nutraceuticals and Food Ingredients: An Overview. *COMPREHENSIVE REVIEWS IN FOOD SCIENCE AND FOOD SAFETY*, 12: 5–23. https://doi.org/10.1111/1541-4337.12005.

Joshi, V. K. (2020). Value addition to fruit and vegetable processing waste- an appraisal. Food Fermentation Technology, 10(2), 35–58.

Juturu, V., & Wu, J. C. (2014). Microbial cellulases: Engineering, production and applications. Renewable and Sustainable Energy Reviews, 33(2014), 188–203.

Kapdan, I. K., & Kargi, F. (2006). Biohydrogen production from waste materials. Enzyme and Microbial Technology, 38(5), 569–582.

Kowalska, H., Czajkowska, K., Cichowska, J., & Lenart, A. (2017). What's new in biopotential of fruit and vegetable by-products applied in the food processing industry. Trends in Food Science & Technology, 67(2017), 150–159.

Kuhad, R. C., & Gupta, R. (2011). Singh A (2011) Microbial cellulases and their industrial applications. *Enzyme Research*, 1–10.

Kumar, K., Yadav, A. N., Kumar, V., Vyas, P., & Dhaliwal, H. S. (2017). Food waste: A potential bioresource for extraction of nutraceuticals and bioactive compounds. Bioresource Bioproducts, 4(1), 18.

Laaksonen, O., Kuldjärv, R., Paalme, T., Virkki, M., & Yang, B. (2017). Impact of apple cultivar, ripening stage, fermentation type and yeast strain on phenolic composition of apple ciders. Food Chemistry, 233, 29–37. doi: 10.1016/j.foodchem.2017.04.067.

Lee, Z. K., Li, S. L., Kuo, P. C., Chen, I. C., Tien, Y. M., & Huang, Y. J. (2010). Thermophilic bio-energy process study on hydrogen fermentation with vegetable kitchen waste. International Journal of Hydrogen Energy, 35(24), 13458–13466.

Levin, D. B., Pitt, L., & Love, M. (2004). Biohydrogen production: Prospects and limitations to practical application. International Journal of Hydrogen Energy, 29(2), 173–185.

Luciana Gabriela Ruiz Rodríguez, Víctor Manuel Zamora Gasga, Micaela Pescuma, Carina Van Nieuwenhove, Fernanda Mozzi, Jorge Alberto Sánchez Burgos. (2021). Fruits and fruit by-products as sources of bioactive compounds. Benefits and trends of lactic acid fermentation in the development of novel fruit-based functional beverages, *Food Research International*, Volume 140, 109854, ISSN 0963-9969, https://doi.org/10.1016/j.foodres.2020.109854.

Malav, A., Meena, S., Sharma, M., Sharma, M., & Dube, P. (2017). A critical review on single cell protein production using different substrates. International Journal of Development Research, 7(11), 16682–16687.

Mann, J. I., & Cummings, J. H. (2009). Possible implications for health of the different definitions of dietary fibre. Nutrition, Metabolism & Cardiovascular Diseases, 19(3), 226–229.

Martins, S., Mussatto, S. I., Martinez-Avila, G., Montanez-Saenz, J., Aguilar, C. N., & Teixeira, J. A. (2011). Bioactive phenolic compounds: Production and extraction by solid-state fermentation review. Biotechnology Advances, 29(3), 365–373.

Metha, D., & Satyanarayana, T. (2016). Bacterial and archaeal a-Amylases: Diversity and Amelioration of the desirable characteristics for industrial applications. Frontiers in Microbiology, 7(2016), 1129.

Mondal, A. K., Sengupta, S., Bhowal, J., & Bhattacharya, D. K. (2012). Utilization of fruits wastes in producing single cell protein. International Journal of Environmental Science and Technology, 1(5), 430–438.

Muniraj, I. K., Uthandi, S. K., Hu, Z., Xiao, L., & Zhan, X. (2015). Microbial lipid production from renewable and waste materials for second-generation biodiesel feedstock. Environmental Technology Review, 4(1), 1–16.

Mushimiyimana, I., & Tallapragada, P. (2016). Bioethanol production from agro wastes by acid hydrolysis and fermentation process. Journal of Scientific and Industrial Research, 75(6), 383–388.

Mushtaq, Q., Irfan, M., Tabssum, F., & Iqbal-Qazi, J. (2017). Potato peels: A potential food waste for amylase production. The Journal of Food Process Engineering, 40(4), e12512.

Palafox-Carlos, H., Ayala-Zavala, F., & González-Aguilar, G. A. (2011). The role of dietary fiber in the bioaccessibility and bioavailability of fruit and vegetable antioxidants. Journal of Food Science, 76(1), R6–R15.

Panda, S. K., Mishra, S. S., Kayitesi, E., & Ray, R. C. (2016). Microbial processing of fruit and vegetable wastes for production of vital enzymes and organic acids: Biotechnology and scopes. Environmental Research, 146(2016), 161–172.

Panesar, R., Kaur, S., & Panesar, P. S. (2015). Production of microbial pigments utilizing agro-industrial waste: A review. Current Opinion in Food Science, 1(2015), 70–76.

Pham, T. P. T., Kaushik, R., Parshetti, G. K., Mahmood, R., & Balasubramanian, R. (2015). Food waste-to-energy conversion technologies: Current status and future directions. Waste Management, 38(2015), 399–408.

Rahman, S. N. A., Masdar, M. S., Rosli, M. I., Majlan, E. H., Husain, T., Kamarudin, S. K. (2016). Overview biohydrogen technologies and application in fuel cell technology. Renewable and Sustainable Energy Reviews, 66(2016), 137–162.

Ramirez-Arias, A. M., Giraldo, L., & Moreno-Pirajan, J. C. (2018). Biodiesel synthesis: Use of activated carbon as support of the catalyst. In Kumar, S., Sani, R. (eds.), Biorefinery of biomass to biofuels. *Biofuel and biorefinery technologies,* (pp. 117–152).

Ravindran, R., & Jaiswal, A. K. (2016). Exploitation of food industry waste for high-value products. Trends in Biotechnology, 34(1), 58–69.

Rispail, N., Morris, P., & Webb, K. J. (2005). Phenolic compounds: Extraction and analysis. In Marquez, A. J. (Ed.), Lotus Japonicus Handbook (pp. 349–354). Dordrecht: Springer.

Sadh, P. K., Duhan, S., & Duhan, J. S. (2018). Agro-industrial wastes and their utilization using solid state fermentation: A review. Bioresource Bioproducts, 5(2018), 1–15.

Said, A., Leila, A., Kaouther, D., & Sadia, B. (2014). Date wastes as substrate for the production of a-amylase and invertase. Iranian Journal of Biotechnology, 12(47), 41–49.

Sarkis, J. R., Boussetta, N., Blouet, C., Tessaro, I. C., Marczak, L. D. F., & Vorobiev, E. (2015). Effect of pulsed electric fields and high voltage electrical discharge on polyphenol and protein extraction from sesame cake. Innovative Food Science and Emerging Technologies, 29(2015), 170–177.

Singh, A., Kuila, A., Adak, S., Bishai, M., & Banerjee, R. (2012). Utilization of vegetable wastes for bioenergy generation. *Agricultural Research*, 1(3), 213–222.

Singh, K., Kumar, T., Prince, Kumar, V., Sharma, S., & Rani, J. (2019). A review on conversion of food waste and by-products into value added products. International Journal of Chemical Studies, 7(2), 2068–2073.

Unakal, C., Kallur, R. I., & Kaliwal, B. B. (2012). Production of a-amylase using banana waste by Bacillus subtilis under solid state fermentation. European Journal of Experimental Biology, 2(2012), 1044–1052.

Verma, N., & Kumar, V. (2020). Utilization of bottle gourd vegetable peel waste biomass in cellulase production by Trichoderma reesei and Neurospora crassa. *Biomass Conversion and Biorefinery.* https://doi.org/10.1007/s13399-020-00727-9.

7

Recent and Novel Technology Used for the Extraction and Recovery of Bioactive Compounds from Fruit and Vegetable Waste

Rohini Dhenge, Massimiliano Rinaldi,
Tommaso Ganino, and Karen Lacey

7.1 Introduction

Food is essential for both ecology and human survival. It can either be consumed in raw form or processed to create goods with added value. The importance of fruits and vegetables in our diets and in human life has led to a large rise in demand for these essential food items (FAO, 2017a, 2017b). Examples of the significant amount of fruits produced globally include 124.73 million metric tons (MMT) of citrus; 114.08 MMT of bananas; 84.63 MMT of apples; 74.49 MMT of grapes; 45.22 MMT of mangoes, mangosteens, and guavas; and 25.43 MMT of pineapples (FAO, 2017a, 2017b). Production of some vegetables includes potato (3820.00 MMT), tomatoes (171.00 MMT), cabbages and other brassicas (71.77 MMT), carrots and turnips (38.83 MMT), cauliflower and broccoli (24.17 MMT), and peas (17.42 MMT) (FAO, 2017a, 2017b). Among all horticulture crops, fruits and vegetables are used the most. Due to the nutrients and health-improving elements they contain, they are eaten raw, minimally processed, and processed. Producing and processing of horticulture commodities, particularly fruits and vegetables, have greatly risen to meet the rising demand as a result of changing diets and expanding populations. A substantial dietary, financial, and environmental issue is being created by significant losses and waste in the fresh fruits and vegetables and processing industries. However, a growing concern has led to an increase in food waste being generated globally as a result of the population's exponential growth, and periodic supply-chain imbalances. Each year,

162 DOI: 10.1201/9781003269199-7

around 1.3 billion tons of the world's produced food are either lost or wasted, and this amount is increasing (Du et al., 2018). The processing of fruits and vegetables alone accounts for 16.5–20.5 MMT of the approximately 55 MMT of food waste produced globally each year (Saini et al., 2019). According to Jha and Sit (2021), India alone accounts for roughly 5.6 million tons of fruits and vegetables lost each year, or about 10% of all agricultural commodities lost after harvest (Figure 7.1). For instance, the FAO of the United Nations estimates that fruit and vegetable losses and waste are the greatest of all food types and may even exceed 60%. The bulk of food loss is caused by improper storage conditions at each level, according to a recent analysis by Rethink Food Loss through Economics and Data (ReFED) (ReFED, 2016). From leftovers at homes, restaurants, and businesses to postharvest garbage at the farming level, this waste is produced and described by Griffin et al. (2009) in great detail. Of the total food waste produced, 60% came from customers and families, with 20% coming from production waste, 1% from processing, and 19% from distribution. Similar problems with the preponderance of food waste created from all municipal solid garbage were observed in emerging nations like China (52.6%), the United States (25%), Bangladesh (75%), and France (32%) (De Clercq et al., 2017).

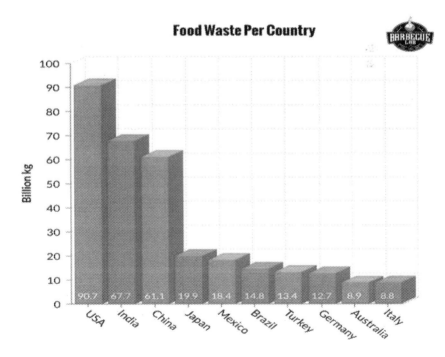

Figure 7.1 *The top 10 countries that produce the most food waste and the amount of total food loss.*

A recent poll conducted in Scandinavia in 2019 projected that households there lost more than 50% of their food due to difficulty choosing between "best before" and "use by" (Filimonau and De Coteau, 2019). The problem has now worsened and affected the broader food system with the onset of the coronavirus illness in 2019 (COVID-19) (Jribi et al., 2020). Fruit and vegetable waste account for a sizeable share (42%) of the waste created out of all the food resources that are wasted. Since methane can be produced as the main by-product from these wastes very cheaply, landfills are typically where they are disposed of. Although methane might be used to produce fuel, greenhouse gas has 25 times more potential to cause global warming than CO_2 (Cantera et al., 2018; Kumar, 2013). While processing co-products, a sufficient amount of waste from fruits and vegetables has also been produced. For example, during harvest or processing, more than 5 million tons of sugar beet pulp, 3.5 million tons of brewer's grain, and 500,000 tons of onion waste are produced, making up over 75% of the waste produced (Gowe, 2015). According to Sagar et al. (2018), the majority of fruits and vegetables yield between 25% and 30% of by-products in the form of peels, seeds, pomace, stems, and other materials. As an illustration, the percentage of waste produced by different produce sections is pomace and rotten fruits (52%) as the most common, followed by rind/skin/peels (42%) and seeds (10%). Fruit and vegetable waste materials are rich sources of bioactive substances such as fibers, polyphenols, flavonoids, tannins, vitamins, essential oils, natural colors, and organic acids (Sridhar et al., 2021). These bioactive compounds have been historically used for therapeutic purposes and offer numerous health benefits. The disposal of such waste poses a significant challenge due to its large quantity (Gowe, 2015). However, discarded peels and seeds, which are often overlooked, are particularly valuable as they contain phytochemical components that can be utilized as food taste enhancers and preservation agents (Sridhar et al., 2021). Additionally, fruits and vegetables with high levels of carotenoids, vitamins, and fiber possess antioxidant and anti-diabetic properties that contribute to the prevention of various human diseases and disorders (Yusuf, 2017). By harnessing these bioactive substances from fruit and vegetable waste, we can not only reduce waste but also tap into a sustainable source of beneficial compounds for various applications.

The utilization of food waste for the creation of bioactive components presents significant opportunities for various industries, including the food business, pharmaceutical and health sectors, and the textile industry. These industries can benefit from incorporating phytochemicals derived from food waste into enriched or functional products (Tostivint et al., 2016). However, the global agricultural and food sector faces obstacles in terms of food safety and waste management. Food waste is a growing concern, as it occupies valuable space in landfills and waste treatment facilities. Both low-income and developing nations experience negative socioeconomic effects from food waste, while middle- and high-income nations generate substantial amounts of household garbage due to consumer behaviors and excessive consumption (Tostivint et al., 2016). To address these challenges, reducing food waste and utilizing it to create value-added products can enhance the effectiveness of the food supply chain (FSC), lower associated costs, and improve food accessibility and security. By valorizing waste products, it is possible to mitigate environmental difficulties and promote sustainable solutions (Tostivint et al., 2016). The efficient valorization of waste products in various applications offers a useful source in lowering environmental difficulties as well as resolving problems in a sustainable manner. The types and makeup of waste derived from fruits and vegetables, their bioactive components, methods of extraction, and potential applications for the resulting bioactive chemicals are all covered in this chapter.

7.2 Bioactive Substances Are Extracted from Fruits and Vegetables

Fruits and vegetables are essential for meeting our daily nutritional requirements. However, the food industry generates a significant amount of waste, which is categorized based on its biological oxygen demand and chemical oxygen demand. This waste contains various bioactive compounds that can be utilized instead of discarded as waste, aligning with the concept of zero waste. These waste materials offer a diverse range of bioactive compounds, such as carotenoids (such as lutein and zeaxanthin), flavonoids (such as hesperetin, quercetin, genistein, and kaempferol), and phenolic acids. These phytochemicals vary in their polarity, solubility, molecular size, bioavailability, and metabolic pathways. Typically, fruit and vegetable waste contain higher levels of bioactive compounds with natural antioxidant properties compared to pulp, indicating the potential of these waste materials as a source of natural antioxidants. "Bioactive compounds" are secondary metabolites that are normally produced in modest levels in addition to primary metabolites like proteins, carbs, and lipids in fruits and vegetables. According to Jha and Sit (2021), these secondary metabolites should help plants grow and increase their capacity for survival (resistance to environmental stress, diseases, and UV radiation). The zero-waste approach involves the effective utilization of agro-industrial waste to produce value-added products, which find multiple applications in the food industry, including as colorants, antioxidants, and preservatives. Recently, these bioactives have been used extensively in nutraceuticals and functional meals.

It has been discovered that fruits or vegetables' meal and processing by-products, such as peels and seeds, contain significant levels of bioactive substances that can be directly incorporated into food products or incorporated using various nanoencapsulation techniques, such as nanoemulsions, to enhance their functional value. Bioactives from fruit and vegetable waste

Extraction and Recovery of Bioactive Compounds   165

have been successfully incorporated into a number of functional food products, including extruded snacks, bakery items, drinks, breakfast cereals, and dairy products (Arya et al., 2021). Furthermore, these compounds find applications in the nutraceutical, cosmetic, and pharmaceutical industries. The risk of developing numerous chronic diseases and the development of age-related issues can be effectively decreased by including five servings of fruits and vegetables each day. It is crucial to have a standardized and reliable approach to screen out compounds that are healthy for people. There are a wide variety of bioactives found in different fruit and vegetable wastes. The only way to unlock several opportunities for further bioactive separation, characterization, and purification that will result in high-quality products for customers is by choosing a suitable extraction procedure. In order to increase their delivery in food applications, functional and bioactive chemicals must be extracted from waste fruits and vegetables in greater quantity, quality, and with better bioactivity or bioavailability. Regarding health, the environment, and the economy, the sustainability of extraction technologies is essential. In addition, green extraction processes should demonstrate their maturity level through technological readiness levels (TRL) on a scale of 1 to 9 (Belwal et al., 2020). Despite being abundant in bioactive compounds, it is typically used as animal feed or left in landfills, posing serious environmental risks due to the release of damaging greenhouse gases (GHG). In order to make use of this bioactive-rich waste, safer and more environmentally friendly extraction methods must be used, such as supercritical fluid extraction, pulsed electric field, hydrodynamic cavitation, etc. Beyond their application as food additives and preservatives, these green-derived bioactives can also be employed in the formulation of functional and nutraceutical foods. Due to their generally recognized as safe (GRAS) status, they are currently employed as food coloring and flavoring additives.

To extract these bioactives that may be utilized as clean labels in food systems, there have been ongoing efforts to find new, easier, safer, more sustainable, environmentally friendly, and cost-effective extraction methods. The current study covers the unique, environmentally friendly methods for extracting bioactives from fruit and vegetable waste, particularly for culinary applications, along with their underlying principles, workings, and operations. The compilation of numerous succinct uses of these green extractives in functional foods has also been attempted. Researchers, academics, and anyone working in the food sector who wants to create goods with natural bioactives and green labeling will find this material useful. In addition, the high moisture and nutrition content promotes the growth of harmful bacteria, which are the main cause of offensive odors and infectious diseases (Sagar et al., 2018). However, many of the methods share the same goal, which is to isolate the specific bioactive from samples that are easy to separate and examine (Azmir et al., 2013). The sample, extraction technology, and extraction solvent are just a few of the variables that affect the extraction of bioactive (Figure 7.2).

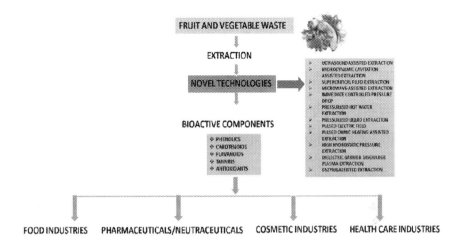

Figure 7.2 A schematic representation of fruit and vegetable waste valorization.

7.3 New and Green Extraction Techniques

According to Chemat et al. (2020), "Green extraction is based on the discovery and design of extraction processes which reduce energy consumption, allow the use of alternative solvents, produce renewable natural products, and ensure a safe and high-quality extract and or products." The term "green extraction" of natural bioactive components is preferable. These methods address some of the drawbacks of conventional methods because they consume less energy, are organic solvents, and are environmentally friendly. Limitations in current procedures spur industrial competitiveness to create more cutting-edge, environmentally friendly, and practical methods. It offers a well-organized chain of low-cost, high-energy processes that starts with the production of the biomatrix and ends with its transformation into bioactive components by extraction and separation.

In recent years, the food, pharmaceutical, and cosmetic industries have heavily utilized new substitutes with strong and environmentally friendly approaches using the concepts of green chemistry and green engineering (Chemat, 2012). Due to their superior practices focused on economic, environmental, and safety issues, green techniques have attracted a lot of attention. Following these guidelines, researchers developed the six green extraction principles and coined the term "green extraction" (Chemat, 2012). The six guiding principles of green extraction processes are:

1. Utilizing sustainable and renewable bio-resources,
2. Using green solvents or water,
3. Using less energy,
4. Producing waste co-products,

5. Using the fewest number of unit operations, and

6. Producing non-denatured and biodegradable extracts.

Academicians, scientists, and researchers are concentrating on extraction process intensification with increased extraction efficacy and good extract quality due to the significance of these six concepts. Recent technologies, such as ultrasound-assisted extraction, hydrodynamic cavitation-assisted extraction, microwave-assisted extraction, liquid-biphasic flotation extraction, cloud point extraction, and pulsed electric field (Gian) assisted extraction, have all demonstrated lower solvent and energy consumption, instantaneous regulated pressure drops, high-voltage electrical discharge (Saini et al., 2019), etc. The following sections provide discussions of these methods. Additionally, the separation of bioactive chemicals is great with the hybridization of extraction, which combines two or more green extraction technologies. In response to the limitations of traditional extraction procedures, innovative techniques have been introduced. Traditional extraction methods often face challenges in achieving high purity, employing costly solvents, requiring lengthy extraction times, potentially degrading heat-labile chemicals, and offering low extraction selectivity. To address these limitations, novel strategies have been developed. Various distinctive and emerging approaches are now being utilized for the extraction process.

7.3.1 Ultrasound-Assisted Extraction (UAE)

Ultrasound-assisted extraction (UAE) is a cost-effective and straightforward technique commonly performed within the frequency range of 20–2000 kHz. This method relies on two key principles: diffusion across cell walls and rinsing of contents after cell disruption. UAE operates by inducing cavitation, which involves the expansion and compression of the matrix, leading to increased permeability of cell walls and improved extraction efficiency. The technique is known to facilitate mass transfer, particle breakdown, and enhanced solvent accessibility to cells. UAE is widely employed for processing liquid–liquid or liquid–solid samples. However, the effectiveness of the UAE is influenced by various factors, including pressure, temperature, frequency, and sonication time (Shen et al., 2017). Among these factors, frequency plays a crucial role in determining both the yield and the properties of the extracted compounds. Studies have demonstrated that using frequencies above 20 kHz can alter the physicochemical properties of phytochemicals, potentially leading to the formation of free radicals. To implement UAE technology in large-scale industrial extraction, it is crucial to prioritize security, sustainability, cost-efficiency, and eco-friendliness. Achieving these goals can be facilitated through process intensification and energy consumption reduction. While UAE has not been reported to be used in large-scale industrial technology, France's Prestation d'Eco-Extraction et de Vectorisation (PEEV) eco-extraction platform is aiming to scale up UAE's industrial extraction method. With further innovation in

168   Wealth out of Food Processing Waste

process design, UAE has the potential to reach a high Technology Readiness Level (TRL) of 9.

7.3.1.1 UAE Applications for Removing Bioactives from Discarded Fruits and Vegetables

Several studies have demonstrated the successful extraction of beneficial compounds such as anthocyanins, polyphenols, flavonoids, and carotenoids from various fruit and vegetable wastes using UAE technology. More and Arya (2021) optimized the pulsed sonication process to extract bioactives from pomegranate peels, resulting in a remarkable extract yield of 0.51 g/g of peel. The extract exhibited exceptional antioxidant properties, containing 160.5 mg GAEAC/g of peel, including 177.5 mg/g of polyphenols and 0.21 mg/g of anthocyanins. The study emphasized that power and duty cycle were the critical factors influencing extraction efficacy. Similarly, using pulsed sonication, the extraction of polyphenols from Kinnow orange peel yielded 36.17 mg GAE/g extract, along with 64.70% DPPH and 57.37% ORAC antioxidant capabilities. A comparison between pomegranate peel and mandarin peel revealed that pomegranate peels required a higher amplitude (80%) to effectively extract the bioactives, potentially due to their increased toughness. The optimal treatment time for pomegranate peel was found to be considerably shorter. Goula et al. (2017) successfully recovered 93.8% of the total carotenoids from pomegranate peels under specific conditions, including a temperature of 51.5°C, an S/S ratio of 0.10, and an amplitude of 58.8%, using sunflower oil as a solvent. This study highlighted the influence of the target molecule and solvent on the effectiveness of UAE.

Another study by Belwal et al. (2019) reported a 25% higher anthocyanin yield for Starkrimson fruit peel using UAE compared to conventional extraction methods. Furthermore, the study identified 13 different anthocyanin molecules. UAE extract also exhibited superior DPPH and FRAP antioxidant capacity, surpassing traditional extraction methods by 28% and 23.2%, respectively.

7.3.2 Hydrodynamic Cavitation-Assisted Extraction (HCAE)

Only a limited number of studies have demonstrated the potential of the hydrodynamic cavitation-assisted extraction (HCAE) system for extracting bioactive compounds from fruit and vegetable waste materials. These studies have explored the effectiveness of HCAE as a standalone method or in combination with other extraction techniques, such as microwave, enzyme, deep eutectic solvents, maceration, and ionic liquids. Panda and Manickam (2019) have acknowledged the effectiveness of HCAE for industrial-scale extraction. Grillo et al. (2019) utilized a pilot-scale HCAE reactor (25 L) to extract various bioactives, including polyphenols, flavanols, methylxanthines, fatty acids, and cocoa butter, from cocoa bean shells. High-performance liquid chromatography (HPLC) was used to quantify the bioactive yield and quality, while liquid and gas chromatography-mass spectrometry (LC-MS and GC-MS) analyses revealed the presence of additional bioactives. Although HCAE is still in the pilot stage and lacks documented industrial applications for bioactive extraction, a "green technologies development platform" in Italy is actively working on developing flow-mode extraction processes using cavitation reactors. This progress could potentially elevate HCAE to a Technology Readiness Level (TRL) of 6–7 and facilitate its transformation into an industrial process (Belwal et al., 2020).

7.3.2.1 Utilizing HCAE to Recover Bioactive from Fruit and Vegetable Waste

The research conducted by Grillo et al. (2019) supported the effectiveness of HCAE by employing semi-continuous techniques with pilot-scale reactors. The overall yield of lipophilic bioactives was 15.8%, and hydrophilic bioactives accounted for approximately 20.5% when compared to the total weight of cocoa bean shells, which predominantly consisted of methylxanthines (theobromine: 160.2 mg/g extract and caffeine: 8.9 mg/g extract), fatty acids (964 mg/g extract), and lipophilic bioactives.

Meneguzzo et al. (2019) proposed regulated HCAE as a novel method for extracting low methoxy pectin from orange peel. They identified a suitable system for processing large quantities of material using a cavitation device with a venturi. The extraction yielded polyphenols (flavanones and derivatives of hydroxycinnamic acid), terpenes (mainly D-limonene), and pectin (with a low degree of esterification). The authors observed that 60% of the bioactive yield was obtained within 10 minutes of extraction, with 53% of that yield achieved within just 2 minutes, indicating the efficacy of HCAE. In the case of bioactive compounds from Silver Fir (Abies alba Mill.) needles, Albanese et al. (2019) obtained a yield of 0.14 mg GAE/mL for polyphenols and 0.55 mg CE/mL for flavonoids using only water as the extraction solvent and a venturi-type hydrodynamic cavitation reactor. Notably, this study demonstrated a significant extraction yield for bioactive compounds from Silver Fir.

7.3.3 Microwave-Assisted Extraction (MAE)

Microwave-assisted extraction (MAE) is a method that utilizes electromagnetic radiation in the frequency range of 300 MHz to 300 GHz, with the common frequency being 2450 MHz (equivalent to approximately 600–700 Watts of energy). This technique operates on the principle that microwave energy absorbed by the sample converts into thermal energy, facilitating the extraction process (Zhang et al., 2011). The mechanism of MAE involves the disruption of hydrogen bonds when a solvent comes into contact with the sample and is heated. This leads to the rotation of dipole molecules and the migration of ions, enabling the diffusion of the solvent and dissolution of the components (Datta et al., 2014). Another possible mechanism involves the evaporation of moisture within the cells, resulting in increased pressure on the cell wall. This, in turn, modifies the physical properties and porosity of the materials, enhancing solvent penetration and improving the yield of biomaterials (Routray and Orsat, 2011a, 2011b).

7.3.3.1 Application of MAE to Extract Bioactives from Fruit and Vegetable Waste

MAE has demonstrated successful extraction of bioactive compounds regardless of the source, solvent, or extraction conditions. Researchers have achieved significant carotenoid extraction yields from carrot, gac fruit, and passion fruit peels using MAE. Optimal extraction conditions, such as 200 W microwave power for 25 minutes with olive oil as the solvent, resulted in superior carotenoid production (86.9% yield) from passion fruit peel (Chutia and Mahanta, 2021). Similarly, flaxseed oil as a solvent with 165 W power for 9.39 minutes successfully recovered approximately 77.48% of carotenoids from carrot waste (Chutia and Mahanta, 2021; Elik et al., 2020). Notably, the use of edible oil as a solvent and the potential application of carotenoid-enriched oil as a source of antioxidant-rich edible oil were intriguing findings (Chutia and Mahanta, 2021; Elik et al., 2020).

Microwave energy played a crucial role in carotenoid extraction in all the mentioned studies. Furthermore, several researchers have reported MAE as an effective method for extracting polyphenolic compounds from various waste materials, including carob fruit peel, pomegranate peel, spinach waste, and banana peel (Vu et al., 2019). Optimum extraction conditions using MAE, such as 600 W power for 60 seconds, resulted in the highest polyphenol yield (87.8 mg/g) from pomegranate peel (Motikar et al., 2021). Conversely, response surface methodology was used to evaluate various MAE treatment settings for polyphenol extraction from spinach waste peel, and the best combination of ethanol, water, and HCl (60/39.9/0.1) produced polyphenols with an approximate concentration of 950 mg/kg fresh weight after 5 minutes at 90°C (Quiles-Carrillo et al., 2019). The study also compared MAE with ultrasound-assisted extraction (UAE) and highlighted UAE as a practical strategy for the spinach matrix. However, the authors did not consider microwave power in the experiment with the spinach matrix, noting that the pomegranate peel is tougher than spinach, which could influence the yield and quality of the extract (Quiles-Carrillo et al., 2019). The complexity of the sample matrix and the power of MAE have been shown to impact the yield and quality of the extract. Additionally, pH can be a significant factor in the effectiveness of MAE. For instance, intensified MAE of banana peel bioactives produced 50.55 mg/g of phenolic compounds after 6 minutes at 960 W and pH 1.0 (Vu et al., 2019)

7.3.4 Supercritical Fluid Extraction (SFE)

The extraction of compounds using this technique involves the use of solvents at temperatures and pressures above their critical points. The critical point refers to the specific temperature (Tc) or pressure (Pc) at which gas and liquid phases can no longer be distinguished from each other. These solvents, known as supercritical fluids, possess unique properties that combine the characteristics of liquids (such as density and solvation power) and gases (including viscosity, diffusion, and surface tension). These properties enable efficient extraction with higher yields in a shorter period of time (Ameer et al., 2017). Supercritical CO_2 extraction has gained prominence as a safe, efficient, and environmentally friendly technology for extracting bioactives from fruit and vegetable waste (Sagar et al., 2018). Although industrial-scale applications of supercritical CO_2 extraction for extracting flavor and aroma chemicals from solid materials are well-established, the valorization of fruit and vegetable waste for bioactive extraction is still at the pilot scale (TRL close to 8), with ongoing research for industrial applicability.

7.3.4.1 Application of SFE to Extract Bioactives from Fruit and Vegetable Waste

Various studies have investigated the extraction parameters and bioactive content of different waste materials. Increasing pressure has been found to enhance carotenoid extraction yield (97%) from carrot peels, but excessive pressures over 250 bar lead to a decline in output (Coelho et al., 2018). Temperature and co-solvents are often adjusted to improve extraction yield by modifying the affinity of CO_2 for bioactives, particularly polar compounds with low solubility. For instance, Fornereto Soldan et al. (2021) demonstrated that increasing temperature and adding ethanol increased oleoresin yield (9.38–10.08%) from capsicum annum waste. The extracted oleoresin contained carotenoids (0.27–2.01 mg/g), phenols (12.30–23.94 mg/g), and flavonoids (0.6–1.52 mg/g).

Modifying the extraction process can also impact the yield of specific compounds. Lima et al. (2019) observed a 40% increase in pipercallosidine yield from pepper leaves when 5% methanol was used as a modifier alongside CO_2. However, Šafranko et al. (2021) noted the production of 5-HMF and chlorogenic acid as by-products of thermal deterioration at temperatures above 175°C. Hesperidin concentration ranged from 0.16 to 15.07 mg/g, while limonin exhibited strong antioxidant activity (13.16 and 30.65% at 100 and 300 bar, respectively) (del Pilar Sanchez-Camargo et al., 2019). Optimizing the extraction process involves considering factors such as solvent and bioactive contact times. Longer contact times generally result in higher extraction yields. Ekinci and Gürü (2019) achieved a 97% recovery of phytosterols from melon seeds with a dynamic extraction time of 1–3 hours, beyond which no significant change was observed, likely due to reaching maximal solubility. Supercritical fluid flow rate also plays a role, affecting residence time and mass transfer. Increasing flow rate enhances extraction potential by reducing mass transfer limitations, but excessively high flow rates may decrease yield due to inadequate contact between solute and supercritical fluid. The addition of high-fat materials, such as ground coconut or peanut, to the raw material can improve solvation power and facilitate the separation of non-polar bioactives and essential oils.

7.3.5 Extraction with Instant Controlled Pressure Drop

7.3.5.1 Applications of DIC for the Removal of Bioactives from Leftover Fruits and Vegetables

Instant controlled pressure drop (DIC) systems have gained attention in contemporary applications for the extraction of active molecules, with several papers highlighting their effectiveness in extracting bioactives from various raw materials under ideal conditions. Mkaouar et al. (2018) demonstrated the successful extraction of phenolic compounds from olive leaves using ethanol, resulting in a 312% increase in extraction yield. The DIC system showed improved bioavailability and antioxidant activity compared to untreated leaf extracts. Louati et al. (2019) also found that DIC processing facilitated the extraction of phenolic compounds and improved drying of orange industry by-products.

Martínez-Meza et al. (2021) observed intensified extraction of anthocyanins, non-extractable proanthocyanidins, and polyphenols from grape pomace using DIC at 2 bar for 60 seconds. However, at higher pressure and longer duration, a decrease in anthocyanin concentration was noted, possibly due to depolymerization of proanthocyanidins and conversion of anthocyanins to phenolic compounds. Allaf et al. (2013) suggested that DIC could enhance the extraction of essential oils, and Rashidi et al. (2018) demonstrated that DIC yielded higher essential oil extraction compared to hydro-distillation, UAE, and Soxhlet extraction methods. Feyzi et al. (2017) supported the efficacy of DIC for extracting bioactives with higher antioxidant activity, such as polyphenols from *Bunium persicum* and essential oils. Efforts are being made to scale up DIC technology from pilot to industrial levels, with the ABCAR-DIC Process Company in La Rochelle, France leading the way (Chemat et al., 2020). However, as the technology is still under development, the TRL for DIC is expected to range from 6 to 8.

7.3.6 Pressurized Hot Water Extraction

Pressurized hot water extraction (PHWE) is an innovative method used to extract bioactive compounds from food and by-product matrices by utilizing hot water as the solvent. One key aspect of this technique is the application of high pressure and temperature conditions, typically above 0.1 MPa and 100°C but below 374°C and 22.1 MPa (Duba et al., 2015). These elevated pressure and temperature levels result in significant changes in properties such as surface tension and viscosity.

The reduction in solvent surface tension during PHWE facilitates the penetration of the solvent into the pores of the peel matrix to a greater extent. These alterations in properties enhance the efficiency of PHWE as an extraction technique, leading to reduced extraction time and increased recovery of phenolic compounds. Several studies have been conducted to investigate the effectiveness of PHWE for extracting bioactive molecules from food and by-products. These studies have identified various factors that influence the efficacy of PHWE, including temperature, pressure, solubility of the target molecule, flow rate, characteristics of modifiers, and the nature of additives used. In summary, PHWE is a promising extraction technique that utilizes hot water under elevated pressure and temperature conditions. By exploiting these conditions, PHWE can effectively extract bioactive compounds from food and by-product matrices, offering advantages such as shorter extraction times and improved recovery rates of phenolic compounds.

7.3.6.1 Application of PHWE

Several studies have examined the potential of PHWE for the recovery of valuable compounds from fruit and vegetable waste peels. Vergara-Salinas et al. (2013) optimized PHWE conditions (10.34 MPa for 5 min at 100°C and 150°C) and achieved the recovery of approximately 98–100% of anthocyanins and tannins from grape peel extract. However, they observed an increase in the degradation of extracted molecules at higher temperatures, leading to the formation of colorless chalcone compounds due to the opening of the pyrilium ring. Tunchaiyaphum et al. (2013) successfully extracted polyphenols and tannins from mango peels using PHWE at 180°C, highlighting its environmental friendliness. In contrast, Ko et al. (2011) suggested PHWE as an excellent alternative for extracting quercetin from onion peel waste but did not identify the optimal pressure value for quercetin extraction. Lopresto et al. (2014) investigated the effect of PHWE on the recovery of D-limonene from citrus peels and found that pre-drying and particle size reduction significantly increased the yield of D-limonene.

Duba et al. (2015) reported an increase in total polyphenol content from 44.3 ± 0.4 to 77 ± 3 mg/g by varying the PHWE temperature from 80 to 120°C. This study also provided the kinetics of PHWE extraction using a two-site kinetic model, which could be a valuable tool for future process development and design. However, it is important to consider the residence time in PHWE, as thermally sensitive compounds may degrade if the extraction time is too long. The effectiveness of PHWE depends on the equilibrium of the desired analytes dispersing into the chosen solvent, which stabilizes after a certain period. Overall, PHWE has shown efficiency in extracting polyphenols, particularly tannins and volatile compounds, from fruit and vegetable waste peels. Its commercialization could facilitate the scaling up of this extraction process in fruit and vegetable processing industries.

7.3.7 Pressurized Liquid Extraction

Pressurized liquid extraction (PLE) is a novel extraction technique that has been widely used for extracting phytochemical compounds from natural food matrices. Initially introduced by Dionex Corporation for the extraction of anthocyanins from seaweed for pharmaceutical purposes (Mustafa and Turner, 2011), PLE has since been successfully applied to extract anthocyanins, flavonoids, and saccharides from various fruits and vegetables, indicating its potential for extracting phytochemicals from waste fruit and vegetable peels. PLE offers advantages in the food industry due to its rapidity and reduced solvent consumption (Joana Gil-Chávez et al., 2013). Several studies have investigated the effectiveness of PLE for extracting phytochemicals from food matrices, identifying factors such as solvent polarity, solvent toxicity, particle size, mass transfer rate, peel moisture content, temperature, pressure, and extraction time as influential parameters.

7.3.7.1 Application of PLE

According to Howard and Pandjaitan (2008), the purity of flavonoids and polyphenolic compounds extracted from spinach stem peels using PLE was approximately 91% and 93%, respectively. The antioxidant activity of the extracts was also found to be 2.1 times and 7.7 times higher compared to conventional extraction methods. Similarly, Cheigh et al. (2012) reported improved extraction yields of narirutin and hesperidin from citrus peels using PLE. These citrus peel extracts demonstrated effective antioxidant properties when added to soybean and peanut oils but had little to no effect or acted as pro-oxidants when incorporated into vegetable oil-in-water systems. However, further studies are needed to determine the antioxidant capacity of these extracts in vivo.

Ju and Howard (2003) found that red grape peel extract obtained using acidified water through PLE at temperatures between 80 and 100°C yielded the highest purity of total anthocyanins (94%) compared to PLE with acidified methanol (90% purity). Machado et al. (2015) investigated the effect of PLE on the extraction of total phenolics (TP) and anthocyanins (AA) from blackberry peels using acidified water as the solvent. They determined that a temperature of 140°C was the optimal PLE condition, resulting in the highest yields of TP (7.36 mg GAE/g) and AA (76.03 mmol TE/g) with purities of 91% and 90%, respectively. Barrales et al. (2018) studied the potential of PLE for retaining phenolic compounds and reducing sugars from citrus peels. They found that PLE at 140°C yielded a two-fold increase in reducing sugars and a five-fold increase in phenolic compounds compared to conventional techniques, with significant purity. While PLE has demonstrated numerous applications and advantages, it is primarily used in systematic settings and requires a significant initial investment. Additionally, in some cases, the mechanical impact of PLE can lead to the inactivation of extracted bioactive molecules. Despite these limitations, the high recoveries (above 90%) of flavonoids, anthocyanins, and reducing sugars obtained through PLE suggest its suitability for quality control analysis and its potential for industrial-scale applications in the food, nutraceutical, and pharmaceutical industries.

7.3.8 Pulsed Electric Field

Pulsed electric field (PEF) is a non-thermal technique that utilizes direct current to process materials. In this method, short-duration, high-voltage pulses are applied to the materials placed between electrodes. The electric current passing through the cell suspension disrupts cell structures, causing the molecules to separate based on their respective charges. PEF can be operated in either batch or continuous mode. Factors such as field strength, energy, pulse number, material properties, and temperature affect the extraction yield and must be considered during process design (Heinz et al., 2003).

7.3.8.1 Application of PEF

El Kantar et al. (2018) examined the impact of PEFAE on the recovery of polyphenols from citrus fruit waste peels (pomelo, orange, and lemon). The results showed that PEF treatment at 3 kV/cm and 10 kV/cm increased polyphenol recovery from the peels, with orange peels reaching up to 22 mg/g DM and citrus peels reaching 16 mg/g DM. However, a slight reduction in polyphenol yield was observed in pomelo peels, which was attributed to their higher density compared to other citrus peels. Koubaa et al. (2016) aimed to extract the red color pigment (betanin) from red prickly pear peel using PEFAE. They found that optimal recovery of the pigment (at 20 kV/cm, 50 min, 20°C) was achieved due to the diffusion of betanin molecules from intracellular to extracellular media facilitated by PEF-induced cell permeabilization. The extraction kinetics of total betanin from pear peel were modeled using the Peleg model. However, it should be noted that microstructural analysis was not considered in these studies, and future research could benefit from investigating the preservation of cell integrity in relation to extraction efficiency.

Hossain et al. (2015) also observed enhanced leaching of polyphenolic compounds from potato peels with increasing electric field magnitude during PEFAE. These studies collectively indicate the feasibility of using PEF-assisted extraction techniques to recover polyphenols, including flavonoids, natural pigments, total phenolic compounds, and proteins, from waste peels. PEF has been employed for the extraction of phenolic compounds and anthocyanins from Merlot skin, where it enhances the stability of bioactive compounds during winemaking and reduces extraction time (Delsart et al., 2012; López et al., 2008). PEF treatment combined with maceration during vinification not only improves color, anthocyanin content, and polyphenol content but also reduces maceration time, thereby enhancing wine quality (López et al., 2008). Additionally, PEF treatment applied to orange peel resulted in a 159% increase in phenolic compound yield at a power of 7 kV/cm (Luengo et al., 2013).

In the extraction of phenolic and flavonoid compounds from onions, PEF treatment significantly increased the yield by 2.2 and 2.7 times, respectively, compared to control samples (Liu et al., 2018). PEF is an environmentally friendly process suitable for extracting valuable components from onions without significant loss of their properties. The electric field intensity and extraction time exert a significant effect on the yield of phenolic and flavonoid compounds (Liu et al., 2018). PEF technology facilitates the extraction of anthocyanins, betanines, carotenoids, and other compounds from natural food sources and by-products, reducing extraction time, solvent quantity, and temperature requirements (Boussetta et al., 2012). One notable advantage of PEF is its non-thermal nature, which preserves the quality of the final product. Moreover, it can be implemented in continuous processes at the pilot or industrial scale, making it a cost-effective emerging technology applicable to large-scale food processing (Boussetta et al., 2012).

7.3.9 Pulsed Ohmic Heating Extraction (POHE)

In many cases, the use of a low electric field at ambient temperature may not be sufficient to extract the desired bioactive molecules, especially polyphenols, which are strongly bonded to the cellular membrane of the peel. Therefore, the application of pulsed ohmic heating extraction (POHE) has emerged as a powerful method to achieve higher yields of phenolic compounds by raising the temperature through ionic movement resulting from the ohmic heating process, while maintaining a moderate electric field (E < 100 V/cm) (Jaeschke et al., 2019). POHE is commonly implemented using an alternating current (AC) power source, making it widely accepted in the food industry due to the accessibility of AC current and reduced risk of electrolysis. Moreover, POHE has been found to effectively inactivate enzymes in foods, preserve organoleptic properties, facilitate product formation and reformation, induce microbial inactivation, and aid in clarification processes (Gavahian et al., 2016; Jittanit et al., 2017; Kim et al., 2017).

Interestingly, the application of pulsed ohmic heating for the extraction of bioactive compounds from food products, residues, and by-products has also been investigated. The effectiveness of POHE in extracting bioactive compounds from waste peel matrices depends on critical factors such as temperature, frequency, field strength, process time, electrical conductivity, and the specific properties of the peel matrix.

7.3.9.1 Application of POHE

In a recent study conducted by Coelho et al. (2019), it was found that POHE treatment applied to tomato peels resulted in a significant increase in the recovery of rutin and polyphenols, with 77% and 58% higher yields compared to the control sample, respectively. Interestingly, the researchers also observed an increase in the bioavailability of lycopene within the temperature range of 70–88°C. However, degradation due to oxidation was observed both below and above this temperature range. Another study by Coelho et al. (2017) demonstrated that POHE extraction at 40°C and 70°C led to a substantial improvement in thenantioxidant activity and phenolic content of tomato

Extraction and Recovery of Bioactive Compounds   177

peels. Nonetheless, further investigations are needed to assess the influence of electrical frequency and field strength on the extraction mechanism.

Pereira et al. (2016) applied POHE for phytochemical extraction from potato peels and found it to be a promising technique for the recovery of anthocyanins and phenolic compounds from vegetable peels while reducing power consumption. El Darra et al. (2013) obtained a 36% higher yield of polyphenolic extracts from grape peel samples when subjected to POHE conditions of 400 V/cm at 50°C. Overall, these studies indicate the potential of POHE as a promising technique for the valorization of waste peels from fruit and vegetable processing industries, eliminating the need for hydroalcoholic solvents.

7.3.10 High Hydrostatic Pressure Assisted Extraction

High hydrostatic pressure (HHP), also known as high pressure processing (HPP), is an environmentally friendly technique that involves processing food at very high pressures ranging from 100 to 300 MPa at ambient temperature for a short duration of 3–5 minutes (Oey et al., 2008). This technique has various applications in food processing, including microbial decontamination, surface modification, enhancement of nutritional properties, and product transformation.

Subsequently, HHP-assisted extraction (HHPE) was recognized as an effective alternative extraction method for plant materials, including foods and their byproducts. HHPE has been shown to be a rapid and efficient process compared to other extraction methods. The extraction mechanism of bioactive compounds from waste peels depends on factors such as peel particle size, presence of interfering molecules, chemical nature of the compounds, extraction conditions, sample moisture, and equilibrium constant. Notably, the US Food and Drug Administration has acknowledged HHP-assisted extraction as an environmentally friendly technology because it requires only power for operation and does not generate any waste (Oey et al., 2008). This recognition highlights the sustainability and potential of HHP as a green extraction technique.

7.3.10.1 Application of HHPE

Strati et al. (2015) conducted a study where HHPE performed at 700 MPa for 10 minutes resulted in a six-fold increase in carotenoid yield from tomato peel compared to solvent extraction. The increase in yield was attributed to the denaturation of carotenoid binding proteins induced by HHP. Similarly, Xi (2006) reported an improved extraction of lycopene from tomato peel after processing at 500 MPa for 12 minutes at 20°C, which was attributed to tissue fissures caused by HHPE, leading to the release of lycopene.

In another study, Xie et al. (2018) investigated the extraction of pectin from potato waste peel using HPPE at 200 MPa for 5 minutes. They found that the pectin extracted through HHPE exhibited enhanced emulsifying properties, making it suitable for various food applications. Naghshineh et al. (2013)

explored the extraction of pectin from lime peel using HHPE and observed a linear relationship between extract yield and applied pressure. However, at 400 MPa, a decrease in extraction yield was observed, which was attributed to the irreversible breakage of intramolecular and intermolecular enzymes caused by the high-pressure treatment, resulting in a decline in activity. Furthermore, Casazza et al. (2012) assessed the potential of HHPE for recovering phenolic molecules from grape peels. The study demonstrated the effectiveness of HHPE in extracting phenolic compounds from the peels.

7.3.11 Dielectric Barrier Discharge Plasma Extraction (DBDE)

Non-thermal plasma, also known as cold plasma, is generated by applying an electric field to a gas. This energy accelerates and ionizes the gas molecules, leading to the production of additional free electrons that further trigger new ionizations. The energized electrons cause the dissociation of molecules, resulting in the formation of radicals and atoms. Excited molecules release excess energy in the form of electromagnetic radiation, including UV radiation. Thus, a plasma mainly consists of energized atoms, free radicals, nitrogen species, ozone, electrons, positive ions, negative ions, and other components (Takamatsu et al., 2015).

In recent years, there has been growing interest in the application of cold plasma extraction techniques in the food industry (Hou et al., 2008). One specific type of non-thermal plasma called dielectric barrier discharge (DBD) plasma has been widely used for enzyme inactivation, microbial inactivation, and other food processing applications. It has also been employed for the extraction of phytochemical compounds from foods and their by-products.

7.3.11.1 Application of DBDE

In a study by S. (2018), the optimization of pectin extraction from ponkan peel using the DBDE technique was conducted with the application of response surface methodology (RSM). The study successfully achieved a significant pectin yield of 27% under the following conditions: pH 2, voltage of 40 V, and solid-to-liquid ratio (S/L) of 1:30 for a duration of 5.5 minutes. It was observed that an extended extraction period (>5.5 min) or a voltage exceeding 40 V led to a decrease in pectin recovery, as prolonged exposure to the plasma system resulted in pectin degradation. However, the precise mechanism underlying this extraction technique still requires further investigation. Despite limited attention given to the use of DBD plasma for the extraction of bioactive compounds from waste peels, it has been recognized for its enhanced selectivity toward targeted molecules (Umair et al., 2019). Moreover, the extraction using DBD plasma requires lower energy consumption and can be conducted without the need for additional chemical compounds. Therefore, it is considered a promising technique for the recovery of bioactive compounds from plant materials and their by-products. However, there are some limitations that need to be addressed for practical implementation of DBDE, such as the limited lifespan of the plasma system, increased operational costs, and potential alteration of physicochemical attributes of the targeted compounds during the extraction process.

7.3.12 Enzyme-Assisted Extraction

Enzyme-assisted extraction is a novel technology that offers an alternative to conventional extraction methods for various valuable components. Unlike organic chemicals, water is used as the solvent in this technique, making it more sustainable and environmentally friendly (Puri et al., 2012). It is particularly beneficial when phytochemicals are dispersed in the cell cytoplasm or when the presence of hydrogen or hydrophobic bonds in the polysaccharide-lignin network hinders extraction using conventional techniques. Due to its eco-friendly nature and ability to achieve higher product yields, enzyme-assisted extraction has gained increased attention in recent years (Puri et al., 2012). The efficiency of enzyme-assisted extraction for different bioactive compounds is influenced by factors such as enzyme concentration, composition, particle size, water-to-solid ratio, and hydrolysis time (Puri et al., 2012). This technique has been successfully employed for the extraction of various components, including carotenoids from pumpkin (Ghosh and Biswas, 2015), anthocyanins from Crocus sativus (Lotf et al., 2015), and anthocyanins from grape skin (Muñoz et al., 2004). For example, pectinase-assisted extraction yielded 18–20 mg/g of phenolics from grape marc seeds (Štambuk et al., 2016), Pectinex® facilitated the extraction of antioxidant phenols (0.152 mg/g) from apple pomace (Oszmianski et al., 2011), achieved a phenolics concentration of 86.6% from grape residues (Gómez-García et al., 2012). Enzyme-assisted extraction using enzymes such as pectinase, cellulase, and protease has proven effective in enhancing compound recovery. However, the cost of enzymes limits the scalability of the process and the yield of the final product. One potential solution is enzyme immobilization, which allows for enzyme reuse without compromising specificity and activity, thereby reducing costs (Puri et al., 2012)

7.4 Conclusion and Future Prospects

The processing industry of fruits and vegetables generates a significant amount of waste that contains valuable bioactive compounds. These compounds can be utilized in various industries as natural colorants, antioxidants, preservatives, and antimicrobial agents. To enhance the functional and textural properties of these bioactive compounds in food systems, nanoencapsulation techniques like nanoemulsions can be employed. Non-conventional extraction methods have been used to extract bioactives from waste, but there is a need for the development of efficient and novel techniques to recover these compounds. Ultrasound-assisted extraction has emerged as a promising method

Table 7.1 Extraction of Various Bioactive Components from Waste Using Non-Thermal Processing Techniques with its Pros and Cons

Technique	Advantages	Disadvantages	Bioactive component	Extraction time-sample size-solvent quantity-power force
Supercritical fluid extraction	- Quick, secure, excellent selectivity, and requires no filtering - Greater penetration and superior mass transfer	- Many variables to optimize - Needs precise temperature and pressure control	Flavonoids, antioxidants, carotenoids, fatty acids, essential oils, terpenes, polyphenols	10–60 min 1–5 g 30–60 mL Moderate
Microwave-assisted extraction	- Quick, simple to use, and moderate consumption of solvent - Reduced extraction time	- Risk of explosion (solvent must absorb microwave energy), costly, necessary filtration step - Not suitable for heat-sensitive and non-polar compounds	Phenolic compounds, glycosides, flavonoids, terpenoids, essential oils, alkaloids, saponins	3–30 min 1–10 g 10–40 mL High
Enzyme-assisted extraction	- Extraction of cell-wall-bound components	- Limited enzymes optimized for extraction efficiency	Anthocyanins, polyphenols, carotene, terpenes, flavonoids	
Pulsed electric field extraction	- Non-thermal technique	- Not suitable for highly electrically conductive products	Phenols, flavonoids, proteins, anthocyanins, carbohydrates	
High-hydrostatic-pressure extraction	- Low energy consumption	- Requires expensive equipment and maintenance	Phenolic compounds, carotenoids, flavonoids, pectin, lutein, lycopene, catechin	
Ultrasound-assisted extraction (acoustic cavitation)	- Simple to use, safe for ambient temperature and atmospheric pressure, uses a moderate amount of solvent, and is reproducible - Low energy consumption	- Filtration is necessary, and high frequency operation may cause some compounds to degrade - Can produce free radicals affecting bioactive compound quality	Phenolic compounds, flavonoids, oils, anthocyanins	10–60 min 1–30 g 50–200 mL Moderate

in the food processing sector due to its higher yield, better quality, simplicity, and environmentally friendly nature.

Nanoemulsions serve as essential delivery systems for various bioactive compounds in the food industry. They offer multiple functions, such as increased bioavailability, controlled ingredient release, modulation of product texture, and improved protection against degradation. Although extensive research is being conducted on different extraction techniques and delivery systems loaded with bioactive compounds, further investigation is required to understand the specific functions of bioactive compounds within food-based delivery systems and their bioavailability in the body (Table 7.1). A comprehensive understanding of the bioaccessibility, metabolism, and absorption of encapsulated compounds through the gastrointestinal tract is desirable, considering the alterations that occur in their properties. Additionally, extensive research should focus on the loading capacity of bioactive compounds in nanoemulsions and their release characteristics. Therefore, it is crucial to extract bioactive compounds from fruit and vegetable waste to reduce environmental pollution. These compounds can potentially replace synthetic antioxidants in various industries, and further research is needed to explore their applications fully.

References

Albanese, L., Bonetti, A., D'Acqui, L. P., Meneguzzo, F., & Zabini, F. (2019). Affordable production of antioxidant aqueous solutions by hydrodynamic cavitation processing of silver fir (Abies alba Mill.) needles. Foods, 8(2), 65.

Allaf, T., Tomao, V., Besombes, C., & Chemat, F. (2013). Thermal and mechanical intensification of essential oil extraction from orange peel via instant autovaporization. Chemical Engineering and Processing: Process Intensification, 72, 24–30. https://doi.org/10.1016/j.cep.2013.06.005.

Ameer, K., Shahbaz, H. M., & Kwon, J. H. (2017). Green extraction methods for polyphenols from plant matrices and their byproducts: A review. Comprehensive Reviews in Food Science and Food Safety, 16(2), 295–315.

Arya, S. S., Venkatram, R., More, P. R., & Vijayan, P. (2021). The wastes of coffee bean processing for utilization in food: A review. Journal of Food Science and Technology, 1–16.

Azmir, J., Zaidul, I. S. M., Rahman, M., Sharif, K., Mohamed, A., Sahena, F., Jahurul, M., Ghafoor, K., Norulaini, N., & Omar, A. (2013). Techniques for extraction of bioactive compounds from plant materials: A review. Journal of Food Engineering, 117(4), 426–436. doi: 10.1016/j.jfoodeng.2013.01.014.

Barrales, L., Escalona-Buendía, H. B., & Aguilar, C. N. (2018). Retention of phenolic compounds and reducing sugars from citrus peel using pressurized liquid extraction. Journal of Food Science and Technology, 55(9), 3643–3650. doi: 10.1007/s13197-018-3253-4.

Belwal, T., Chemat, F., Venskutonis, P. R., Cravotto, G., Jaiswal, D. K., Bhatt, I. D, & Luo, Z. (2020). Recent advances in scaling-up of non-conventional extraction techniques: Learning from successes and failures. TrAC Trends in Analytical Chemistry, 127, 115895.

Belwal, T., Huang, H., Li, L., Duan, Z., Zhang, X., Aalim, H., & Luo, Z. (2019). Optimization model for ultrasonic-assisted and scale-up extraction of anthocyanins from Pyrus communis 'Starkrimson' fruit peel. Food Chemistry, 297, 124993.

Boussetta, N., Vorobiev, E., & Reess, T. (2012). Pulsed electric field-assisted extraction of bioactive compounds. In Pulsed Electric Fields Technology for the Food Industry: Fundamentals and Applications (pp. 239–268). Springer.

Cantera, S., Muñoz, R., Lebrero, R., López, J. C., Rodríguez, Y., García-Encina, P. A. (2018). Technologies for the bioconversion of methane into more valuable products. Curr. Opin. Biotechnol., 50, 128–135. https://doi.org/10.1016/j.copbio.2017.12.021.

Casazza, A. A., Aliakbarian, B., Mantegna, S., Cravotto, G., & Perego, P. (2012). Extraction of phenolic compounds from grape skins and defatted grape seeds using power ultrasound: A study on the mechanisms of ultrasound-assisted extraction. Ultrasonics Sonochemistry, 19(3), 582–590.

Cheigh, C. I., Chung, M. S., & Chung, H. J. (2012). Effect of pressurized liquid extraction on extraction yield and antioxidant activity of narirutin and hesperidin from citrus peel. Journal of Agricultural and Food Chemistry, 60(19), 4756–4761. doi: 10.1021/jf2050515.

Chemat, F. (2012). Microwave-Assisted Extraction for Bioactive Compounds. (Vol. 4). Springer Science & Business Media.

Chemat, F., Vian, M. A., Fabiano-Tixier, A.-S., Nutrizio, M., Jambrak, A. R., Munekata, P. E., Lorenzo, J. M., Barba, F. J., Binello, A., & Cravotto, G. (2020). A review of sustainable and intensified techniques for extraction of food and natural products. Green Chemistry, 22(8), 2325–2353. doi: 10.1039/C9GC03878G.

Chutia, H., & Mahanta, C. L. (2021). Green ultrasound and microwave extraction of carotenoids from passion fruit peel using vegetable oils as a solvent: Optimization, comparison, kinetics, and thermodynamic studies. Innovative Food Science & Emerging Technologies, 67, 102547.

Coelho, J. P., Filipe, R. M., Robalo, M. P., & Stateva, R. P. (2018). Recovering value from organic waste materials: Supercritical fluid extraction of oil from industrial grape seeds. The Journal of Supercritical Fluids, 141, 68–77.

Coelho, M. S., Saldaña, E., Branco, R., Pereira, S. A., Saraiva, J. A., & Alves, V. D. (2019). Pulsed ohmic heating as an innovative technology to extract high-value compounds from tomato peels. Food and Bioprocess Technology, 12(10), 1716–1726.

Coelho, M. S., Saraiva, J. A., & Alves, V. D. (2017). Valorization of tomato peel using pulsed ohmic heating: Effect on antioxidant activity and phenolic content. Food Chemistry, 221, 1372–1380.

Datta, A. K., Sumnu, G., & Raghavan, G. S. V. (2014). Dielectric properties of foods. In Engineering Properties of Foods (pp. 523–588). CRC Press.

De Clercq, D., Wen, Z., Gottfried, O., Schmidt, F., & Fei, F. (2017). A review of global strategies promoting the conversion of food waste to bioenergy via anaerobic digestion. Renew. Sustain. Energy Rev., 79, 204–221. https://doi.org/10.1016/j.rser.2017.05.047.

del Pilar Sanchez-Camargo, A., Gutierrez, L. F., Vargas, S. M., Martinez-Correa, H. A., Parada-Alfonso, F., & Narvaez-Cuenca, C. E. (2019). Valorisation of mango peel: Proximate composition, supercritical fluid extraction of carotenoids, and application as an antioxidant additive for an edible oil. The Journal of Supercritical Fluids, 152, 104574.

Delsart, C., Ghidossi, R., Poupot, C., Cholet, C., Grimi, N., Vorobiev, E., ... & Peuchot, M. M. (2012). Enhanced extraction of phenolic compounds from Merlot grapes by pulsed electric field treatment. *American journal of Enology and Viticulture, 63*(2), 205–211.

Du, C., Abdullah, J., Greetham, D., Fu, D., Yu, M., Ren, L., Li, S., & Lu, D. (2018). Valorization of food waste into biofertilizer and its field application. J. Clean. Prod., 1–49.

Duba, K. S., Casazza, A. A., Mohamed, H. B., Perego, P., & Fiori, L. (2015). Extraction of polyphenols from grape skins and defatted grape seeds using subcritical water: Experiments and modeling. Food and Bioproducts Processing, 94, 29–38. doi: 10.1016/j.fbp.2015.01.001.

Duba, K. S., Shrestha, S., Ambigaipalan, P., Head, M., Adhikari, B., & Johnson, S. K. (2015). Pressurized Hot Water Extraction (PHWE) of bioactive compounds from plants and algae: A review. Food Research International, 73, 84–92. doi: 10.1016/j.foodres.2015.03.008.

Ekinci, M. S., & Gürü, M. (2019). Extraction of phytosterols from melon (Cucumis melo) seeds by supercritical CO2 as a clean technology. Green Processing and Synthesis, 8(1), 677–682.

El Darra, N., Rajha, H. N., Boussetta, N., Maroun, R. G., Louka, N., Vorobiev, E., & Louka, M. L. (2013). Effect of pulsed electric field on polyphenol profile and cell viability of grape peel extracts. Innovative Food Science & Emerging Technologies, 17, 95–102.

El Kantar, S., Delsart, C., Perrier, A., & Vorobiev, E. (2018). PEF-assisted extraction of polyphenols from citrus fruit waste peels: Impact of electric field intensity. Journal of Food Engineering, 238, 135–143. doi: 10.1016/j.jfoodeng.2018.06.007.

Elik, A., Yanık, D. K., & Göğüş, F. (2020). Microwave-assisted extraction of carotenoids from carrot juice processing waste using flaxseed oil as a solvent. Lwt, 123, 109100.

FAO. (2017a). QC. www.fao.org/faostat/en/#data. Accessed January, 2018.

FAO. (2017b). FAO Statistics Data 2014. www.fao.org/faostat/en/≠data. Accessed June 26, 2017.

Feyzi, E., Eikani, M. H., Golmohammad, F., & Tafaghodinia, B. (2017). Extraction of essential oil from Bunium Persicum (Boiss.) by instant controlled pressure drop (DIC). Journal of Chromatography A, 1530, 59–67. https://doi.org/10.1016/j.chroma.2017.11.033.

Filimonau, V., & De Coteau, D. A. (2019). Food waste management in hospitality operations: A critical review. Tourism Manag., 71, 234–245. https://doi.org/10.1016/j.tourman.2018.10.009.

Gavahian, M., Jafari, S. M., & Rezaei, K. (2016). Pulsed electric field-assisted extraction of bioactive compounds from plants. In S. M. Jafari, A. Koocheki, A. A. Mohebbi, & L. Khazaei (Eds.), Pulsed Electric Fields Technology for the Food Industry: Fundamentals and Applications (pp. 179–209). Springer.

Ghosh, D., & Biswas, M. (2015). Enzyme-assisted extraction of bioactive compounds from pumpkin (Cucurbita maxima) pulp. Food Science and Biotechnology, 24(6), 2207–2216.

Gómez-García, M., Oliete, B., Roselló, C., & Blanco, C. A. (2012). Enzymatic extraction of phenolics from red grape pomace of cv. Monastrell: Optimization and modeling. Food and Bioproducts Processing, 90(4), 671–677.

Goula, A. M., Ververi, M., Adamopoulou, A., & Kaderides, K. (2017). Green ultrasound-assisted extraction of carotenoids from pomegranate wastes using vegetable oils. Ultrasonics Sonochemistry, 34, 821–830.

Gowe, C. (2015). Review on potential use of fruit and vegetables by-products as a valuable source of natural food additives. Food Science and Quality Management, 45(5), 47–61.

Griffin, M., Sobal, J., & Lyson, T. A. (2009). An analysis of a community food waste stream. Agric. Hum. Val., 26, 67–81. https://doi.org/10.1007/s10460-008-9178-1.

Grillo, Giorgio, Luisa Boffa, Arianna Binello, Stefano Mantegna, Giancarlo Cravotto, Farid Chemat, Tatiana Dizhbite, Liga Lauberte, and Galina Telysheva. (2019). Cocoa bean shell waste valorisation; extraction from lab to pilot-scale cavitational reactors. Food Research International, 115, 200–208.

Heinz, V., Toepfl, S., & Knorr, D. (2003). Impact of temperature on lethality and energy efficiency of apple juice pasteurization by pulsed electric fields treatment. Innovative Food Science & Emerging Technologies, 4(2), 167–175.

Hossain, M. B., Tiwari, B. K., & Gangopadhyay, N. (2015). Effect of pulsed electric field processing parameters on the extraction of polyphenols from potato peels. Food and Bioprocess Technology, 8(5), 1070–1080. doi: 10.1007/s11947-015-1489-1.

Hou, Y. M., Dong, X. Y., Yu, H., Li, S., Ren, C. S., & Xiu, Z. L. (2008). Disintegration of biomacromolecules by dielectric barrier discharge plasma in helium at atmospheric pressure. *IEEE transactions on plasma science, 36*(4), 1633–1637.

Howard, L. R., & Pandjaitan, N. (2008). Pressurized liquid extraction of flavonoids from spinach. Journal of Food Science, 73(3), C151–C157.

Jaeschke, D. P., Lee, J., & Reineke, K. (2019). Pulsed electric field-assisted extraction of bioactive compounds. In S. M. Jafari & M. R. Mozafari (Eds.), Advanced Extraction Techniques: Food, Agriculture, and Environmental Applications (pp. 69–92). CRC Press.

Jittanit, W., Jittanit, S., Sa-Nguanpeng, S., & Apichartsrangkoon, A. (2017). Bioactive compounds extraction from plants using pulsed electric field extraction: A review. In R. Fuchs & E. Miltner (Eds.), Pulsed Electric Fields Technology for the Food Industry: Microbial Inactivation and Enhancement of Extraction Processes (pp. 169–194). Springer.

Jha, A. K., & Sit, N. (2022). Extraction of bioactive compounds from plant materials using combination of various novel methods: A review. *Trends in Food Science & Technology, 119,* 579–591.

Joana Gil-Chávez, G., Villa, J. A., Ayala-Zavala, J. F., Heredia, J. B., Sepulveda, D., Yahia, E. M., & González-Aguilar, G. A. (2013). Technologies for extraction and production of bioactive compounds to be used as nutraceuticals and food ingredients: An overview. Comprehensive Reviews in Food Science and Food Safety, 12(1), 5–23. doi: 10.1111/1541-4337.12005.

Jribi, S., Ben Ismail, H., Doggui, D., & Debbabi, H. (2020). COVID-19 virus outbreak lockdown: What impacts on household food wastage? Environ. Dev. Sustain., 22, 3939–3955. https://doi.org/10.1007/s10668-020-00740-y.

Ju, Z. Y., & Howard, L. R. (2003). Effects of solvent and temperature on pressurized liquid extraction of anthocyanins and total phenolics from dried red grape skin. Journal of Agricultural and food Chemistry, 51(18), 5207–5213.

Kim, B. G., Seo, J. W., & Kang, D. H. (2017). Extraction of bioactive compounds using pulsed electric fields. In M. R. Mozafari & G. R. Ziaee (Eds.), Pulsed Electric Fields Technology for the Food Industry: Basic Principles and Applications (pp. 43–67). CRC Press.

Ko, J.-H., Ahn, B.-N., Kang, M.-C., Kang, S.-M., Ahn, D.-H., & Byun, H.-G. (2011). Simultaneous extraction of quercetin, kaempferol, and isorhamnetin from onion solid wastes using subcritical water extraction. Journal of Food Science, 76(9), C1335–C1339. doi: 10.1111/j.1750-3841.2011.02395.x.

Koubaa, M., Gavahian, M., & Vorobiev, E. (2016). Red prickly pear (Opuntia ficus-indica) peel extraction using pulsed electric fields: Impact on phenolic content, antioxidant activity and color. *Innovative Food Science & Emerging Technologies*, 38(Part B), 245–251. doi: 10.1016.

Kumar, S. S. (2013). Eco-friendly practice of utilization of food wastes. International Journal of Pharmaceutical Science, 2319–6718, ISSN (Online 2).

Luengo, E., Álvarez, I., & Raso, J. (2013). Improving the pressing extraction of polyphenols of orange peel by pulsed electric fields. *Innovative Food Science & Emerging Technologies, 17,* 79–84.

Lima, R. N., Ribeiro, A. S., Cardozo-Filho, L., Vedoy, D., & Alves, P. B. (2019). Extraction from leaves of Piper klotzschianum using supercritical carbon dioxide and co-solvents. The Journal of Supercritical Fluids, 147, 205–212.

Liu, J., Zhou, Y., Wang, Y., & Wang, J. (2018). Non-thermal pulsed electric field enhances the extraction of phenolic compounds from onions. Food Chemistry, 253, 1–6.

López, N., Puértolas, E., Condón, S., Álvarez, I., & Raso, J. (2008). Application of pulsed electric fields for improving the maceration process during vinification of red wine: influence of grape variety. *European Food Research and Technology, 227,* 1099–1107.

Lopresto, C. G., Tine, M. R., Gómez, C. L., & Riba, J. P. (2014). Recovery of D-limonene from citrus peel waste by hydrodistillation and extraction with compressed CO2. Waste Management, 34(5), 870–875. doi: 10.1016/j.wasman.2014.02.009.

Lotf, A., Ahmadi, N., Zakerin, A., & Badii, F. (2015). Enzyme-assisted extraction of anthocyanins from Crocus sativus petals. Food Chemistry, 169, 218–224.

Louati, I., Bahloul, N., Besombes, C., Allaf, K., & Kechaou, N. (2019). Instant controlled pressure-drop as texturing pretreatment for intensifying both final drying stage and extraction of phenolic compounds to valorize orange industry by-products (Citrus sinensis L.). Food and Bioproducts Processing, 114, 85–94.

Machado, A. P. D. F., Pasquel-Reátegui, J. L., Barbero, G. F., & Martínez, J. (2015). Pressurized liquid extraction of bioactive compounds from blackberry (Rubus fruticosus L.) residues: A comparison with conventional methods. Food Research International, 77, 675–683.

Martínez-Meza, Y., Pérez-Jiménez, J., Rocha-Guzmán, N. E., Rodríguez-García, M. E., Alonzo-Macías, M., & Reynoso-Camacho, R. (2021). Modification on the polyphenols and dietary fiber content of grape pomace by instant controlled pressure drop. Food Chemistry, 360(November 2020), Article 130035. https://doi.org/10.1016/j.foodchem.2021.130035.

Meneguzzo, F., Brunetti, C., Fidalgo, A., Ciriminna, R., Delisi, R., Albanese, L., Zabini, F., Gori, A., dos Santos Nascimento, L.B., De Carlo, A. and Ferrini, F (2019). Real-scale integral valorization of waste orange peel via hydrodynamic cavitation. *Processes*, 7(9), 581.

Extraction and Recovery of Bioactive Compounds 185

Mkaouar, S., Krichen, F., Bahloul, N., Allaf, K., & Kechaou, N. (2018). Enhancement of bioactive compounds and antioxidant activities of olive (Olea europaea L.) leaf extract by instant controlled pressure drop. Food and Bioprocess Technology, 11, 1222–1229.

More, P. R., & Arya, S. S. (2021). Intensification of bio-actives extraction from pomegranate peel using pulsed ultrasound: Effect of factors, correlation, optimization and antioxidant bioactivities. Ultrasonics Sonochemistry, 72, 105423.

Motikar, P. D., More, P. R., & Arya, S. S. (2021). A novel, green environment-friendly cloud point extraction of polyphenols from pomegranate peels: A comparative assessment with ultrasound and microwave-assisted extraction. Separation Science and Technology, 56(6), 1014–1025.

Muñoz, O., Sepúlveda, E., Lutz, M., Peña-Neira, Á., & Rubio, P. (2004). Enzyme-assisted extraction of anthocyanins from grape skin: Kinetics and mathematical modeling. Journal of Food Engineering, 64(3), 295–301.

Mustafa, A., & Turner, C. (2011). Pressurized liquid extraction as a green approach in food and herbal plants extraction: A review. Analytica Chimica Acta, 703(1), 8–18.

Naghshineh, M., Olsen, K., & Georgiou, C. A. (2013). High hydrostatic pressure extraction of pectin from lime peel. Journal of Food Engineering, 116(2), 466–471.

Oey, I., Van der Plancken, I., Van Loey, A., & Hendrickx, M. (2008). Does high pressure processing influence nutritional aspects of plant based food systems?. *Trends in Food Science & Technology, 19*(6), 300–308.

Oszmianski, J., Wojdylo, A., Gorzelany, J., & Kapusta, I. (2011). Identification and characterization of low molecular weight polyphenols in berry leaf extracts by HPLC-DAD and LC-ESI/MS. Journal of Agricultural and Food Chemistry, 59(24), 12830–12835.

Panda, D., & Manickam, S. (2019). Cavitation technology—The future of greener extraction method: A review on the extraction of natural products and process intensification mechanism and perspectives. Applied Sciences, 9(4), 766.

Pereira, R. N., Vicente, A. A., & Rodrigues, A. E. (2016). Optimization of anthocyanin extraction from potato peels using pulsed electric fields. Journal of Food Engineering, 169, 204–210.

Puri, M., Sharma, D., & Barrow, C. J. (2012). Enzyme-assisted extraction of bioactives from plants. *Trends in biotechnology, 30* (1), 37–44.

Quiles-Carrillo, L., Mellinas, C., Garrigos, M. D. C., Balart, R., & Torres-Giner, S. (2019). Optimization of microwave-assisted extraction of phenolic compounds with antioxidant activity from carob pods. Food Analytical Methods, 12, 2480–2490.

Rashidi, S., Eikani, M. H., & Ardjmand, M. (2018). Extraction of Hyssopus officinalis L. essential oil using instant controlled pressure drop process. Journal of Chromatography A, 1579, 9–19. https://doi.org/10.1016/j.chroma.2018.10.020.

ReFED. (2016). A roadmap to reduce U.S. Food waste by 20 percent. J. Chem. Inf. Model., 53, 1689–1699. https://doi.org/10.1017/CBO9781107415324.004.

Routray, W., & Orsat, V. (2011a). Blueberries and their anthocyanins: Factors affecting biosynthesis and properties. Comprehensive Reviews in Food Science and Food Safety, 10(6), 303–320.

Routray, W., & Orsat, V. (2011b). Microwave-assisted extraction of flavonoids: A review. Food Research International, 44(4), 1089–1099.

S., Z. (2018). Study on the extraction and degradation of citrus pectin by dielectric barrier discharge plasma. Vol. 12.

Šafranko, S., Ćorković, I., Jerković, I., Jakovljević, M., Aladić, K., Šubarić, D., & Jokić, S. (2021). Green extraction techniques for obtaining bioactive compounds from mandarin peel (Citrus unshiu var. Kuno): Phytochemical analysis and process optimization. Foods, 10(5), 1043.

Sagar, N. A., Pareek, S., Sharma, S., Yahia, E. M., & Lobo, M. G. (2018). Fruit and vegetable waste: Bioactive compounds, their extraction, and possible utilization. Comprehensive Reviews in Food Science and Food Safety, 17(3), 512–531.

Saini, A., Panesar, P. S., & Bera, M. B. (2019). Valorization of fruits and vegetables waste through green extraction of bioactive compounds and their nanoemulsions-based delivery system. Bioresources and Bioprocessing, 6(1), 1–12.

Shen, X., Zhu, X., Li, P., & Chen, X. (2017). Recent advances in ultrasonic-assisted extraction of bioactive compounds from plant materials. Trends in Food Science & Technology, 67, 160–172.

Soldan, A. C. F., Arvelos, S., Watanabe, E. O., & Hori, C. E. (2021). Supercritical fluid extraction of oleoresin from Capsicum annuum industrial waste. Journal of Cleaner Production, 297, 126593.

Sridhar, A., Ponnuchamy, M., Kumar, P. S., Kapoor, A., Vo, D. V. N., & Prabhakar, S. (2021). Techniques and modeling of polyphenol extraction from food: A review. In Environmental Chemistry Letters. Springer International Publishing. https://doi.org/10.1007/s10311-021-01217-8.

Štambuk, P., Tomašković, D., Tomaz, I., Maslov, L., Stupić, D., & Karoglan Kontić, J. (2016). Application of pectinases for recovery of grape seeds phenolics. 3 Biotech, 6, 1–12.

Strati, I. F., Gogou, E., & Oreopoulou, V. (2015). Extraction of carotenoids from tomato processing waste streams using high-pressure technologies. Food and Bioprocess Technology, 8(3), 592–602.

Takamatsu, T., Uehara, K., Sasaki, Y., Hidekazu, M., Matsumura, Y., Iwasawa, A., ... & Okino, A. (2015). Microbial inactivation in the liquid phase induced by multigas plasma jet. PloS one, 10(7), e0132381.

Tostivint, C., Bergheaud, V., & Couvert, A. (2016). Valorization of fruit and vegetable wastes. In Fruit and Vegetable Waste (pp. 3–24). Academic Press.

Tunchaiyaphum, S., Eshtiaghi, M. N., & Yoswathana, N. (2013). Extraction of bioactive compounds from mango peels using green technology. International Journal of Chemical Engineering and Applications, 4(4), 194.

Umair, M., Jinap, S., & Maaruf, A. G. (2019). Plasma-assisted extraction: An innovative and sustainable technique for the recovery of bioactive compounds from plants. Food Chemistry, 276, 180–191.

Vergara-Salinas, J.R., Bulnes, P., Zúñiga, M.C., Pérez-Jiménez, J., Torres, J.L., Mateos-Martín, M.L., Agosin, E. and Pérez-Correa, J.R. (2013). Effect of pressurized hot water extraction on antioxidants from grape pomace before and after enological fermentation. Journal of Agricultural and Food Chemistry, 61(28), 6929–6936.

Vu, H. T., Scarlett, C. J., & Vuong, Q. V. (2019). Maximising recovery of phenolic compounds and antioxidant properties from banana peel using microwave assisted extraction and water. Journal of Food Science and Technology, 56, 1360–1370.

Xi, J. (2006). Effects of high hydrostatic pressure on the stability and antioxidant activity of lycopene in tomato. Food Chemistry, 97(4), 601–605.

Xie, J., Li, X., Wang, Q., Liu, Y., Wang, L., Zou, Y., & Liu, C. (2018). High pressure processing assisted extraction and emulsifying properties of pectin from potato waste peel. Journal of Food Engineering, 235, 70–76.

Yusuf, M. (2017). Agro-Industrial waste materials and their recycled value-added applications: Review. Handb. Ecomater., 1–12. https://doi.org/10.1007/978-3-319-48281-1.

Zhang, W., Yao, Y., Sullivan, N., & Chen, Y. (2011). Modeling the primary size effects of citrate-coated silver nanoparticles on their ion release kinetics. Environmental Science & Technology, 45(10), 4422–4428.

8

Value Addition of Spices Processing Industrial Waste

Sabbu Sangeeta, Sweta Rai, Preethi Ramachandran, Poonam Yadav, and Gaurav Chandola

8.1 Introduction

The term "spices" applies to natural plant or vegetable products, or a mixture thereof, used in whole and ground form, mainly for imparting flavor, aroma, and pungency to food. They are low-volume, high-value crops that play a vital role in the national economy of India and provide a strong footing in the international market (Peter, 2001). Usually, spices are dried roots, bark, or seeds that are used whole, crushed, or powdered (Pruthi, 1999). Spices are divided into different groups based on their plant components, such as floral, fruits, berries, seeds, rhizomes, roots, leaves, kernel, aril, bark, bulb, etc.

India is considered the "Land of Spices" and is one of the major spice-producing and -exporting countries of the world, contributing about 20–25% of the world trade in spices. It is a well-known fact that spices have played a significant role in food flavoring and medicine since ancient times. Important spices in India include pepper, cardamom, chili, ginger, turmeric, coriander, cumin, fennel, fenugreek, celery, saffron, tamarind, and garlic. Other spices produced and exported in small quantities are aniseed, bishop's weed (*Ajawajn*), dill seed, poppy seed, Bay leaves (*Tejpat*), curry leaves, cinnamon, *Kokam*, and a few other culinary herbs. Tree spices like clove, nutmeg, mace, star anise, allspice, and some herbal spices like rosemary, thyme, marjoram, oregano, chive, parsley, sage, savory, tarragon, and basil are produced in small quantities and are mainly used in domestic cooking. India produced 11,125,010 tons of total spices in an area of 4,388,953 ha (hectares) during the year 2021–22 (SBI, 2023). The production statistics of important spices in India during 2021–22 is given in Table 8.1. The gradual increase in spice production (Figure 8.1) and the development of spice processing industries have increased the need for investigating the utilization of by-products and waste generated to develop value-added products. The waste generated in the spice industry can often be utilized in various ways, promoting sustainability and minimizing environmental impact.

188

DOI: 10.1201/9781003269199-8

Table 8.1 Production of Spices in India during 2021–22

Spices	Area (ha)	Production (tons)
Pepper	283,962	70,000
Cardamom (small)	69,190	23,340
Cardamom (large)	45,039	8,812
Chili	882,000	1,836,222
Ginger	431,218	3,523,436
Tamarind	40,345	152,409
Cloves	1,924	1,209
Nutmeg	23,353	18,429

Source: Spice Board, Cochin, India

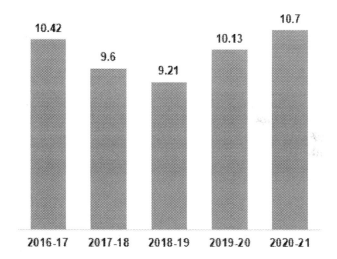

Figure 8.1 *Growth rate of spices production (million tons).*
Source: **Spices Board of India**

The global market size of spices was valued at USD 5.86 billion in 2019 and is expected to expand at a compound annual growth rate (CAGR) of 6.5% from 2020 to 2027 (Annonyms, 2023). Increasing global demand for authentic cuisines is one of the foremost factors driving the consumption of spices. The development of spice processing industries and the production of value-added products thereof has led to the generation of spice-based by-products and inedible waste (Table 8.2 and Figure 8.2). Although the waste generated from the spice industry does not cause much serious damage, its disposal poses concerns regarding environmental pollution. Hence, strategies are needed for the effective utilization of industrial wastes to prevent such issues.

8.2 Waste Utilization of Major Spices

The spice processing industry produces around 80–90% of residues as waste. At present these waste materials are not properly utilized for commercial use and are disposed of into the environment, causing adverse effects (Sowbhagya, 2019). The spent residues from the processing of different spices such as chili, cumin, coriander, and pepper are an excellent source of dietary fibers, which can be an inexpensive functional ingredient in various products. Mustard meal or its residue can be used to prevent soil-born pathogens and prevent nematodes. Similarly, some other spice wastes can be effectively utilized as organic fertilizers and pesticides in agriculture (Sowbhagya, 2019). The highly priced value-added product of the spice processing industry is spice oil and oleoresin. These products generate 80–90% spice spent, which are rich in various nutrients. Spice crops and their by-products have export prospects and significant internal demand in most industries. The by-products of the spice industry could be used as food ingredients, pharmaceuticals, cosmetics, and functional components in value-added products (Panak-Balentic et al., 2018). This chapter provides insight on the utilization of various spice wastes for the development of value-added products.

Table 8.2 Value-Added Products and Waste Generated by Various Spices

Spice	Value-added products	Waste
Chilli	Essential oils, oleoresins, chili paste, pickles, sauces	chili seeds and chili spent
Turmeric	Essential oils, oleoresins, curcumin powder, turmeric powder, spice mixes	Spent, immature turmeric, turmeric cake after extraction of oil
Pepper	Essential oils, oleoresins, white pepper, canned green pepper, green pepper in brine	Spent, pinheads, light berries, immature berries
Ginger	Ginger candy, ginger oil, ginger powder, paste, dried ginger, oleoresins, pickles, syrups	Peel, spent, stems, leaves
Cardamom	Essential oils, oleoresins, Bleached cardamom, dried cardamom, cardamom powder	Outer pericarp, spent, immature cardamom
Coriander	Essential oils, oleoresins, coriander powder, curry powder, spice mixes	Spent, stalk, empty coriander seeds
Tamarind	Pickles, sauces, candy, tamarind pulp–based beverages	Pod husk, seed, seed husk, fibers
Onion	Frozen onion rings, paste, dehydrated onion flakes, onion powder, onion oil, onion vinegar, onion sauce, onion wine	Peel, top and bottom part
Garlic	Garlic powder, garlic salt, garlic vinegar, garlic cheese croutons, garlic potato chips, garlic bread	Peel, stalk
Cinnamon	Quills, quillings, featherings, chips, cut pieces, powder, cinnamon bark oil, cinnamon leaf oil, cinnamon extracts	Cinnamon chip, spent, bark

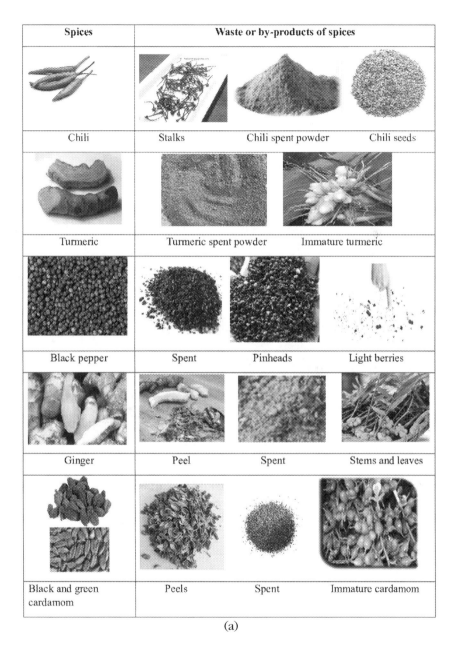

Figure 8.2 Waste generated during value addition of different spices.

8.3 Chili (*Capsicum annum*)

Chili (*Capsicum annum*) is an annual plant belonging to the family Solanaceae. There are more than 400 chili varieties in the world. It is an important cash

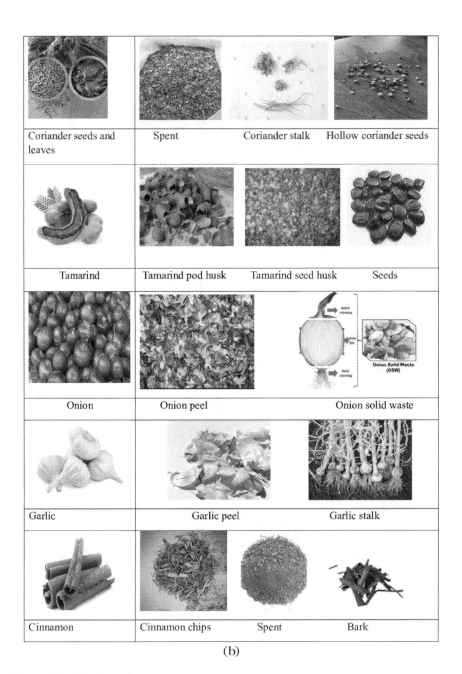

(b)

Figure 8.2 *(Continued)*

crop. Originally, chili belonged to tropical America and the West Indies from where it was introduced to India in the 17th century by the Portuguese (Geetha and Selvarani, 2017). Since then, it has become an integral part of Indian cuisine and is valued for its characteristic pungency, color, and aroma.

Chili fruit is commonly known for its pungency, which is due to the presence of alkaloids called capsaicin ($C_{18}H_{27}NO_3$) that possess a high medicinal value. Capsaicin helps in directly reducing various free radicals (Bhattacharya et al., 2010; Chakrabarty et al., 2017). Another bioactive compound, capsaicinoids, contributes the antioxidant, anti-cancer, anti-arthritic, antibacterial, and analgesic properties of chili (Ahuja and Ball, 2006; Prasad et al., 2006). Green chili is rich in vitamins (A and C) and contains rutin, which has high medicinal value. After maturation, the chilli turns red. The intensity of the red color is dependent on two carotenoid pigments, i.e., capsanthin (30–60%) and capsorubin (6–18%) (Peter, 2001). It is used as a flavoring and coloring agent in various food products. Oleoresins extracted from chili seeds have great pharmacological uses, e.g., for pain relieving balms. Chili is also a good source of vitamins (A, B, C, K) and many minerals (Ca, Mg, K, Fe, Cu, etc.) (Chakrabarty et al., 2017). Vitamin C and carotenoids found in chili act as an antioxidant, which might help to improve insulin regulation (Ahuja et al., 2006).

Chili is subjected to curing, grading, packing, and storage during processing. A number of variable products like chili powder, chili oleoresins, chili seed oil, etc. are processed commercially. During processing, chili seeds and chili spent are generated as waste that can be utilized for the generation of various components and the effluent treatment of industrial wastewater. At the same time, nearly 6.11% of the annual chili production in India (approximately 112,194 tons) go to waste due to the lack of an adequate postharvest management infrastructure (NABCONS, 2022). Majority of the postharvest losses in chili (5.2%) occur during farm operations like sorting and grading (2.14%), harvesting (1.07%), and storage (0.93%) (NABCONS, 2022).

Chili seeds can be utilized for the extraction of chili seed oil, which has been found to comprise triglycerides (60%) in which linoleic and other unsaturated fatty acids are predominant. They can be added in the pickle-making process for added pungency (Pruthi, 2003). The solvent-extracted chili seed has an appreciable amount of protein (27–29%) which can be utilized as fertilizer or animal feed (Pruthi, 2003). It is reported that chili spent after the extraction of oleoresins contains about 28% protein and 36% fiber, which can make good animal feed (Tandon, 1964). Chili spent left after oleoresin extraction has good antioxidant properties as well (Florence and Sowbhagya, 2006). Previous studies revealed that aqueous extract of chili spent exhibited 72% free radical scavenging activity, while methanolic extract exhibited 63% free radical scavenging activity. The phenolic content of chili spent was found to be 45 mg/g, which indicates that it has great potential to be used as a source of nutraceuticals (Joshi and Sharma, 2011).

A recent study was conducted for recycling chili stems obtained during the processing of chilli powder. Cookies were prepared with 7% nanocellulose, which was extracted from the chili stem, and it was observed that the cookies were acceptable in overall quality and appearance and increased the total dietary fiber content in the diet of mice, which inhibited weight gain (Ma et al., 2023). Waste chili stem may also be utilized to produce bio-oil through a catalytic process at 450°C with Ni-Ca/SiO$_2$ catalysts (Ma et al., 2021). It is also revealed from the research that chili stalk waste is an effective tool to adsorb acid violet 49 dye from water and textile industrial effluent (Yakuth et al., 2021). Previous research has shown chili waste to have moderate antifungal activity against two common Ascomycetes fungi (*Sphaeropsis sapinea* and *Leptographium procerum*). An increase in fungal inhibition was also noticed when chili and *Lactobacillus* sp. isolated from the chili juice were combined with the chili stems (Singh and Chittenden, 2008). The lactic acid bacterium that grows on chili waste has potent antifungal activity, especially for wood products (O'Callahan et al., 2011). Chili waste, mainly stalks, may be considered a good source of non-dairy lactic acid bacteria in yogurt making. Inoculation of 8–10 chili stalks in a sample of 200 ml of milk is sufficient to produce yogurt with desirable pH and acidity (Olatide et al., 2019). The major phenolic compounds rutin, epicatechin, and catechin comprised 90% of the total compounds detected by HPLC from the waste of jalapeño peppers after the scalding process (Sandoval-Castro et al., 2017).

8.4 Turmeric

Turmeric (*Curcuma longa* L.) is an herbaceous perennial native to the Indo Malayan region and belongs to the family Zingiberaceae. It is a tropical commercial crop that has several uses in the food and pharmacological industries. Turmeric is commonly used as a condiment in vegetables, meat, fish, and pickle preparation. The yellow color compound, i.e., curcumin, found in turmeric is gaining popularity as a natural color pigment in various foods. In pharmaceuticals, it is valued as having anti-cancer, anti-inflammatory, antiseptic, anti-diabetic, and antiviral properties (Bhowmik et al., 2009). Turmeric contains 8.92% moisture, 2.85% ash, 4.60% crude fiber, and 6.85% fat. It also contains 9.40% crude protein, 67.38% carbohydrate, and 3.1–3.4% curcumin (Verma et al., 2018). Turmeric is commonly processed for the manufacturing of turmeric powder and the extraction of essential oils, oleoresins, and curcumin. Turmeric spent oleoresins (TSO) obtained after the extraction of curcuminoids and de-flavored and depigmented turmeric (DDT) are the three major wastes from turmeric processing that may be utilized for the extraction of several beneficial components, such as polyphenols and starch, and utilized for the production of biodegradable films and nanocomposite powders (Verma et al., 2021).

Processing of turmeric spent powder to develop value-added products in the food and pharmaceutical industries is usually adopted to reduce industrial waste. Dosa, a popular and healthy food of southern India was prepared by Vedashree et al. (2016) by replacing rice with turmeric spent flour. CoFe$_2$O$_4$ magnetic nanocomposite powder was developed from discarded turmeric by chemical co-precipitation of CoCl$_2$•4H$_2$O and Fe•Cl$_3$•6H$_2$O by utilizing NaOH solution as a precipitant followed by drying and fine grinding (Mehran et al., 2016). Similarly, waste turmeric such as depigmented turmeric may be utilized for the extraction of essential oils (Sehgal et al., 2016). Turmeric cake after the extraction of oil and oleoresins can be used for the recovery of starch, as it contains an appreciable amount of dietary fiber (45%), with 43% insoluble fiber and 2% soluble fiber, which has better water-holding capacity and microstructural properties (Kuttigounder et al., 2011). Such starch can be used in the preparation of biodegradable films (Maniglia et al., 2015). Partial hydrolysis using pressurized hot water contributed to the polymer modification of DDT waste, which can attribute to this material's application as a complementary starch source to conventional ones like corn and rice. DDT can also be a sustainable alternative for the recovery of bioactive constituents (Santana et al., 2018). Similarly, turmeric waste generated from the extraction process is used for the recovery of mixed biopolymers composed of starch and curcuminoids with relevant quality through supercritical fluid and pressurized liquid extraction processes (Santana et al., 2017). The antidiabetic property of spent turmeric powder was investigated in streptozotocin-induced diabetic rats by feeding at a 10% level, which improved the status of rats by 18% (Kumar et al., 2005). Turmeric spent can also be subjected to extraction with acetone to remove residual curcuminoids (Joshi and Sharma, 2011). Oil (3–5%) extracted from the waste turmeric comprises oxygenated sesquiterpenes, sesquiterpenes hydrocarbons, monoterpenes hydrocarbon, and oxygenated monoterpenes (Purseglove et al., 1981), which is effective against many microorganisms, such as *Staphylococcus albus*, *Staphylococcus aureus*, and *bacillus typhus* (Negi et al., 1999).

8.5 Pepper (*Piper nigrum* L.)

Black pepper of commerce is the dried, matured, and unripened fruit of *Piper nigrum*. It is known as the "King of Spices" or "Black gold of India" and is a member of the family Piperaceae. It is a tropical, perennial climbing plant and native to the Malabar seashore of India. It is cultivated in India, Indonesia, Malaysia, Brazil, Madagascar, Sri Lanka, Vietnam, Thailand, China, and the Federated State of Micronesia. India is one of the leading producers, consumers, and exporters of black pepper in the world. Among all the spice crops, pepper has the highest contribution to foreign exchange and trade turnover. The alkaloid piperine is the major constituent responsible for black pepper's biting taste. The other pungent alkaloids present in black pepper in smaller amounts are chavicine, piperidine, and piperetine (Hammouti et al., 2019). Pepper's characteristic aromatic odor is due to the presence of volatile oils in the cells of the pericarp. Several value-added and diversified products (white pepper, decorticated black pepper, pepper oil, pepper oleoresin, piperine, green pepper in brine, dehydrated green pepper, frozen green pepper, etc.) are produced from the pepper.

During processing, a large amount of pepper spent, outer pericarp, ungraded pepper, light berries, pinheads, spikes, and *Varagu* (unfertilized buds) is generated as waste. Among these trade wastes of pepper, pepper pinheads and light berries contain a good amount of piperine (4.68–5.96%), followed by spikes (0.93%) and Varagu (0.48%) (Rao et al., 1960). The utilization of pepper waste for the recovery of oleoresin and essential oil has been reported (Pruthi, 1968). After the extraction of oleoresin, spent residue is utilized for the recovery of starch. Pepper pericarp has good fiber content (23–26%) that can be extracted (Dewin, 1955). The green skin of the pepper that is obtained as waste during the manufacturing of dehydrated green pepper and white pepper can be used to produce a natural green colorant that has great food applications (Sampathu et al., 2001). A recent study revealed that pepper pericarp contained high amounts of total phenol (1421.95 ± 22.35 mg GAE/100 g), total flavonoid (983.82 ± 8.19 mg CE/100 g), and piperine (2352.19 ± 68.88 mg/100 g) (Lee et al., 2020). The principal monoterpene compounds found in the pepper pericarp are α-pinene (9.2%), 2-β-pinene (14.3%), δ-3-carene (21.5%), and dl-limonene (18.8%), and the primary sesquiterpenes are α-copaene (5.1%) and caryophyllene (17.2%), which imparts a potent odor to the pepper (Lee et al., 2020). The ethanol extract of pepper waste can be used to produce hot red pepper oil (8%) and no-heat red pepper oil (4%), while capsaicinoids (0.8%) and capsaicin (0.4%) have been obtained through the flash column chromatography of pepper waste material (Shu et al., 2023).

8.6 Ginger (*Zingiber officinale*)

Ginger is one of the earliest known oriental spices and is being cultivated in India both as a fresh vegetable and as a dried spice. It is a tropical crop belonging to the family Zingiberaceae and it is indigenous to Southeast Asia. India is the largest producer and consumer of ginger, contributing about 33.6% of the total global production, followed by Nigeria, China, and Indonesia (FAO, 2021). It is one of the earliest known treasured spices esteemed for its pungency and aroma, and it is viewed as a medicinal plant in Indian Ayurvedic systems. It is commonly applied as a digestive aid and spiritual beverage and is known for its aphrodisiac, antiemetic, anti-cancerous, anti-platelet, antimicrobial, antiparasitic, antioxidant, anti-inflammatory, analgesic, hepato-protective, and immune-stimulating properties (Malhotra and Singh, 2003). Ginger processing includes cleaning, sorting, peeling, drying, grading, and packaging (Bag, 2018). Commercially, ginger is used in the soft drink manufacturing, baking, alcohol, and meat processing industries to a great extent. Ginger brine is most popular and preferred by the Japanese (Jiang et al., 2005). Ginger oil itself is a value-added product (around USD 100 per kg) that is expected to grow at 9.41% over the period to reach a total market size of USD 189.431 million by 2025 (Inthalaeng et al., 2023).

Ginger peel, spent, stem, and leaves are the major constituents of ginger waste that may be generated after processing. Ginger stem and leaves generated after harvesting are considered a good source of protein (6.02%), sugar (52.14%), cellulose (48.48%), and hemicellulose and lignin (31.50%) (Chen et al., 2021; Zendrato et al., 2021), while the ginger spent after extraction of oleoresin contains 9.37% ash, 11.74% protein, 50.79% starch, 2.09% pectin, 3.84% cellulose, and 4.52% sugar (Konar et al., 2013). The residual essential oils and oleoresins present in ginger spent contain various valuable compounds (zingiberene, zingerone, gingerols, shogaols, α-curcumene, β-bisabolene, and β-sesquiphellandrene) that can be extracted by applying supercritical carbon dioxide, subcritical water extraction, microwave-assisted extraction, and ultrasound-assisted extraction (Gao et al., 2021; Inthalaeng et al., 2023). Ginger spent subjected to column chromatography and HPLC of butanol extract resulted in the isolation of a potential antioxidant, i.e., diacetoxy-7-(3, 4 dihydroxy phenyl)-1-(3, 4 dihydroxy phenyl) heptane (Sajjad Khan et al., 2006). Natural oils and materials derived from ginger wastes have great antibacterial and antifungal activities. The inclusion of ginger essential oils into chitosan-carboxymethyl cellulose films improved their antifungal properties (Noshirvani et al., 2017), and a similar report has been given in polyvinylidene fluoride membranes (Fahrina et al., 2020). The incorporation of ginger cellulosic fibers (100–200 nm width) by 5% into chitosan-PVA composites improved the mechanical strength and antibacterial activity against *Bacillus cereus, Escherichia coli, Staphylococcus aureus*, and *Salmonella typhimurium* (Jacob et al., 2018). Starch extracted (5%) from ginger waste may also be utilized to produce a keratin-starch biocomposite film with reduced moisture ingress and improved antimicrobial activity (Oluba et al., 2021). The defibrillated waste ginger cellulose (100–200 nm width) blended into magnetic ginger nanofiber composites helps in improving the absorption of copper ions from contaminated water supply in comparison to their nonmagnetic counterpart composites (Jacob et al., 2019).*Figure 8.2 Waste generated during value addition of different spices.*

8.7 Cardamom (*Elettaria cardamomum*)

Cardamom is known as the "Queen of Spices", and it is one of the most expensive and exotic spices in the world. It is a perennial tropical herb plant belonging to the ginger family Zingibaraceae. Cardamom is indigenously grown in the evergreen forests of the Western Ghats in South India. There are two main types of cardamom: Small green cardamom (*Eletteria cardamomum*) and large red/black cardamom (*Amomum subulatum* Roxb). The most common type is the small green cardamom, which is commercially grown in Guatemala, Tanzania, Sri Lanka, El Salvador, Vietnam, Laos, and Cambodia, while large cardamom is mainly grown in India, Nepal, Malaysia, Pakistan, Singapore, and Bhutan. India is the main exporter of dried cardamom (Rajathi et al., 2017). Cardamom possesses various medicinal properties, such as anti-inflammatory, hepato-protective, and anti-ulcer, and is used for making antidotes to snake venom. The volatile oil of cardamom contains cineole, terpinene, terpiniol, sabinine, α-pinene, and limonene, which act as a tonic for the heart and liver, an appetizer, promote the elimination of bile, and help reduce congestion of the liver (Rajathi et al., 2017). Processing small cardamom includes harvesting, cleaning, drying, grading, grinding, and packaging. It is also processed for various products such as cardamom seeds, cardamom oil, and oleoresin. During the processing of small cardamom pod husks, spent free from oil and flavor, burned seeds, immature seeds, etc. are produced as waste, whereas from large cardamom, false spike (obtained after removing capsules), shoots, old rhizomes, dark-colored pericarp or husk, mucilaginous sweet pulp, spent free from oil and flavor, etc. waste is generated.

Generally, the seeds-to-husk ratio in small cardamom pods are 70:30, respectively. The husk of small cardamom contains 0.2–1.0% oil with similar properties as seed oil that can be extracted from the husk (Govindarajan et al., 1982). After the extraction of oleoresins from small cardamom, the spent cake that is recovered contains starch, carbohydrate, protein, fiber, etc. in an appreciable amount, which can be used for animal feed, fuel for a boiler, and as a source of manure (Ashokkumar et al., 2020). Small cardamom spent can also be utilized in various food and cosmetic industries for flavor purposes. Scented sticks, commonly known as "*Agarbatti*" can also be formulated from the cardamom spent.

The husk generated from large cardamom processing contains a total anthocyanin percentage of 0.99% on a dry weight basis, which can be a good source of natural water-soluble red colorant. Among the total anthocyanin, two major pigments are predominant, i.e., cyanidin 3-glucoside and cyanidin 3,5-diglucoside (Puru Naik et al., 1999). Various value-added products (jam, jelly, alcohol, etc.) can be made from the mucilaginous sweet pulp, and it also contains a good amount of sugar, i.e., 10–12% (Shankaracharya et al., 1997). The husk or the outer cover of cardamom is considered good for healing tooth pain, headaches, and stomach pain. The spent of large cardamom has a good quantity of protein (8.36–10.04%), mucilage (3.75–4.77%), starch (41.8–47.2%), and sugars (5.2–13.02%). The total color content in fresh husk ranged from 0.49% to 1.16%, while in dry husk it was from 0.22% to 0.4%. The spent and pericarp of large cardamom are a good source of antioxidants (Puru et al., 2006), and both can be utilized for the extraction of volatile oil and the encapsulation of flavor compounds, i.e., 1,8-cineole (Puru et al., 2004). Nowadays, large cardamom pericarp is used as an inexpensive and environmentally friendly adsorbent for sequestering the Congo red dye from water by activation of black cardamom waste with potassium hydroxide (KOH) at 600°C (Aftab et al., 2023). A natural dye with good fastness on the gray scale may also be obtained from black cardamom peel by mordanting with alum in both alkali and aqueous medium (Singh and Srivastava, 2017). Liquid smoke is generated from cardamom by-products (husk) through pyrolysis at 550°C, which can be applied as a preservative, while biochar may produce from the cardamom husk at 350°C and can be used as an energy source (Rahmawati et al., 2022).

8.8 Waste Utilization of Minor Spices

8.9 Coriander (*Coriandrum sativum* L.)

Coriander (*Coriandrum sativum* L.) is a multipurpose herb mainly cultivated for its foliage and seeds, which have great significance in international trade. India is the largest producer, consumer, and exporter of coriander in the world, with an annual production volume of around three lakh tons (SBI, 2023). Coriander has a long history of therapeutic use for various health problems, such as infusing coriander seeds thrice a day to lower blood sugar. A combination of coriander, cardamom, and caraway seeds is a useful medicine in digestion, flatulence, vomiting, dysentery, insomnia, anxiety, and cardio-vascular and urinary disorders (Patel et al., 2012). Coriander seeds contain thalides, which increase the amount of anti-cancer protective enzymes. Coriander seeds contain a wide array of health beneficial compounds, such as minerals, phenolics, fatty acids, and essential oils, among others (Laribi et al., 2015). However, seed shattering and spillage of coriander seed during harvesting and threshing operations, lead to significant postharvest losses (4.72%) in farm-level operations (NABCONS, 2022). Commercially, coriander is processed mainly to produce ground spice, essential oils, oleoresins, and curry powder (Bhat et al., 2014). Coriander seed oil is used in various industries for flavoring products like pastries, cookies, cakes, buns, candy, cocoa, chocolate, puddings, cordial syrup, gelatin desserts, tobacco, and alcoholic beverages, particularly gin.

Coriander spent and residual coriander seeds after the extraction of essential oils are considered the main waste of this industry. Traditionally this waste is used to feed animals or sent to landfills. But nowadays residual coriander seeds and spent can be utilized to extract various phytochemicals and dietary fibers, to produce fuel, and for bioremediation. Rojas-Flores et al. (2023) generated 0.882–0.154 V Voltage and 2.287–0.072 mA current at pH 3.9–0.16 and with an electrical conductivity of 160.42–4.54 mS/cm by the manufacturing of the single-chamber microbial fuel cells using zinc and copper as an electrode, and coriander waste was used as a substrate to observe their power for the generation of bioelectricity. Coriander straw fiber has been successfully incorporated as a reinforcing filler in polypropylene and biobased low-density polyethylene composite materials, which enhance 50% of flexural (from 19 to 28 MPa) and tensile strength (from 12 to 17 MPa) through twin-screw extrusion compounding and injection molding (Uitterhaegen et al., 2018). Coriander seed spent can also be employed as an eco-friendly and cheap adsorbent for the bioremediation of acid black 52 dye from the aqueous solution and textile industrial effluent with maximum adsorption capacity (432.0 mg/g) at pH 2.0 (Taqui et al., 2019).

8.10 Tamarind

Tamarind (*Tamarindus indica* L.) is an important nutritionally rich, under-utilized crop that belongs to the family Leguminosae. It is native to tropical

Africa but has been cultivated in the Indian subcontinent for so long that it is sometimes said to be native to that country (Bhadoriya et al., 2011). The pods are left to ripen on the tree until the outer shell is dry and easily detached from the pulp (Shankaracharya, 1997). A typical pod of tamarind contains around 55% pulp, 34% seeds, and 11% shell (Caluwe et al., 2010). It is one of the most important multipurpose tropical trees whose almost every part is utilized in one way or another. It is a valuable timber species used in making furniture, tool handles, charcoal, and fuelwood (Kumar and Bhattacharya, 2008). The leaves of the tamarind are used for medicinal purposes, and seeds are exploited for the extraction of different polysaccharides. The most important and valuable part of tamarind is its acidic pulp, which is used as a favorite ingredient in culinary preparations such as curries, chutneys, sauces, and sherbets (Little and Wadsworth, 1964). Tamarind has been used as medicine since ancient times. It has been known to be useful for treating constipation, gall disorders, and liver diseases, among other health problems (Aida et al., 2001). Extract from the tamarind pulp is used as a therapeutic drink in feverish conditions, digestion, convalescence, bowel complaints, biliary disorders, dysentery, and rheumatism. The pectin extracted from tamarind pulp has higher antioxidant potential than apple pectin, commercial pectin, guar gum, derivatives sulfates, oligosaccharides, and xanthan, demonstrating that the physicochemical and rheological potential may be used as an excipient in pharmaceutical and food products (Sharma et al., 2015). Pulp is the main raw material for all tamarind value-added products, such as tamarind juice concentrate, toffee, pulp powder, jam, pickles, and more. Tamarind pod shells and tamarind seeds are the by-products or waste usually obtained during tamarind processing.

The chief use of tamarind seed powder is in the preparation of jute and in some cotton yarn or as a creaming agent for rubber latex or as a soil stabilizer (Bhattacharya et al., 1991). Tamarind waste, such as tamarind husk, pulp, seeds, fruit, and the effluent generated during tartaric acid extraction, can be used as supplements for the production of ethanol (9.7%, w/v) from 22.5% reducing sugars of molasses through a yeast fermentation process because the tamarind aqueous effluent released after tartaric acid extraction contains 7.5% total reducing sugar, which could be directly fermented to yield ethanol (Patil et al., 1998). Tamarind fruit contains 35–40% seed by weight, which can be subjected to roasting, decortication, and sorting to produce tamarind seed powder. This powder mainly comprises gums and is capable of forming a gel, thus it can also be applied as a rheology modifier in food products (Prabhu and Teli, 2011). Tamarind seed gum/tamarind xyloglucan is used as a thickening, stabilizing, and gelling agent in food (Gupta and Devendra, 1988). Tamarind seed contains 7–9% oil, which can be extracted and used as an illuminant and varnishing agent (El-Sidding, 2006). Tamarind gum extracted from seeds is composed of glucose, xylose, and galactose at a ratio of 3:2:1 that are subjected to chemical modification via carbamation and sonication and are used in the preparation of biopolymers, which can be used as thickeners for textile printing (Hebeish et al., 2017). Roasted tamarind seed powder has more technical application in a habitually consumed food product, such as cookies, by replacing 15% of wheat flour (Silva et al., 2022). Tamarind seed powder is also a natural coagulant to reduce chemical oxygen demand (COD) for the treatment of dairy wastewater by 81.81% (Magar and Jadhav, 2018). Reduction in COD (90.95%) and BOD (94.87%) in dairy wastewater is also recorded by Meghana et al. (2020). A similar reduction in BOD value (from 71 mg/L to 53.83 mg/L) is also observed in the wastewater of the tofu industry (Nurika et al., 2007). Tamarind seed powder is applied to purify water, as it reduced turbidity by 98.12% when it was used at 300 mg/l (Raju et al., 2018). Tamarind seed powder also has been found to be effective in treating industrial waste-water containing detergent by reducing turbidity by 97.78% and COD by 43.5% at a dosage of 400 mg/l of tamarind seed powder (Ronke et al., 2016). Judith et al. (2017) also reported that extracting tamarind seeds with 0.5 M NaCl facilitates maximum turbidity reduction during wastewater treatment. It is, therefore, recommended that local industries consider using tamarind waste for wastewater treatment as an alternative to chemical coagulants because it is biological, affordable, and readily available.

Tamarind fruit contains 30% shell by weight that can be used as biomass mate-rial for manufacturing briquettes. Tamarind shell's calorific value is 16.3 MJ/kg, with 99% combustion efficiency (Rao et al., 2015). Tamarind fruit shells are an effective absorbent for the removal of Congo red dye from agricultural solid waste (Reddy, 2006). Similarly, tamarind pod shells act as an effective absorbent for the removal of methylene blue and amaranth dyes from aque-ous solutions (Ahalya et al., 2012). Tamarind pod shells can also be used as an alternative low-cost absorbent for the removal of Ni (II) and Cr (VI) ions in the remediation of wastewater (Pandharipande and Rohith, 2013). Tamarind peel flour also has great importance in food processing industries by incorporating peel flour (15%) with wheat flour (85%) to produce cookies acceptable for a consumer market (Silva et al., 2022).

8.11 Cumin

Cumin (*Cuminum cyminum* Linn.) is an important commercial seed spice belong-ing to the family Apiaceae. The plant is native from the East Mediterranean to South Asia where it is cultivated extensively. In India, it is commonly known as *Zeera/Jeera*. It is an annual herb and has been used since ancient times as a medicine and spice in food. India is the world's largest producer and con-sumer of cumin. Many value-added products of cumin seeds, like oleoresins and cumin oil, are also exported from India (Srinivasan, 2018; Sowbhagya, 2019). Two types of cumin seeds are available in India, i.e., black cumin (*Nigella sativa*) and *Shahi Jeera* (*Bunium persicum*). Black cumin belongs to the family Ranunculaceae, while *Shahi Jeera* belongs to the Apiaceae family. Cumin seeds have mostly been used in the cuisines of many different food

Value Addition of Spices Processing Industrial Waste   201

cultures, in both whole and ground forms. In India, cumin seeds have been used for thousands of years as a traditional ingredient of innumerable dishes, including kormas and soups, and form an ingredient of several other spice blends. Besides food application, it also has many uses in traditional medicine. In the Ayurvedic system of medicine in India, cumin seeds have immeasurable medicinal values (digestive stimulant and antidiabetic, neuroprotective, cardioprotective, chemopreventative, and antioxidant properties). Cumin is an important component of most of the Ayurvedic decoctions, specifically used for conditions like bloating, vomiting, diarrhea, dyspepsia, dysentery, malabsorption syndrome, fever, and skin diseases (Sharma et al., 2016). Cumin spent and immature seeds are the main waste of the cumin processing industry and have great potential to produce essential oils and fiber and to be incorporated in various value-added food products.

Cumin spent residue generated after oil extraction has been reported to be rich in dietary fiber, which exhibits significant water-retention properties. Cumin spent residue contained approximately 6% moisture, 5% crude fiber, 5% fat, 20% protein, and 8% starch (Sowbhagya et al., 2007). It is also reported that cumin spent residue is a good source of dietary fiber (62.1%), which is much higher than that of many fruits and vegetables. Cumin spent contains an appreciable amount of soluble dietary fiber (5.5%), thiamine (0.05 mg/100g), riboflavin (0.28 mg/100g), niacin (2.7 mg/100g), zinc (6.5 mg/100g), and iron (6.0 mg/100g) (Milan et al., 2008). Thus, spent cumin has great potential for producing healthy and functional products that might also strengthen its industrial use as a protein supplement. Black cumin cake can be utilized to produce Amphisin (an antifungal agent) by *Pseudomonas fluorescens* DSS73, which is effective against *Aspergillus* species (Ciurko et al., 2023). Another study also revealed that methanol extract from cumin waste is rich in p-coumaric acid, ferulic acid, ellagic acid, and cinnamic acid, which can inhibit the cell cycle and enhance apoptosis in HT29 cancer cells and prevent further promotion of colon cancer (Arun et al., 2016). Cumin spent residue subjected to 150 ppm methanolic extract and 250 ppm aqueous extract exhibited 90% and 80% of antioxidant activity, respectively (Sowbhagya, 2006). Cold-pressed black cumin waste may be utilized for the development of starch bread with reduced hardness, elasticity, and chewiness of the starch bread crumb. Starch bread enriched with black cumin seed and black cumin waste contains a higher quantity of 2-hydroxybenzoic acid, 2-hydroxyphenyl acetic acid, and 4-hydroxyphenyl acetic acid (Rozylo et al., 2021). Waste generated from cumin processing has good nutrient composition and can be a good substitute for conventional foodstuffs in health food formulations and can find application in therapeutic foods.

8.12 Clove (*Syzygium aromaticum*)

Clove is an evergreen spice plant native to Indonesia and the Philippines, and it belongs to the family Myrtaceae. It is used in many cuisines of Indian,

African, and other Middle Eastern countries due to its aroma. Clove is largely used in the preparation of foods, spice mixes, as well as in many medicines. Clove generally contains huge medicinal components, such as acetyl eugenol; beta-caryophyllene; vanillin; crategolic acid; tannins such as bicornin, gallotannic acid, methyl salicylate, the flavonoids eugenin, kaempferol, rhamnetin, and eugenitin; triterpenoids such as oleanolic acid, stigmasterol, and campesterol; and several sesquiterpenes, which increase the demand of clove in both flavor and pharmaceutical industry (Bendre et al., 2016). The essential oil extracted from the clove has great antimicrobial activity that is used for coating various food products to increase their shelf life (Goni et al., 2009). Clove leaf, clove spent after extraction of essential oil, immature clove, etc. are the main by-products of clove processing that can be used in different ways to reduce environmental pollution.

Clove has a good nutritional composition in terms of carbohydrates (51.3%), fat (16.6%), fiber (11.5%), protein (6.9%), ash (6%), potassium (0.6 g/100 g), calcium (0.27 g/100 g), and magnesium (0.09 g/100 g). So, after the extraction of essential oil and oleoresin, clove spent can be utilized as a good source of fiber, protein, carbohydrates, and various minerals (Al-Jasass and Al-Jasser, 2012). Clove buds waste has a great potential to develop commercially eco-friendly packaging material (cardboard) by mixing clove buds waste powder (seven parts) with recycled paper (three part), tapioca (20%), and chitosan (1%) in acid solution through a casting process in order to improve its mechanical properties, which are degraded at 286.58°C (Ratri et al., 2020). Eco-enzymes are produced from the organic wastes of nutmeg, clove, and eucalyptus leaves to control the growth of *E. coli* and *S. aureus* (Rijal, 2022). Wastewater from clove oil distillation contains many valuable aromatic compounds, such as eugenol and eugenol acetate that can be recovered to 79.44% and 54.05%, respectively, by using coffee husks as a biosorbent with adsorption efficiency of 37.96%. Clove oil (0.31%) can also be recovered from the wastewater of the clove distillation process (Maimulyanti et al., 2019).

Currently, low-cost and biodegradable biomaterials are prepared from clove leaf waste to remove noxious pollutants (methylene blue) from water bodies and the ecosystem with a biosorption capacity of approximately 9.80 mg/g (Kusuma et al., 2023). Biopellet (an alternative fuel solution in the co-firing system with coal) is prepared from clove leaf refining waste after drying the leaves and then mixing with glue or molasses in a ratio of 80% powder and 20% molasses followed by heating (at 90°C), molding, and pressing (93 kg/cm²) (Sijabat and Siregar, 2021). Clove leaves are utilized as a biosorbent for the sequestration of malachite green from a water-soluble solution through activation using potassium hydroxide (Sudarni, et al., 2021). Clove leaf waste is also utilized for the development of a natural antioxidant, i.e., 3-(4-allyloxy-3-methoxyphenyl)-1-phenyl prop-2-en-1-one (Chalcone derivative) (Eden et al., 2021).

8.13 Mustard

Mustard is an annual plant that belongs to the *Brassicaceae* family (formerly *Cruciferae*). Several varieties of mustard are available in India, such as yellow-white mustard (*Sinapsis alba* L.), black mustard (*Brassica nigra* L.), brown mustard (*Brassica juncea* L.), and wild mustard (*Sinapsis arvensis* L.). India ranks first in mustard production among Asian countries. Mustard seeds are rich in lipid content (23–47%), and the lipid profile varies according to the species and climacteric conditions. Usually, the fatty acid of mustard seeds contains erucic acid (26.5–36.5%), oleic acid (22%), linoleic acid (19.5–22%), and linolenic acid (9–15%) in significant amounts depending on the species (Sawicka et al., 2007). Mustard seed is a good source of protein, fiber, minerals, vitamins, antioxidants, and phytonutrients, such as alpha-linolenic acid, erucic acid, palmitic acid, tocopherols, tocotrienols, carotene, oryzanol, squalene, and thiamine (Billman, 2013). The plant has several health benefits, including antimicrobial (due to the presence of allyl isothiocyanate), anti-diabetic, antimalarial, etc. properties (Sahu et al., 2020).

Mustard oil is the primary product of mustard seeds. It can be extracted by either cold pressing or mechanical expelling and solvent extraction. After extracting the oil, mustard cake is generated as a by-product, which constitutes about 60% of the mustard seed (Sehwag and Das, 2015). The nutritional quality of mustard cake is very high in terms of protein content (33–40%), lignin (6.35%), total extractable polyphenol (2.1%), sulfur-containing amino acids (cysteine and methionine), and carbon-to-nitrogen ratio (1:4). It is also a good source of total phenols (77–81 mg/kg), glucosinolates, and phytates (Boscariol et al., 2019). Mustard cake utilized as animal feed and the partial replacement of soybean meal by mustard cake at 80 gm/kg in the diet of rabbits had no adverse effect on the animals' growth and health (Tripathi et al., 2003). Mustard cake is rich in nitrogen (4.8%), phosphorus (2% as P_2O_5), and potassium (1.3% as K_2O) which is very necessary for maintaining soil fertility and the proper growth of plants (Balesh et al., 2005). The presence of isothiocyanates in mustard cake has pesticidal properties and weed suppression capacity after enzymatic action on glucosinolates. Several researchers have documented that mustard cake can control weeds and act as biopesticides (Meyer et al., 2011). Mustard cake is a good option to produce enzymes (lipase and phytase) and lactic acid through solid-state fermentation (Bano et al., 1993). It can also be utilized as a nitrogen source for rice straw substrate and has been shown to increase the yield of oyster mushrooms by 50–100% (Bano et al., 1993). Several researchers have already evaluated the use of defatted mustard cake as a suitable source of biopolymer with good water vapor permeability, water solubility, tensile strength, and elongation of film (Hendrix et al., 2012; Kim et al., 2012). After extracting the oil, polar antioxidants like phenolics and lignans or flavonoids remain in the mustard meal and can be further extracted by using polar organic solvents that possess moderated levels of antioxidant activity and can be successfully applied to special foods (Swati et al., 2015).

8.14 Celery (*Apium graveolens* L.)

Celery (*Apium graveolens* L.) is an important aromatic plant grown mostly for its fresh herbs as salad crops, but the dried fruits are used as spices. It belongs to the family Apiaceae, and in India it is commonly grown in Punjab, Haryana, and Uttar Pradesh. The postharvest operations of fresh celery include the removal of small lateral branches and damaged leaves, packaging, pre-cooling, and then storing at 0°C and 95% relative humidity for a short period of time (Malhotra, 2006). However, celery seeds are cured in the sun for 7–10 days, followed by threshing, cleaning, screening, grading, and packaging (Malhotra, 2005). Celery seeds are used for the extraction of volatile oil (1.5–3.0%) which mainly contains 60–70% limonene, phthalides and β-salinene, coumarins, furanocoumarins (bergapten), flavonoids (apiin), 10–20% selinene, 2.5–3.0% sedanolid, and 0.5% sedanonic anhydride (Farrell, 1999). Fresh celery and its seed oil contain various medicinal properties, including antirheumatic, antispasmodic, diuretic, hypotensive, and anti-inflammatory (Prajapati et al., 2003). Many types of processed products are prepared from fresh celery (juice blends, freeze-dried celery, canned celery, pickling celery, etc.) as well as from celery seeds (essential oils, oleoresins, celery powder, celery salt, celery seed vinegar, etc.). Normally, 60% of celery is wasted on consumer-level and on-farm food loss (Qin and Horvath, 2021). Around 1,200 tons of spent residue of celery are disposed of as waste annually in India (Mathew, 2004). During the production of different types of value-added products, various by-products and waste are generated, such as stalks, leaves, seed residue, andcelery spent residue after oil extraction, that can be further utilized for the extraction of valuable components.

A huge amount of celery spent is obtained annually from spice oleoresin industries, out of which a maximum portion is utilized as fuel for boilers and some portion goes to animal feed. Celery spent is also considered an alternative source of cereal dietary fiber, as it contains 57.6% total dietary fiber. Aqueous methanolic extract of celery spent possessed 33% of free radical scavenging activity at 50 ppm, whereas 100 and 200 ppm exhibited 63% and 92% radical scavenging activity, respectively (Sowbhagya, 2006). Ethanol (85%) extract of waste celery stalk produced 45.5% total sugar and 15.2% mannitol, while stalk plus leaf residues produced 33.9% total sugar and 13.3% mannitol, which can be further used for the preparation of dietary fiber–rich food supplements (Ruperez and Toledano, 2003). Celery spent residue after oil and oleoresin extraction contained 61% total dietary fiber, 53.5% insoluble dietary fiber, 7.5% soluble dietary fiber, 19% protein, 7.9% starch, and 5% fat. So, the incorporation of celery spent residue (7.5%) resulted in a 7.8% increase in the dietary fiber of cookies compared to the control sample (3.5% dietary fiber) (Sowbhagya et al., 2011). Essential oil extracted from waste celery seeds mainly contained limonene (75.06%), which exhibited great antimicrobial activities against many bacteria (*Bacillus subtilis*, *Staphylococcus aureus*, *Escherichia coli*, and *Pseudomonas aeruginosa*) as well as fungi (*Candida vini, Aspergillus niger,* and *Penicillium expansum*) and the oil also showed high radical scavenging activity for DPPH (IC_{50} = 89.11 g/l) (Dąbrowska et al., 2020).

Celery by-products also act as natural fermentation media for the growth of yeast (*Yarrowia lipolytica*), which further enhanced the production of citric acid (Yalcin, 2012). Lignin-rich nano cellulose (43.9%) is isolated from depectinized celery by sulfuric acid treatment and then high-pressure homogenization. This lignin-rich nano cellulose mixed with 20% xylonic acid is cast into a flexible and transparent film with better antibacterial properties (Luo et al., 2020). Celery residue after modification by acid and cetyltrimethylammonium bromide, act as a bio-adsorbent for Congo red dye from polluted water with maximum monolayer adsorption capacities of 238.09 and 526.32 mg/g for acid-modified celery residue and cetyltrimethylammonium bromide-acid-modified celery residue, respectively (Mohebali et al., 2019). Celery root peel waste, which has high nutritional value and is beneficial for healthy nutrition, has valorization potential to re-formulate *Ayran* (a traditional Turkish drink) with good organoleptic quality and increased shelf life by at least 50% compared to the control drink (Aşkın Uze, 2023). Hybrid liquid fertilizer was prepared from celery waste, and it was found that this fertilizer with 2 g of ammonia after seven days of fermentation could be used to grow a mung bean plant (Zakaria et al., 2023). Currently, celery root peel is combined with pea pods and mixed vegetable waste for the production of bacterial cellulose from *Komagataeibacter hansenii* (Bozdag et al., 2023).

8.15 Cinnamon (*Cinnamomum zeylanicum*)

Cinnamon is a valuable spice that is obtained from the bark of an evergreen tree (*Cinnamomum zeylanicum*) that belongs to the Laurel family. It is native to Sri Lanka and now it is grown all over the world. Cinnamon is mainly used in the aroma and essence industries due to its fragrance, which can be incorporated into different varieties of foodstuffs, perfumes, and medicinal products. Cinnamaldehyde and trans-cinnamaldehyde are the main valuable components of cinnamon, which contribute to the fragrance and are commonly present in essential oil (Jakhetia, 2010). Procyanidins and catechins are also found in cinnamon bark that possess the antioxidant activities (Peng et al., 2008; Maatta-Riihinen et al., 2005). Cinnamon bark oil has various important pharmaceutical activities due to its antimicrobial, antifungal, antioxidant, and antidiabetic properties (Rao and Gan, 2014). Cinnamon also contains a variety of resinous components, including cinnamaldehyde, cinnamate, cinnamic acid, and numerous types of essential oils (trans-cinnamaldehyde, cinnamyl acetate, eugenol, L-borneol, caryophyllene oxide, b-caryophyllene, L-bornyl acetate, E-nerolidol, α-cubebene, α-terpineol, terpinolene, and α-thujene) (Tung et al., 2008; Tung et al., 2010). Cinnamon has been used in everyday applications for different purposes and in various forms. Numerous cinnamon products in the form of bark, essential oils, bark powder, spice mixes, phenolic compounds, flavonoids, and isolated components are generated after processing (Rao and Gan, 2014).

Cinnamon bark is utilized as spice and its leaves, fruit, root, stalk, and flowers are considered as waste. During processing, inferior quality bark, broken quills, sticks, and inferior feathering are produced as waste. The leaf of the cinnamon plant is used for the production of essential oils through steam distillation, which is rich in eugenol content (80–85%) and also used for culinary purposes (Pruthi, 1980; Pruthi et al., 1974). Cinnamon bark oil can be extracted from the cinnamon waste after first grinding the waste material and then macerating it in a saturated salt solution for two days and distilling it. The yield of oil depends on the maturity, quality, and age of the leaves and bark. The wastewater effluent of cinnamon bark oil distillery is rich in cinnamaldehyde (66.2%), which can be used as a fungicide in the postharvest treatment of various fruits, such as bananas (Ranasinghe et al., 2003). Researchers also revealed that the aqueous extract of cinnamon leaves consists of various valuable compounds, such as flavonols and flavones, phenolic acids, lactones, terpenoids, phenylpropanoids, and flavanols (Yang et al., 2022).

Nowadays cinnamon biowaste is introduced as a new adsorbent for various chemicals that can be used for the effluent or wastewater treatment generated from the textile, food, and pharmaceutical industries. The activated carbon is developed from the cinnamon sticks (a low-cost farming by-product) with high surface area (894 m^2/g) and high pore volume (0.49 cm^3/g), which enhances the adsorption of chlorpyrifos from agricultural wastewater (Ettish et al., 2021). Cinnamon wood biochar is a by-product of bioenergy plants and has efficient capacity (113 mg/g) to adsorb sulfamethoxazole (a sulfonamide antibiotic used to treat urinary infections) from aqueous solution (Keerthanan et al., 2021). Similarly, the C-Cinnamal Calix [4] Resorsinarene compound is extracted from cinnamon oil waste and has great potential to adsorb harmful dyes, such as methanil yellow, with an absorption capacity of 2.1375 mg/g for 150 minutes of contact time (Etika and Nasra, 2020). Cinnamon logged waste can be utilized as a biopesticide to suppress pest attacks on mangrove seedlings when it is applied at 75% concentration five times a day for seven days (Salatalohy and Baguna, 2022). Cinnamon waste material after polyphenol extraction subjected to enzymatic hydrolysis can be used as feedstock for the yeast *Rhodosporidium toruloides* to produce carotenoids up to 2.00 (±0.23) mg/L (Bertacchi et al., 2021). The waste of cinnamon bark biomass is also investigated as a low-cost and eco-friendly biosorbent for removing cationic methylene blue dye from wastewater effluent with maximum adsorption capacities of 123.25 mg g–1 for 80 minutes, whereas cinnamon bark biomass modified with MnO_2, which enhances the high surface charge density to increase the adsorption capacity up to 394.56 mg g–1 for 20 minutes (Yardimci and Kanmaz, 2023).

8.16 Fenugreek

Fenugreek (*Trigonella foenum graecum*) is an annual herb belonging to the Leguminosae family that originates from the regions of West Africa (Balasubramanian et al., 2016). Presently, more than 260 different species of fenugreek are distributed across various parts of the world, including Asian countries like India, Pakistan, China, and Iran, and parts of Australia, the United Kingdom, Mediterranean Europe, and Russia (Hilles and Mahmood, 2021). Owing to the characteristic flavor, color, and medicinal properties, fenugreek seeds and leaves hold significant economic importance as a culinary ingredient and medicinal herb. Fenugreek seeds have a sharp, bitter taste and contain several bioactive compounds, such as pyridine alkaloids, mainly choline (0.5%), trigonelline (0.2–0.38%), carpaine, and gentianine; free amino acids, including histidine, 4-hydroxyisoleucine (0.09%), lysine, and arginine; flavonoids, mainly isovitexin, apigenin, vitexin, orientin, quercetin, and luteolin (Wani and Kumar, 2018). Whole or ground form seeds of fenugreek are used for flavoring in many foods mostly in spice blends, curry powder, and teas. Moreover, it is an excellent source of soluble dietary fiber (26.8%), chemically identified as galactomannans, and therefore used as an emulsifying agent, food stabilizer, and adhesive (Dhull et al., 2020). Fresh fenugreek leaves contain β-carotene (19 mg/100 g) and ascorbic acid (220.97 mg/100 g) alongside several vitamins and minerals, such as riboflavin, thiamine, carotene, niacin, phosphorous, calcium, zinc, and iron. Having better retention of nutrients, fenugreek leaves are used as vegetables in some diets (Yadav and Sehgal, 1997).

Commercially, fenugreek seeds are subjected to several processing operations, including threshing, grading, drying, pressing, and/or grinding, whereas the leaves undergo washing, blanching, trimming, chopping, drying, and/or grinding. During processing, the inedible portions like fenugreek spent, stems, roots and rootlets, damaged leaves, and broken seeds are generated as industrial waste, which accounts for 45–50% of the initial weight (Gomez et al., 2003). However, being exceptionally rich in bioactive compounds, these inedible portions have the potential to be utilized effectively for the generation of biogas, bioactive compounds extraction, as well as fertilizer and soil amendments rather than being disposed of as waste. The vegetable residue of fenugreek (stem and root) has been utilized for the synthesis of effective biosorbent for eliminating Basic Violet 14 (BV 14) from wastewater (Jain et al., 2020). Fenugreek seed husk incorporated at a 10% level into wheat flour for the preparation of good-quality and high-fiber (13.36 %) muffins (Srivastava et al., 2012). Unprocessed food waste, including raw vegetable waste (fenugreek) and banana waste, is the best source of biogas (methane) generation due to its biodegradable capacity (Shukla et al., 2010). Trigoxazonane, a monosubstituted trioxazonane compound found in fenugreek root exudate, demonstrates significant inhibition ranging from 27% to 54% in the germination of the parasitic achlorophyllous weed *Orobanche crenata*, commonly known as Broomrapes (Evidente et al., 2007).

8.17 Onion (*Allium cepa*) and Garlic (*Allium sativa*)

Onions and garlic are largely used as vegetables as well as spices due to their characteristic flavor in various types of nourishment items around the world. Both belong to the same family, i.e., Liliaceae and are commonly used for culinary purposes. Onions originated in Central Asia, whereas garlic originated in China. Globally, 99,968,016 tons of onion are produced from an area of 5,192,651 ha (FAO, 2021). China is the biggest producer of onions, with 18,122,435 tons per year, followed by India with 18,122,435 tons (FAO, 2021).

Other than culinary reasons, both garlic and onions are also utilized for remedial purposes all around the globe. They are utilized for curing many diseases, such as colds, coughs, fevers, diabetes, rheumatism, intestinal worms, colic, and flatulence. Onions are believed to help in congestion of the lungs, expanding the airways, and stimulating appetite. The mixture of rue and onion is used to get rid of parasites in the digestive system (Rane and Gaikwad, 2019). Natural antioxidants found in onions, such as cyanidin peonin, quercetin, kaempferol, isorhamnetin, and their glycosides, helps in reducing the risk of many forms of cancer, ulcers, cataract formation, and the prevention of vascular and heart disease by the inhibition of lipid peroxidation and lowering of low-density lipoprotein (LDL) cholesterol levels (Corzo-Martínez et al., 2007; Campos et al., 2003). Garlic is traditionally used for the treatment of dysentery, liver disorders, tuberculosis, facial paralysis, high blood pressure, and bronchitis. Onions contain essential oils and organic sulphide, i.e., allyl propyl disulphide (a sulfur-containing compound) that are responsible for their pungent odors and for many of their health-promoting effects, while garlic contains almost 33 sulfur compounds. Sulfur compounds such as alliin, allicin, ajoene, allylpropyl disulfide, diallyl trisulfide (DATS), S-allylcysteine (SAC), vinyldithiins, S-allylmercaptocysteine, etc. have shown to possess great antifungal properties (Rane and Gaikwad, 2019). In recent years, the consumption and processing of onions and garlic has increased due to their flavor and abundance of health benefits. Onions are processed mainly into dehydrated onion, canned onion, and pickled onion, while garlic is processed in the form of canned garlic, pickled garlic, garlic paste, mixed ginger and garlic paste, etc. Onion and garlic skin is the main waste of these spices that is not used during culinary preparation. But the skins contain many medicinal and nutritional components that can be further utilized.

A rise in the extent of processing onions results in huge processing waste, where the skins contribute most of the waste (Narashans et al., 2021). A previous study revealed that the outer skin of an onion has much higher antioxidant activity than the inner, edible parts due to the presence of free and glycosidically bonded quercetin (2–10% w/w) and oxidized quercetin derivatives, such as minor flavonols and phenolic compounds (Kotenkova and Kupaeva, 2019; Skerget et al., 2009; Griffiths et al., 2002; Suh et al., 1999). Onion skin is also rich in dietary fiber and bio sugars that can be used to prepare many functional foods (Choi et al., 2015). A study of the phytochemical profile of onion skin extract revealed that flavonols and anthocyanins are major metabolites. Quercetin, quercetin glucosides, and their dimer and trimer derivatives, and, among anthocyanins, cyanidin 3-glucoside, are the most abundant bioactive compounds present in onion skin and enhance several biological activities, like oxygen radical absorbance capacity and in vitro alpha-glucosidase assays (Celano et al., 2021). Moreover, onion skin extract has also been reported as an anticarcinogenic, hypocholesterolemic, good cardiovascular agent and has an antiasthmatic effect (Sagar et al., 2020). Acrylamide is a common human carcinogen found in many thermally processed foods that produces cytotoxicity and immunotoxicity in liver cells. This can be effectively treated with onion peel powder supplementation through reduced acrylamide-stimulated immunotoxin and cytotoxic effects because of its direct ROS (reactive oxygen species) scavenging activity (Shabir et al., 2022). Different parts of the onion that constitute waste include the top and bottom parts and the outer scale (skin). However, these inedible portions have great nutritional value. Previous analysis reported that the top and bottom parts are rich in protein (8.76%) and ash content (11.46%), whereas the outer scale is rich in carbohydrates (66.12%), fat (15.71%), and fiber (26.84%). X-ray spectrometry also indicated that the outer scale contains the highest level of calcium (3.05%), followed by the onion bulb and the top and bottom parts (Bello et al., 2013). Oil extracted from the outer scale of the onion bulb also contains 52.87% linoleic acid (Bello et al., 2013). Another study also revealed that onion peel is rich in carbohydrates (88.56%) and low in protein content (0.88%), while the ethanolic extract of onion peels yielded 98.52 µg QUE ml^{-1} total flavonoid, 664.30 µg ml^{-1} GAE total phenol, total antioxidant property (1338.15 µg ml^{-1}) and scavenged DPPH radical 27.76 µg ml^{-1} (Ifesan, 2017). Hence, onion skin powder may be incorporated into many bakery products, like bread and pizza, as good sources of energy that are widely consumed as well (Sagar et al., 2020).

Before the incorporation of onion waste into the food ingredient, it should be stabilized for safety purposes. Some common stabilizing techniques are sterilization, pasteurization, and freezing, presenting crucial steps for the valorization of onion waste as a safe food ingredient (Sharma et al., 2016). Wheat bread fortified with 3% onion skin powder enhanced the antioxidant as well as the sensory properties of the bread (Dziki et al., 2014). Onion skin powder up to 6% can be utilized as a good supplement of antioxidant and dietary fiber during manufacturing of dried and fried noodles that led to a significant increase in total phenols (171.76 GAE mg/100g in dry and 123.18 GAE mg/100g in fried noodles) and dietary fibers (7.98% in dry and 8.08% in fried noodles) (Sayed et al., 2014). Onion peel powder acts as a prebiotic due to the presence of fructooligosaccharides, which can enhancs the growth of microbes during the fermentation process (Cebin et al., 2020; Kimoto-Nira et al., 2019) and is also incorporated with basic ingredients for manufacturing various food products, such as gluten-free bread (Bedrnicek et al., 2020), wheat pasta (Michalak-Majewska et al., 2020), Hanwoo Tteokgalbi (Chung et al., 2018), etc. Extract of onion peel has been utilized to produce wheat bread (Piechowiak et al., 2020), bean paste (Sęczyk et al., 2015), and wheat flour extrudates (Tonyali and Sensoy, 2017), with higher antioxidant activity and phenolic content and film from *Artemisia sphaerocephala* Krasch gum with good potential for intelligent packaging materials and gas-sensing labels with pH indicator (Liang et al., 2018).

Water extract from onion skin has potential antioxidant activity due to the presence of two main phenolic compounds, i.e., quercetin and quercetin monoglucoside, which help in the curing of meat products (Bedrnicek, et al., 2019) and have shown inhibitory activities against urease and xanthine oxidase enzymes (Nile et al., 2017). Ethanolic water extract of onion peel possesses antiproliferative properties by arresting the cell cycle, intrinsic apoptosis (mitochondrial membrane potential modification), and caspase 3 activation. In addition to this, onion waste also increased intracellular ROS with possible NF-kB activation causing a proteasome down regulation, which helps in reducing cancer proliferation (Paesa et al., 2022). Production of bioethanol and bio-piezoelectric material from onion juice residue and onion skin, respectively, is an efficient use of waste material (Kim et al., 2017; Maiti et al., 2017). According to Mondal et al. (2019), onion peel dust can be effectively utilized as an adsorbent for the removal of nitrate from aqueous solutions. Modification of onion skin waste with aluminum oxide increased the removal percentage of lead (91.23%) and cadmium (94.10%) depending on concentration, contact time, adsorbent dosage, and pH (Yusuff, 2022). A two-step batch fermentation process is developed in which *Saccharomyces cerevisiae* AHU3532 and *Acetobacteraceti* TUA549B are used for the fermentation of alcohol and acetic acid, respectively, from onion waste to produce vinegar. It was found that the concentration of total amino acids (2100 mg l⁻¹) and total organic acid (2830 mg l⁻¹) in onion vinegar was higher than those in other types of vinegar (Horiuchi et al., 1999). The outer skin of red onion contains a good amount of anthocyanins and phenolics that can be used as a natural antioxidant and colorant in many food products, such as jellies and gums, sugar-enrobed confectionery, foamed commodities, stiff-boiled candies, and fat-based coatings (Ali et al., 2016). It is also reported that pectin from onion peels can be recovered through hydrolysis (semicontinuous) of onion-peel waste with subcritical water treatment (Compaore et al., 2017)

During the commercial production of peeled garlic, approximately 22–27% w/w garlic waste (skin) is generated (Reddy and Rhim, 2014). According to the researchers garlic peel waste contains an appropriate amount of dietary fiber (62.10%), total sugars (6.51%), dry matter (80.8%), total ash (7.37%), and a good source of many minerals such as B (18.0 mg/kg), Al (826 mg/kg), S (1635 mg/kg), K (9081 mg/kg), Ca (20610 mg/kg), Cr (18.40 mg/kg), Mn (35.4 mg/kg), Fe (682 mg/kg), Zn (12.9 mg/kg), Se (0.058 mg/kg), and Mo (1.480 mg/kg) (Zhivkova, 2021). So, this can be utilized for variable purposes. Garlic skin waste with the combination of agave waste (a medicinal plant that belongs to the family Asparagaceae) is availed for the production of cellulose nano particles (Varela, 2021). Garlic peel has adequate potential to eliminate the methylene blue dye from aqueous solution (Hameed and Ahmad, 2009), and it can also be used as the precursor for carbons via hydrothermal carbonization and KOH activation for CO_2 adsorption (Huang et al., 2019). Low-cost antimicrobials have been produced form garlic skin, which exhibited maximum inhibition against a varied range of microbes, including *Escherichia coli, Bacillus megaterium, Bacillus cereus, Staphylococcus aureus, Colletotrichum falcatum, Fusarium moniliforme,* and *Rhizoctonia solani* (Naqvi et al., 2020). Similarly, aqueous methanolic extracts of garlic skin (0.4, 0.8, and 1.6 g/L), and garlic stem (0.8 and 1.6 g/L) have successfully been used to prevent *Saprolegnia parasitica* infestation on rainbow trout eggs that enhanced the chances of utilization of these waste in aquaculture (Ozcelik et al., 2020).

The compost prepared from a mixture of onion, garlic, and cabbage waste is found to be effective against *Sclerotium cepivorum*, which causes white rot in the Allium species (Shalaby and El-Kot, 2009). Negi et al. (2012) also observed the heavy metals (Pb^{2+}, Sn^{2+}, Fe^{2+}, Hg^{2+}, As^{3+}, and Cd^{2+}) adsorption capacity of onion and garlic waste powder from both synthetic and industrial effluents is very high and the technique appears industrially applicable and viable. Another study on the quality of duck meat indicates that feeding ducks onion husk powder and garlic husk powder significantly increased the protein and decreased cholesterol contents of duck meat because of the presence of antioxidant compounds such as allinin, allisin, and flavonoid, which are able to decrease fat and cholesterol levels in the body (Putri et al., 2016). The water extract of onion and garlic peels biowaste can be used as a natural growth enhancer in a sustainable way for growing microgreens of fenugreek and falooda seeds (Patil et al., 2021). The peels of garlic and onion are exploited to develop flexible bio-composite film with good biocompatibility and antimicrobial activity (99.95% for garlic and 99.65% for onion skin film) through an acylation process followed by the application of poly lactic acid as a matrix (Salunkhe, 2022).

References

Aftab, R. A., Zaidi, S., Khan, A. A. P., Usman, M. A., Khan, A. Y., Chani, M. T. S., and Asiri, A. M. (2023). Removal of congo red from water by adsorption onto activated carbon derived from waste black cardamom peels and machine learning modeling. *Alexandria Engineering Journal, 71*, 355–369.

Ahalya, N., Chandraprabha, M. N., Kanamandi, R. D., and Ramachandra, T. V. (2012). Adsorption of methylene blue and amaranth on to tamarind pod shells. *Journal of Biochemcal Technology, 3*(5), 189–192.

Ahuja, K. D., and Ball, M. J. (2006). Effects of daily ingestion of chilli on serum lipoprotein oxidation in adult men and women. *British Journal of Nutrition, 96*(2), 239–242.

Ahuja, K. D., Robertson, I. K., Geraghty, D. P., and Ball, M. J. (2006). Effects of chili consumption on postprandial glucose, insulin, and energy metabolism. *The American Journal of Clinical Nutrition, 84*(1), 63–69.

Aida, P., Rosa, V., Blamea, F., Tomas, A., and Salvador, C. (2001). Paraguayan plants used in traditional medicine. *Journal of Ethnopharmacology, 16*(1), 93–98.

Ali, O. H., Al-sayed, H., Yasin, N., and Afifi, E. (2016). Effect of different extraction methods on stablity of anthocyanins extracted from red onion peels (*Allium cepa*) and its uses as food colorants. *Bulletin of National Nutritional Institute of Arab Republic, 47,* 1–24.

Al-Jasass, F. M., and Al-Jasser, M. S. (2012). Chemical composition and fatty acidcontent of some spices and herbs under Saudi Arabia conditions. *The Scientific World Journal, 2012,* 859892.

Anonymous. (2023). Spices Market Size, Share & Trends Analysis Report By Product (Pepper, Turmeric), By Form (Powder, Whole, Chopped & Crushed), By Region (North America, Europe, APAC, CSA, MEA), And Segment Forecasts, 2020–2027. https://www.grandviewresearch.com/industry-analysis/spices-market#:~:text=Report%20Overview,driving%20the%20consumption%20of%20spices. (Assessed on 16/03/2023).

Arun, K. B., Aswathi, U., Venugopal, V. V., Madhavankutty, T. S., and Nisha, P. (2016). Nutraceutical properties of cumin residue generated from Ayurvedic industries using cell line models. *Journal of Food Science and Technology, 53*(10), 3814–3824.

Ashokkumar, Kaliyaperumal; Murugan, Muthusamy; Dhanya, M.K.; Warkentin, Thomas D. (2020). Botany, traditional uses, phytochemistry and biological activities of cardamom [Elettaria cardamomum (L.) Maton] – A critical review. *Journal of Ethnopharmacology, 246*(), 112244–. doi:10.1016/j.jep.2019.112244.

Aşkın Uzel, R. (2023). "Sustainable green technology for adaptation of circular economy to valorize agri-food waste: celery root peel as a case study", *Management of Environmental Quality, 34*(4), pp. 1018–1034.

Bag, B. B. (2018). Ginger processing in india (Zingiber officinale): A review. *International Journal of Current Microbiology and Applied Science, 7*(4), 1639–1651.

Balasubramanian, S., Roselin, P., Singh, K. K., Zachariah, J., and Saxena, S. N. (2016). Postharvest processing and benefits of black pepper, coriander, cinnamon, fenugreek, and turmeric spices. *Critical Reviews in Food Science and Nutrition, 56*(10), 1585–1607.

Balesh, T. F., Zapata, B. I., Aune, L., and Sitaula, B. (2005). Evaluation of mustard meal as organic fertilizer on tef (Eragrostis tef (Zucc) Trotter) under field and greenhouse conditions. *Nutrient Cycling in Agro Ecosystems, 73,* 49–57.

Bano, Z., Shashirekha M. N., and Rajarathnam, S. (1993). Improvement of the bioconversion and biotransformation efficiencies of oyster mushroom by supplementation of it rice straw substrate with oil seed cake. *Enzyme Microbial Technology, 15*(11), 985–989.

Bedrnicek, J., Jirotkova, D., Kadlec, J., Laknerova, I., Vrchotova, N., Triska, J., Samkova, E., and Smetana, P. (2020). Thermal stability and bioavailability of bioactive compounds after baking of bread enriched with different onion by-products. *Food Chemistry, 319,* 126562.

Bedrnicek, J., Laknerova, I., Linhartova, Z., Kadlec, J., Samkova, E., Barta, J., Bártova, V., Mraz, J., Pesek, M., Winterova, R., Vrchotova, N., Triska, J., and Smetana, P. (2019). Onion waste as a rich source of antioxidants for meat products. *Czech Journal of Food Sciences, 37*(4), 268–275.

Bello, M. O., Olabanji, I. O., Abdul-Hammed, M., and Okunade, T. D. (2013). Characterization of domestic onion wastes and bulb (Allium cepa L.): Fatty acids and metal contents. *International Food Research Journal, 20*(5), 2153–2158.

Bendre, R. S., Rajput, J. D., Bagul, S. D., and Karandikar, P. S. (2016). Outlooks on medicinal properties of eugenol and its synthetic derivatives. *Natural Products Chemistry and Research, 4*(3), 1–6.

Bertacchi, S., Pagliari, S., Cantù, C., Bruni, I., Labra, M., and Branduardi, P. (2021). Enzymatic hydrolysate of cinnamon waste material as feedstock for the microbial production of carotenoids. *International Journal of Environmental Research and Public Health, 18,* 1146.

Bhadoriya, S. S., Ganeshpurkar, A., Narwaria, J., Rai, G., and Jain, A. P. (2011). Tamarindus indica: Extent of explored potential. *Pharmacognosy Review, 5*(9), 73–81.

Bhat, S., Kaushal, P., Kaur, M., and Sharma, H. K. (2014). Coriander (*Coriandrum sativum* L.): Processing, nutritional and functional aspects. *African Journal of Plant Science, 8*(1), 25–33.

Bhattacharya, A., Chattopadhyay, A., Mazumdar, D., Chakravarty, A., and Pal, S. (2010). Antioxidant constituents and enzyme activities in chili peppers. *International Journal of Vegetable Science, 16*, 201–211.

Bhattacharya, S., Bal, S., Mukjerji, K. R., and Bhattacharya, S. (1991). Rheological behaviour of tamarind seed. *Journal of Food Engineering, 13*(2), 151–158.

Bhowmik, D., Chiranjib, K. P., Kumar, S., Chandira, M., and Jayakar, B. (2009). Turmeric: A herbal and traditional medicine. *Archives of Applied Science Research, 1*(2), 86–108.

Billman, G. E. (2013). The effects of omega-3 polyunsaturated fatty acids on cardiac rhythm; a critical assessment. *Pharmacology and Therapy, 140*(1), 53–80.

Boscariol, R. G., Hilkner, M. H., de Alencar, S. M., and de Castro, R. J. S. (2019). Biologically active compounds from white and black mustard grains: An optimization study for recovery and identification of phenolic antioxidants. *Indian Crop Production, 135*, 294–300. DOI: 10.1016/j.indcrop.2019.04.059.

Bozdağ, G., Pinar, O., Gündüz, O. et al. (2023). Valorization of pea pod, celery root peel, and mixed-vegetable peel as a feedstock for biocellulose production from Komagataeibacter hansenii DSM 5602. *Biomass Conv. Bioref. 13*, 7875–7886.

Caluwe, E. D., halamov, K., and Damme, P. V. (2010). Tamarind (Tamarindus indica L.): A review of traditional uses, phytochemistry and pharmacology. *African Focus, 23*(1), 85–110.

Campos, Z., Coutinho, M. and Magnusson, M. (2003). Terrestrial activity of caiman in the Pantanal, Brazil. Copeia; *American Society of Ichthyologists and Herpetologists (ASIH),* 2003(3) (Sep. 5, 2003), pp. 628–634.

Cebin, A. V., Seremet, D., Mandura, A., Martinic, A., and Komes, D. (2020). Onion solid waste as a potential source of functional food ingredients. *Engineering Power, 15*(3), 7–14.

Celano, R., Docimo, T., Piccinelli, A. L., Gazzerra, P., Tucci, M., Sanzo, R. D., Carabetta, S., Campone, L., Russo, M., and Rastrelli, L. (2021). Onion peel: Turning a food waste into a resource. *Antioxidant, 10*(304), 1–17.

Chakrabarty, S., Islam, A. M., and Islam, A. A. (2017). Nutritional benefits and pharmaceutical potentialities of chili: A review. *Fundamental and Applied Agriculture, 2*(2), 227–232.

Chen, X., Wang, Z., and Kan, J. (2021). Polysaccharides from ginger stems and leaves: Effects of dual and triple frequency ultrasound assisted extraction on structural characteristics and biological activities. *Food Bioscience, 42*, 101166.

Choi, I. S., Cho, E. J., Moon, J. H., and Bae, H. J. (2015). Onion skin waste as a valorization resource for the by-products quercetin and biosugar. *Food Chemistry, 188*, 537–542.

Chung, Y. K., Choi, J. S., Yu, S. B., and Choi, Y. I. (2018). Physicochemical and storage characteristics of hanwoo tteok-galbi treated with onion skin powder and blackcurrant powder. *Korean Journal of Food Science and Animal Resources, 4*(38), 737–748.

Ciurko, D., Łaba, W., Kancelista, A., John, L., Gudina, E. J., Lazar, Z., and Janek, T. (2023). Efficient conversion of black cumin cake from industrial waste into lipopeptide biosurfactant by Pseudomonas fluorescens. *Biochemical Engineering Journal, 197*, 108981.

Compaore, A., Dissa, A. O., Rogaume, Y., Putranto, A., Chen, X. D., Mangindaan, D., Zoulalian, A., Remond, R., and Tiendrebeogo, E. (2017). Application of the reaction engineering approach (REA) for modeling of the convective drying of onion. *Drying Technology, 35*, 500–508.

Corzo-Martínez, M., Corzo, N., and Villamiel, M. (2007). Biological properties of onions and garlic. *Trends in Food Science and Technology, 18*, 609–625.

Dąbrowska, J. A., Styczyńska, A. K., and Śmigielski, K. B. (2020). Biological, chemical, and aroma profiles of essential oil from waste celery seeds (*Apium graveolens* L.). *Journal of Essential Oil Research, 32*(4), 24–32.

Dewin, H. (1955). Natural and synthetic pepper substances (problems of constitution synthesis in relation to pepper flavour). *SeifenOele. Fette, Wachse, 81*, 489.

Dhull, S. B., Sandhu, K. S., Punia, S., Kaur, M., Chawla, P., and Malik, A. (2020). Functional, thermal and rheological behavior of fenugreek (Trigonella foenum—graecum L.) gums from different cultivars: A comparative study. *International Journal of Biological Macromolecules, 159*, 406–414.

Dziki, D., Rozyło, R., Gawlik-Dziki, U., and Swieca, M. (2014). Current trends in the enhancement of antioxidant activity of wheat bread by the addition of plant materials rich in phenolic compounds. *Trends in Food Science & Technology*, *40*(1), 48–61.

Eden, W. T., Alighiri, D., Wijayati, N., and Mursiti, S. (2021). Synthesis of chalcone derivativefromclove leaf waste as a natural antioxidant. *Pharmaceutical Chemistry Journal*, *55*(3), 269–274.

El-Siddig, K. (2006). Fruits for the future 1 revised edition Tamarind *Tamarindus indica L.* United Kingdom: Southampton Centre for Underutilized Crops; p. 188.

Etika, S. B., and Nasra, E. (2020). Utilization of C-Cinnamal Calix [4] resorcinarene as adsorbent for methanil yellow. *Journal of Physics: Conference Series*, 1788, doi:10.1088/1742-6596/1788/1/012012.

Ettish, M. N., El-Sayyad, G. S., Elsayed, M. A., and Abuzalat, O. (2021). Preparation and characterization of new adsorbent from cinnamon waste by physical activation for removal of Chlorpyrifos. *Environmental Challenges*, 5, 100208.

Evidente, A., Fernández-Aparicio, M., Andolfi, A., Rubiales, D., and Motta, A. (2007). Trigoxazonane, a monosubstituted trioxazonane from Trigonella foenum-graecum root exudate, inhibits Orobanche crenata seed germination. *Phytochemistry*, *68*(19), 2487–2492.

Fahrina, A., Arahman, N., Mulyati, S., Aprilia, S., Nawi, N. I. M., Aqsha, A., Bilad, M. R., Takagi, R., and Matsuyama, H. (2020). *Development of polyvinylidene fluoride membrane by incorporating bio-based ginger extract as additive Polymers*, *12*, 1–11.

FAO. (2021). Onion production data. Food and Agriculture Organization of the United Nations. www.fao.org/faostat/en/#data/QC/visualize.

Farrell, K. T. (1999). Spices, condiments and seasonings. Westport, CT: The AVI Publishing Company, Inc.; pp. 60–63.

Florence, S. P., and Sowbhagya, V. (2006). Antioxidant potency of cumin and chili spent. Proceeding of the 18th Convention of Food Scientist and Technologist (ICFOST), India.

Gao, Y., Ozel, M. Z., Dugmore, T., Sulaeman, A., and Matharu, A. S. (2021). A biorefinery strategy for spent industrial ginger waste. *Journal of Hazardous Materials*, *401*, 123400.

Geetha, R., and Selvarani, K. (2017). A study of chilli production and export from India. *International Journal of Advance Research and Innovative Ideas in Education*, *3*(2), 205–210.

Gomez, S., Roy, S. K., and Pal, R. K. (2003). Primary processing of fenugreek (Trigonella foenum graecum L.)—An eco-friendly approach for convenience and quality. *Plant Foods for Human Nutrition*, *58*, 1–10.

Goni, P., López, P., Sánchez, C., Gómez-Lus, R., Becerril, R., and Nerín, C. (2009). Antimicrobial activity in the vapour phase of a combination of cinnamon and clove essential oils. *Food Chemistry*, *116*(4), 982–989.

Govindarajan, V. S., Shanthi, N., Raghuveer, K. G., and Lewis, Y. S. (1982). Cardamom production technology, chemistry, and quality. *CRC Critical Reviews in Food Science and Technology*, *16*, 229–326.

Griffiths, G., Trueman, L., Crowther, T., Thomas, B., and Smith, B. (2002). Onions-a global benefit to health. *Phytotheraphy Research*, *16*, 603–615.

Gupta, B. S., and Devendra, C. (1988). Availability and utilization of non-conventional feed resources and their utilization by non-ruminants in South Asia. Non-conventional feed resources and fibrous agricultural residues: Strategies for expanded utilization. Proceedings of a Consultation held in Hissar, India, 21–29 March, 62–75.

Hameed, B. H., and Ahmad, A. A. (2009). Batch adsorption of methylene blue from aqueous solution by garlic peel, an agricultural waste biomass. *Journal of Hazardous Materials*, *164*, 870–875.

Hammouti, B., Dahmani, M., Yahyi, A., Ettouhami, A., Messali, M., Asehraou, A., Bouyanzer, A., Warad, I., and Touzani, R. (2019). Black pepper, the "king of spices": Chemical composition to applications. *Arabian Journal of Chemical and Environmental Research*, *6*(1), 12–56.

Hebeish, A., Ragheb, A. A., Abdel-Thalouth, I., Yousef, M. A., and Mahmoud, R. M. (2017). Benign printing paste from the waste of tamarind seeds for textile printing. Part I: Basic Data. *Egyptian Journal of Chemistry*, *60*, 117–127.

Hernández-Varela, J.D., Chanona-Pérez, J.J., Resendis-Hernández, P., Gonzalez Victoriano, L., Méndez-Méndez, J.V., Cárdenas-Pérez, S., & Calderón Benavides, H.A. (2022). Development and characterization of biopolymers films mechanically reinforced with garlic skin waste for fabrication of compostable dishes. *Food Hydrocolloids*, 124, Part A, 107252. https://doi.org/10.1016/j.foodhyd.2021.107252.

Hendrix, K., Mathew, J., Hahn, L., and Sea, C. (2012). Defatted mustard seed meal-based biopolymer film development. *Food Hydrocolloids*, *26*(1), 118–125.

Hilles, A. R., and Mahmood, S. (2021). Historical background, origin, distribution, and economic importance of fenugreek. In: Naeem, M., Aftab, T., Masroor, M., and Khan, A., Fenugreek: Biology and Applications. Singapore: Springer; pp. 3–11.

Horiuchi, J. I., Kanno, T., and Kobayashi, M. (1999). New vinegar production from onions. *Journal of Bioscience and Bioengineering*, *88*(1), 107–109.

Huang, G., Liu, Y., Wu, X., and Cai, J. (2019). Activated carbons prepared by the KOH activation of a hydrochar from garlic peel and their CO2 adsorption performance. *New Carbon Materials*, *34*(3), 247–257.

Ifesan, B. O. T. (2017). Chemical composition of onion peel (*Allium cepa*) and its ability to serve as a preservative in cooked beef. *International Journal of Science and Research Methodology*, *7*(4), 25–34.

Inthalaeng, N., Gao, Y., Remon, J., Dugmore, T. I. J., Ozel, M. Z., Sulaeman, A., and Matharu, A. S. (2023). Ginger waste as a potential feedstock for a zero-wasteginger biorefinery: A review. *RSC Sustainability*, *1*, 213–223.

Jacob, J., Haponiuk, J. T., Thomas, S., Peter, G., and Gopi, S. (2018). Use of ginger nanofibers for the preparation ofcellulose nanocomposites and theirantimicrobial activities. *Fibers*, *6*(79), 2–11.

Jacob, J., Peter, G., Thomas, S., Haponiuk, J. T., and Gopi, S. (2019). In-situ synthesis and characterization of biocompatible magnetic ginger nanofiber composites for copper (II) removal from water. *Materials Today Communications*, *21*, 100690.

Jain, S. N., Sonawane, D. D., Shaikh, E. R., Garud, V. B., and Dawange, S. D. (2020). Vegetable residue of fenugreek (Trigonella Foenum-Graecum), waste biomass for removal of Basic Violet 14 from wastewater: Kinetic, equilibrium, and reusability studies. *Sustainable Chemistry and Pharmacy*, *16*, 100269.

Jakhetia, V., Patel, R., and Khatri P. (2010). Cinnamon: A pharmacologicalreview. *Journal of Advanced Scientific Research*, *1*(2), 19–12.

Jiang, X., Williams, K. M., Liauw, W. S., Ammit, A. J., Roufogalis, B. D., Duke, C. C., and McLachlan, A. J. (2005). Effect of ginkgo and ginger on the pharmacokinetics and pharmacodynamics of warfarin in healthy subjects. *British Journal of Clinical Pharmacology*, *59*(4), 425–432.

Joshi, V. K., and Sharma, S. K. (2011). Food processing waste management: Treatment and utilization technology. New Delhi: New India Publishing Agency; pp. 194–226.

Judith, E. C. S. S., Anantharaj, R., Ambedkar, B., and Dhanalakshmi, J. (2017). Treatment of synthetic turbid water using natural tamarind seeds atatmospheric conditions. *Research Journal of Pharmaceutical, Biological and Chemical Sciences*, *8*, 352–360.

Keerthanan, S., Jayampathi, T., Jayasinghe, C., and Vithanage, M. (2021). Cinnamon wood drived biochar for detoxifying sulfamethoxazole from aqueous solutions. Proceedings of the 8th international symposium on water quality and human health: Challenges ahead. Postgraduate Institute of Science.

Kim, H. M., Song, Y., Wi, S. G., and Bae, H.-J. (2017). Production of D-tagatose and bioethanol from onion waste by an intergrating bioprocess. *Journal of Biotechnology*, *260*, 84–90.

Kim, I. H., Yang, H. J., Noh, B. S., Chung, S. J., Min, S. C. (2012). Development of defatted mustard meal-based composed film and its application to smoked salmon to retard lipid oxidation. *Food Chemistry*, *133*(4), 1501–1509.

Kimoto-Nira, H., Ohashi, Y., Amamiya, M., Moriya, N., Ohmori, H., and Sekiyama, Y. (2019). Fermentation of onion (*Allium cepa* L.) peel by lactic acid bacteria for production of functional food. *Journal of Food Measurement and Characterization*, *1*(14), 142–149.

Konar, E. M., Harde, S. M., Kagliwal, L. D., and Singhal, R. S. (2013). Value-added bioethanol from spent ginger obtained after oleoresin extraction. *Industrial Crops and Products, 42*, 299–307.

Kotenkova, E. A., and Kupaeva, N. V. (2019). Comparative antioxidant study of onion and garlic waste and bulbs. IOP Conf. *Series: Earth and Environmental Science, 333*, 012031. doi:10.1088/1755-1315/333/1/012031.

Kumar, C. S., and Bhattacharya, S. (2008). Tamarind seed: Properties, processing and utilization. *Critical Reviews in Food Science and Nutrition, 48*(1), 1–20.

Kumar, S. G., Shetty, A. K., Sambaiah, K., and Salimath, P. V. (2005). Antidiabetic property of fenugreek seed mucilage and spent turmeric in streptozotocin-induced diabetic rats. *Nutrition Research, 25*(11), 1021–1028.

Kusuma, H. S., Aigbe, U. O., Ukhurebor, K. E., Onyancha, R. B., Okundaye, B., Simbi, I., Ama, O. M., Darmokoesoemo, H., Widyaningrum, B. A., Osibote, O. A., and Balogun, V. A. (2023). Biosorption of methylene blue using clove leaves waste modified withsodium hydroxide. *Results in Chemistry, 5*, 1–15.

Kuttigounder, D., Lingamallu, J. R., and Bhattacharya, S. (2011). Turmeric powder and starch: Selected physical, physicochemical, and microstructural properties. *Journal of Food Science, 76*(9), 1284–1291.

Laribi, B., Kouki, K., M'Hamdi, M., and Bettaieb, T. (2015). Coriander (*Coriandrum sativum L.)* and its bioactive constituents. *Fitoterapia, 103*, 9–26.

Lee, J. G., Chae, Y., Shin, Y., and Kim, Y. J. (2020). Chemical composition and antioxidant capacity of black pepper pericarp. *Applied Biological Chemistry, 63*(1), 1–9.

Liang, T. Q., Sun, G. H., Cao, L. L., Li, J., and Wang, L. J. (2018). Rheological behaviour of film-forming solutions and film properties from Artemisia sphaerocephalaKrasch gum and purple onion peel extract. *Food Hydrocolloids, 82*, 124–134.

Little, E. L., and Wadsworth, F. W. (1964). Common trees of Puerto Rico and the Virgin Islands. Washington: US Department of Agriculture; 557 p.

Luo, J., Huang, K., Zhou, X., and Xu, Y. (2020). Preparation of highly flexible and sustainable lignin-rich nanocellulose film containing xylonic acid (XA), and its application as an antibacterial agent. *International Journal of Biological Macromolecules, 163*, 1565–1571.

Ma, Y., Bao, H., Hu, X., Wang, R., and Dong, W. (2021). Productions of phenolic rich bio-oil using waste chilli stem biomass by catalytic pyrolysis: Evaluation of reaction parameters on products distributions. *Journal of the Energy Institute, 97*(8), 233–239.

Ma, Y., Chai, X., Bao, H., Huang, Y., and Dong, W. (2023). Study on nanocellulose isolated from waste chilli stems processing as dietary fiber in biscuits. *PLoS One, 18*(1), e0281142.

Maatta-Riihinen, K. R., Kahkonen, M. P., Torronen, A. R., and Heinonen, I. M. (2005). Catechins and procyanidins in berries ofvaccinium species and their antioxidant activity. *Journal of Agricultural and Food Chemistry, 53*(22), 8485–8491.

Magar, S. B., and Jadhav, M. V. (2018). Use of Herbal coagulants for treatment of dairy wastewater. *International Journal for Research Trends and Innovation, 3*(12), 24–34.

Maimulyanti, A., Prihadi, A. R., Rosita, T., and Safrudin, I. (2019). Adsorption and recovery of aroma compounds fromwastewater of clove oil distillation using coffee huskbiosorbent. *Science Asia, 45*(5), 446–451.

Maiti, S., Kumar Karan, S., Lee, J., Kumar Mishra, A., Bhusan Khatua, B., and Kon Kim, J. (2017). Bio-waste onion skin as an innovative nature-driven piezoelectric material with high energy conversion efficiency. *Nano Energy, 42*, 282–293.

Malhotra, S. K. (2005). Celery cultivation practices. (In Hindi). *NRCSS, Ajmer: Extension Folder* No. 8; pp. 1–4.

Malhotra, S. K. (2006). Celery. In: Handbook of Herbs and Spices. Cambridge: Woodhead Publishing; pp. 317–336.

Malhotra, S. K., and Singh, A. P. (2003). Medicinal properties of ginger. *Natural Product Radiance, 2*(6), 296–301.

Maniglia, B. C., de-Paula, R. L., Domingos, J. R., and Tapia-Blacido, D. R. (2015). Turmeric dye extraction residue for use in bioactive film production: Optimization of turmeric film plasticized with glycerol. *LWT–Food Science and Technology, 64*(2), 1187–1195.

Mathew, A. G. (2004). Future of spices and floral extract. *Indian Perfumer, 48,* 35–40.

Meghana, M., Nagarajappa, D. P., and Kumar, P. S. K. (2020). Treatment of dairy wastewater using tamarind kernel powder as a low-cost adsorbent. *International Research Journal of Engineering and Technology, 7*(8), 3810–3816.

Mehran, E., Farjami-Shayesteh, S., and Sheykhan, M. (2016). Structural and magnetic properties of turmeric functionalized $CoFe_2O_4$ nano composite powder. *Chinese Physics B, 25*(10), 107504.

Meyer, S. L., Zasada, I. A., Orisajo, S. B., and Morra, M. J. (2011). Mustard seed meal mixtures: Management of meloidogyne incognita on pepper and potential phytotoxicity. *Journal of Nematology, 43*(1), 7–15.

Michalak-Majewska, M., Teterycz, D., Muszyński, S., Radzki, W., and Sykut-Domańska, E. (2020). Influence of onion skin powder on nutritional and quality attributes of wheat pasta. *PLoS One, 1*(15), e0227942.

Milan, K. S. M., Dholakia, H., Kaul Tiku, P., and Prakash, V. (2008). Enhancement of digestive enzymatic activity by cumin (*Cuminum cyminum L.*) and role of spent cumin as a bio-nutrient. *Food Chemistry, 110*(3), 678–683.

Mohebali, S., Bastani, D., and Shayesteh, H. (2019). Equilibrium, kinetic and thermodynamic studies of a low-costbiosorbent for the removal of Congo red dye: Acid and CTAB-acidmodified celery (*Apium graveolens*). *Journal of Molecular Structure, 1176*, 181–193.

Mondal, N. K., Ghosh, P., Sen, K., Mondal, A., and Debnath, P. (2019). Efficacy of onion peel towards removal of nitrate from aqueous solution and field samples. *Environmental Nanotechnology, Monitoring & Management, 11*, 100222.

NABCONS. (2022). Study to determine post-harvest losses of agri produces in India. New Delhi: Ministry of Food Processing Industries; pp. 120–155.

Naqvi, S. A. Z., Irfan, A, Zahoor, A. F., Zafar, M., Maria, A., Chand, A. J., and Ashfaq, S. (2020). Determination of antimicrobial and antioxidant potential of agro-waste peels. *Annals of the Brazilian Academy of Sciences, 92*(2), e20181103. DOI: 10.1590/0001-3765202020181103.

Negi, P. S., Jayaprakasha, G. K., Jagan Mohan Rao, L., and Sakariah, K. K. (1999). Antibacterial activity of turmeric oil: A byproduct from curcumin manufacture. *Journal of Agricultural and Food Chemistry, 47*(10), 4297–4300.

Negi, R., Satpathy, G., Tyagi, Y. K., and Gupta, R. K. (2012). Biosorption of heavy metals by utilizing onion and garlic wastes. *International Journal of Environment and Pollution, 49*, 179–199.

Nile, S. H., Nile, A. S., Keum, Y. S., and Sharma, K. (2017). Utilization of quercetin and quercetin glycosides from onion (*Allium cepa L.*) solid waste as an antioxidant, urease, and xanthine oxidase inhibitors. *Food Chemistry, 235*, 119–126.

Noshirvani, N., Ghanbarzadeh, B., Gardrat, C., Rezaei, M. R., Hashemi, M., Le Coz, C., and Coma, V. (2017). Cinnamon and ginger essential oils to improve antifungal, physical and mechanical properties of chitosan-carboxymethyl cellulose films. *Food Hydrocolloids, 70*, 36–45.

Nurika, I., Mulyarto, A. R., and Afshari, D. K. (2007). Utilization of tamarind (tamarindus indica) seed for coagulation of tofu industry wastewater (study on tamarind seed powder concentration andstirring time). *Jurnal Teknologi Pertanian, 8*(3), 215–220.

O'Callahan, D. R., Singh, T., and McDonald, I. R. (2011). Evaluation of lactic acid bacterium from chilli waste as a potential antifungal agent for wood products. *Journal of Applied Microbiology, 112*(3), 436–442.

Olatide, M., Arawande, J. O., and George, O. O. (2019). Pilot study on chilli stalks as a source of non-dairy lactic acid bacteria in yogurt making. *Applied Food Science Journal, 3*(1), 5–8.

Oluba, O. M., Obi, C. F., Akpor, O. B., Ojeaburu, S. I., Ogunrotimi, F. D., Adediran, A. A., and Oki, M. (2021). Fabrication and characterization of keratin starch biocomposite film from chicken feather waste and ginger starch. *Scientific Reports, 11*(1), 8768.

Ozcelik, H., Tastan, Y., Terzi, E., and Sonmez, A. Y. (2020). Use of onion (*Allium cepa*) and garlic (*Allium sativum*) wastes for the prevention of fungal disease (*Saprolegnia* parasitica) on eggs of rainbow trout (*Oncorhynchus mykiss*). *Journal of Fish Diseases*. DOI: 10.1111/jfd.13229.

Paesa, M., Nogueira, D. P., Velderrain-Rodriguez, G., Esparza, I., Jiménez-Moreno, N., Mendoza, G., Osada, J., Martin-Belloso, O., Rodríguez-Yoldi, M. J., and Ancín-Azpilicueta, C. (2022). Valorization of onion waste by obtaining extracts rich in phenolic compounds and feasibility of its therapeutic use on colon cancer. *Antioxidants*, *11*, 2–18.

Pandharipande, S. L., and Rohith, P. (2013). Tamarind fruit shell adsorbent synthesis, characterization, and adsorption studies for removal of Cr (VI) & Ni(II) ions from aqueous solution. *International Journal of Engineering Science & Emerging Technologies*, *4*(2): 83–89.

Panak Balentić J, Ačkar Đ, Jokić S, Jozinović A, Babić J, Miličević B, Pavlović N. (2018). Cocoa shell: A by-product with great potential for wide application. *Molecules 23*(6), 1404.

Patel, D. K., Desai, S. N., Gandhi, H. P., Devkar, R. V., and Ramachandran, A. V. (2012). Cardio protective effect of Coriandrum sativum L. on isoproterenol induced myocardial necrosis in rats. *Food and Chemical Toxicology*, *50*(9), 3120–3125.

Patil, B. G., Gokhale, D. V., Bastawde, K. B., Puntambekar, U. S., and Patil, S. G. (1998). The use of tamarind waste to improve ethanol production fromcane molasses. *Journal of Industrial Microbiology & Biotechnology*, *21*, 307–310.

Patil, M., Jana, P., and Murumkar, C. (2021). Effect of onion and garlic biowaste on germination and growth of microgreens. *International Journal of Scientific Reports*, *7*(6), 302–305.

Peng, X., Cheng, K. W., and Ma, J. (2008). Cinnamon bark proanthocyanidinsas reactive carbonyl scavengers to prevent the formationof advanced glycation endproducts. *Journal of Agricultural and Food Chemistry*, *56*(6), 1907–1911,

Peter, K. V. (2001). Handbook of herbs and spices. New York, Washington, DC: Woodhead Publishing Limited, CRC Press; pp. 1–15.

Piechowiak, T., Grzelak-Blaszczyk, K., Bonikowski, R., and Balawejder, M. (2020). Optimization of extraction process of antioxidant compounds from yellow onion skin and their use in functional bread production. *LWT- Food Science and Technology*, *117*, 108614.

Prabhu, K. H., and Teli, M. D. (2011). Eco-dyeing using *Tamarindus indica* L. seed coat tannin as a natural mordant fortextiles with antibacterial activity. *Journal of Saudi Chemical Society*, *18*(6), 864–872.

Prajapati, N. D., Purohit, S. S., Sharma, A., and Kumar, T. (2003). A handbook of medicinal plants. Jodhpur, India: Agribios India; pp. 362–363.

Prasad, N. B. C., Gururaj, H. B., Kumar, V., Giridhar, P., Parimalan, R., Sharma, A., and Ravishankar, G. A. (2006). Influence of 8-methylnonenoic acid on capsaicin biosynthesis invivo and invitro cell cultures of Capsicum spp. *Journal of Agricultural Food Chemistry*, *54*(5), 1854–1859.

Pruthi, J. S. (1968). Review on the chemistry & quality evaluation of spices. *Part II. Black pepper. Journal of Indian Spices*, *5*, 11.

Pruthi, J. S. (1980). Spices and condiments: Chemistry, microbiology, technology. New York: Academia Press; pp. 1–499.

Pruthi, J. S. (1999). Quality assurance in spices and spice products, modern methods of analysis. New Delhi: Allied Publishers Ltd; pp. 72–89.

Pruthi, J. S. (2003). Advances in post-harvest processing technologies of Capsicum: Fixed chili seed oil. In De, A. K. (ed.), Advances in Post-Harvest Processing Technologies of Capsicum: Fixed Chili Seed Oil. New York: Taylor and Francis; pp. 175–213.

Pruthi, J. S., Bhat, A. V., and Varkey, A. G. (1974). Cinnamon leaf oil, Annual report CFTRI, India.

Puru, N. J., Ramalakshmi, and Sampathu, S. R. (2006). Antioxidant potential of large cardamom husk and spent powder. In: Proceedings of the 18th Convention of Food Science and Technologists (ICFOST) Dec.

Puru Naik, J., Jagan Mohan Rao, L., Mohan Kumar, T. M., and Sampathu, S. R. (2004). Chemical composition of the volatile oil from the pericarp (husk) of large cardamom (*Amomum subulatumRoxb.*). *Flavour and Fragrance Journal*, *19*(5), 441–444.

Puru Naik, J., Jagan Mohan Rao, L., and K N. Gurudutt. (1999). Anthocyanin Pigments of Large Cardamom (Amomum subulatum Roxb.) Pods. *J. Food Sci. Technol.*, *36*(4), 358–360.

Purseglove, J. W., Brown, E. G., Green, C. L., and Robin, S. R. J. (1981). In: Spices Vol. 2. New York: Longman Group Ltd.; pp. 606–608.

Putri, P. E., Mangisah, I., and Suthana, N. (2016). The effect of dietary supplementation of onion and garlic husk powder on protein, cholesterol and fat of duck meat. Proceedings of International Seminar on Livestock Production and Veterinary Technology, pp. 422–427. DOI: http://dx.doi.org/10.14334.

Qin, Y., and Horvath, A. (2021). Contribution of food loss to greenhouse gas assessment of high-value agricultural produce: California production, US consumption. *Environmental Research Letters*, *16*(1), 014024.

Rahmawati, D., Mansur, D., Sulaswatty, A., and Diastuti, H. (2022). Conversion of cardamom by-product into liquid smoke andbiochar by pyrolysis. *Malaysian Journal of Chemistry*, *24*(2), 293–301.

Rajathi, A. A., Sundarraj, A. A., Leslie, S., and Shree, M. P. (2017). Processing and medicinal uses of cardamom and ginger—a review. *Journal of Pharmaceutical Sciences and Research*, *9*(11), 2117–2122.

Raju, T. D., Reji, A. K., Raheem, N., Sasikumar, S., Vikraman, V., Shimil C. P., and Sneha, K. M. (2018). Role of *Moringa oleifera* and tamarind seed in water treatment. *International Journal of Engineering Research and Technology*, *7*(4), 453–462.

Ranasinghe, L. S., Jayawardena, B., and Abeywickrama, K. (2003). Use of waste generated from cinnamon bark oil (*Cinnamomum zeylanicum* Blume) extraction as a postharvest treatment for Embul banana. *Food, Agriculture &Enviroment*, *1*(2), 340–344.

Rane, P. S., and Gaikwad, S. T. (2019). Medicinal properties of onion and garlic: A review. *JETIR*, *6*(5), 50–58.

Rao, A. S., Kumar, A. A., and Ramana, M. V. (2015). Tamarind seed processing and by-products. *Agricultural Engineering International: CIGR Journal*, *17*(2), 200–204.

Rao, P. V., and Gan, S. H. (2014). Cinnamon: A multifaceted medicinal plant. *Evidence-Based Complementary and Alternative Medicine*. http://dx.doi.org/10.1155/2014/642942

Rao, T. N. R., Dwarkanath, C. T., and Johar, D. S. (1960). Analysis of piperine by spectrophotometric methods: Part I. *Journal and Proceedings of the Institution of Chemists, India*, *32*.

Ratri, P. J., Ayurini, M., Khumaini, K., and Rohbiya, A. (2020). Clove oil extraction by steam distillation and utilization of clove buds waste as potential candidate for eco-friendly packaging. *Jurnal Bahan AlamTerbarukan*, *9*(1), 47–54.

Reddy, J. P., and Rhim, J. W. (2014). Isolation and characterization of cellulose nanocrystals from garlic skin. *Materials Letters*, *129*, 20–23. https://doi.org/10.1016/j.matlet.2014.05.019.

Reddy, M. C. S. (2006). Removal of direct dye from aqueous solutions with an adsorbent made from tamarind fruit shell, an agricultural solid waste. *Journal of Scientific and Industrial Research*, *65*, 443–446.

Rijal, M. (2022). Application of eco-enzymes from nutmeg, clove, and eucalyptus plant waste in inhibiting the growth of *E. coli* and *S. aureus* In Vitro. *Journal of Biology Science & Education*, *11*(1), 31–44.

Rojas-Flores, S., De La Cruz-Noriega, M., Cabanillas-Chirinos, L., Nazario-Naveda, R., Gallozzo-Cardenas, M., Diaz, F., and Murga-Torres, E. (2023). Potential use of coriander waste as fuel for the generation of electric power. *Sustainability*, *15*(2), 896.

Ronke, R. A., and Saidat, O. G. A. G. (2016). Coagulation-flocculation treatment of wastewater using tamarind seed powder. *International Journal of Chemistry Technology and Research*, *9*(5), 771–780.

Rozylo, R., Piekut, J., Wójcik, M., Kozłowicz, K., Smolewska, M., Krajewska, M., and Bourekoua, H. (2021). Black cumin pressing waste material as a functional additive for starch bread. *Materials*, *14*(16), 4560.

Rupérez, P. and Toledano, G. (2003). Celery by-products as a source of mannitol. *Eur Food Res Technol*, *216*, 224–226. https://doi.org/10.1007/s00217-003-0663-x.

Narashans Alok Sagar, Anil Khar, Vikas, Ayon Tarafdar, Sunil Pareek, "Physicochemical and Thermal Characteristics of Onion Skin from Fifteen Indian Cultivars for Possible Food

Applications", *Journal of Food Quality*, vol. 2021, Article ID 7178618, 11 pages, 2021. https://doi.org/10.1155/2021/7178618.

Sagar, N. A., Pareek, S., and Gonzalez-Aguilar, G. A. (2020). Quantification of flavonoids, total phenols and antioxidant properties of onion skin: A comparative study of fifteen Indian cultivars. *Journal of Food Science & Technology*, 57(7), 2423–2432.

Sahu, M., Devi, S., Mishra, P., and Gupta, E. (2020). Mustard is a miracle seed to human health. In Ethnopharmacological Investigation of Indian Spices. USA: IGI Global; pp. 154–162.

Sajjad Khan, M., Salma, K., Deepak, M., Shivananda, B. G. (2006). Antioxidant activity of a new diarylheptanoid from Zingiber officinale. *Pharmacognosy magazine*, 2, 254–257.

Salatalohy, A., and Baguna, F. L. (2022). Potential and utilization of cinnamon logged waste as biopesticide. AGRIKAN—*JurnalAgribisnisPerikanan*, 15(2), 413–418.

Salunkhe, S., Chaudhary, B. U., Tewari, S., Meshram, R., and Kale, R. D. (2022). Utilization of agricultural waste as an alternative for packaging films. *Industrial Crops & Products*, 188, 115685.

Sampathu, S. R., Lakshminarayana, S., Sowbhagya, H. B., and Krishnamurthy, N. (2001). A process for the preparation of stable natural green colorant from the skin of fresh green pepper berries. Patent 386/DEL/2001.

Sandoval-Castro, C.J., Valdez-Morales, M., Oomah, B.D. et al. (2017). Bioactive compounds and antioxidant activity in scalded Jalapeño pepper industrial byproduct (Capsicum annuum). *J Food Sci Technol*, 54, 1999–2010. https://doi.org/10.1007/s13197-017-2636-2.

Santana, Á. L. D., Osorio-Tobón, J. F., Cárdenas-Toro, F. P., Steel, C. J., and Meireles, M. A. D. A. (2018). Partial-hydrothermal hydrolysis is an effective way to recover bioactives from turmeric wastes. *Food Science and Technology*, 38, 280–292.

Santana, Á. L., Zabot, G. L., Osorio-Tobón, J. F., Johner, J. C., Coelho, A. S., Schmiele, M., and Meireles, M. A. A. (2017). Starch recovery from turmeric wastes using supercritical technology. *Journal of Food Engineering*, 214, 266–276.

Sawicka, B., Kotiuk, E., Bienia, B., Krochmal-marczak, B., and Wójcik, S. (2007). The importance of mustard (*sinapis alba*) indian mustard (*brassica juncea* var. Sareptana) and black mustard (*brassica nigra*) in nutrition and phytotherapy. *Acta Scientiarum Polonorum. Agricultura*, 2(6), 17–27.

Sayed, H. S., Hassan, N. M. M., and Khalek, M. H. A. (2014). The effect of using onion skin powder as a source of dietary fiber and antioxidants on properties of dried and fried noodles. *Current Science International*, 3(4), 468–475.

SBI. (2023). Spice wise area and production, Spice Board of India. www.indianspices.com/. Accessed on July 28, 2023.

Sęczyk, L., Swieca, M., and Gawlik-Dziki, U. (2015). Nutritional and health-promoting properties of bean paste fortified with onion skin in the light of phenolic—food matrix interactions. *Food and Function*, 11(6), 3560–3566.

Sehgal, H., Jain, T., Malik, N., Chandra, A., and Singh, S. (2016). Isolation and chemical analysis of turmeric oil from rhizomes. In *Proceedings of the Chemical Engineering towards Sustainable Development, Chennai, India*, 27–30.

Sehwag, S., and Das, M. (2015). A brief overview: Present status on utilization of mustard oil and cake. *Indian Journal of Traditional Knowledge*, 14(2), 244–250.

Shabir, I., Pandey, V. K., Dar, A. H., Pandiselvum, R., Manzoor, S., Mir, S. A., Shams, R., Dash, K. K., Fayaz, U., Khan, S. A., Jeevarathinam, G., Zhang, Y., Rusu, A. V., and Trif, M. (2022). Nutritional profile, phytochemical compounds, biological activities, and utilisation of onion peel for food applications: A review. *Sustainability*, 14, 2–15.

Shalaby, M. E., and El-kot, G. A. (2009). Management of allium white rot caused by *Sclerotium cepivorum* by using compost of certain plant wastes. *Journal of Agricultural Science Mansoura University*, 34(5): 4255–4267.

Shankaracharya, N. B., Puru Naik, J., and Narayan, M. S. (1997). Physico-chemical characteristics and quality of large cardamom *Amomum subulatumRoxb. Beverage and Food World*, 24(1), 33–38.

Sharma, K., Mahato, N., Nile, S. H., Lee, E. T., and Lee, Y. R. (2016). Economical and environment-friendly approaches for usage of onion (*Allium cepa* L.) wastes. *Food & Function, 8*(7), 3354–3369.

Sharma, L. K., Agarwal, D., Rathore, S. S., Malhotra, S. K., and Saxena, S. N. (2016). Effect of cryogenic grinding on volatile and fatty oil constituents of cumin (Cuminum cyminum L.) genotypes. *Journal of Food Science and Technology, 53*, 2827–2834.

Sharma, R., Kamboj, S., Khurana, R., Singh, G., and Rana, V. (2015). Physicochemical and functional performance of pectin extracted by QbD approach from Tamarindus indica L. pulp. *Carbohydr. Polym, 134*(10), 364–374.

Shu, J., Yin, Y., and Liu, Z. (2023). Integrated processes turning pepper sauce waste into valuable by-products. *Foods, 12*(1), 67.

Shukla, P. V., Bhalerao, T. S., and Ingle, S. T. (2010). Comparative study of biogas production from different food wastes. *Journal of Environmental Research and Development, 4*(4), 958–963.

Sijabat, P. S., and Siregar, Y. (2021). Study of distillation waste by clove for alternative fuel power plant: A review. In: IOP Conference Series: Materials Science and Engineering. Bristol: IOP Publishing; p. 12081.

Silva, E. I. G., Silva, J. B. D., Albuquerque, J. M., and Messias, C. M. B. D. O. (2022). Utilizing tamarind residues in the São Francisco valley: Food and nutritional potential. *Ciência Rural, 52*, e20210708.

Singh, R., and Srivastava, S. (2017). Application of natural dye obtained from peel of black cardamom on silk fabric. *International Journal of Home Science, 3*(2), 94–96.

Singh, T., and Chittenden, C. (2008). In-vitro antifungal activity of chilli extracts in combination with Lactobacillus casei against common sapstain fungi. *International Biodeterioration and Biodegradation, 62*(4), 364–367.

Skerget, M., Majheniè, L., Bezjak, M., and. Knez, Z. (2009). Antioxidant, radical scavenging and antimicrobial activities of red onion (allium cepa l) skin and edible part extracts. *Chemical and Biochemical Engineering Quarterly, 23*(4), 435–444.

Sowbhagya, H. B. (2006). Newer chemical and technological approaches for the preparation of spice flavoring from selected spices. Ph.D. thesis, University of Mysore, India.

Sowbhagya, H. B. (2019). Value-added processing of by-products from spice industry. *Food Quality and Safety, 3*(2), 73–80.

Sowbhagya, H. B., Mahadevamma, S., Indrani D., and Srinivas, P. (2011). Physicochemical and microstructural characteristics of celery seed spent residue and in fluence of its addition on quality of biscuits. *Journal of Texture Studies, 42*(5), 369–376.

Sowbhagya, H. B., Suma, P. F., Mahadevamma, S., and Tharanathan, R. N. (2007). Spent residue from cumin—a potential source of dietary fiber. *Food Chemistry, 104*(3), 1220–1225.

Srinivasan, K. (2018). Cumin (*Cuminum cyminum*) and black cumin (*Nigella sativa*) seeds: Traditional uses, chemical constituents, and nutraceutical effects. *Food Quality and Safety, 2*(1), 1–16.

Srivastava, D., Rajiv, J., Mahadevamma, M. M. N., Puru Naik, J., and Srinivas, P. (2012). Effect of fenugreek seed husk on the rheology and quality characteristics of muffins. *Food and Nutrition, 3*, 1473–1479.

Sudarni, D. H. A., Aigbe, U. O., Ukhurebor, K. E., Onyancha, R. B., Kusuma, H. S., Darmokoesoemo, H., and Widyaningrum, B. A. (2021). Malachite green removal by activated potassium hydroxide clove leaf agrowaste biosorbent: Characterization, kinetic, isotherm, and thermodynamic studies. *Adsorption Science and Technology, 2021*, 1–15.

Suh, H. J., Lee, J. M., Cho, J. S., Kim, Y. S., and Chung, S. H. (1999). Radical scavenging compounds in onion skin. *Food Research International, 32*, 659–664.

Swati, S., Sehwag, S., and Das, M. (2015). A brief overview: Present status on utilization of mustard oil and cake. *Indian Journal of Traditional Knowledge, 14*(2), 244–250.

Tandon, G.L., Dravid, S.V., and Siddappa, G.S. (1964). Oleoresin of Capsicum (red chillies) – some technological and chemical aspects. *Journal of Food Science, 29*, 1–5.

Taqui, S. N., Yahya, R., Hassan, A., Khanum, F., and Syed, A. A. (2019). Valorization of nutraceutical industrial coriander seed spent by the process of sustainable adsorption system of acid black 52 from aqueous solution. *International Journal of Environmental Research, 13*, 639–659.

Tonyali, B., and Sensoy, I. (2017). The effect of onion skin powder addition on extrudate properties. *Acta Horticulturae, 1152*, 393–398.

Tripathi, M. K., Mishra, A. S., Misra, A. K., and Prasad, R. (2003). Effect of graded levels of high glucosinolate mustard (brassica júncea) meal inclusion on nutrient utilization, growth performance, organ weight, and carcass composition of growing rabbits. *World Rabbit Science, 11*(4), 211–226.

Tung, Y. T., Chua, M. T., Wang, S. Y., and Chang, S. T. (2008). Antiinflammation activities of essential oil and its constituents from indigenous cinnamon (*Cinnamomum osmophloeum*) twigs. *Bioresource Technology, 99*(9), 3908–3913.

Tung, Y. T., Yen, P. L., Lin, C. Y., and Chang, S. T. (2010). Anti-inflammatory activities of essential oils and their constituents from different provenances of indigenous cinnamon (*Cinnamomum osmophloeum*) leaves. *Pharmaceutical Biology, 48*(10), 1130–1136.

Uitterhaegen, E., Parinet, J., Labonne, L., Mérian, T., Ballas, S., Véronèse, T., and Evon, P. (2018). Performance, durability and recycling of thermoplastic biocomposites reinforced with coriander straw. *Composites Part A: Applied Science and Manufacturing, 113*, 254–263.

Vedashree, M., Pradeep, K., Ravi, R., and Madhava, N. M. (2016). Turmeric spent flour: Value addition to breakfast food. *International Journal of Nutritional Sciences, 1*(2), 1–5.

Verma, M., Pandey, J., Joshi, S., and Mitra, A. (2021). Turmeric production, composition and its non-conventional uses-a review. *The Pharma Innovation Journal, 10*(12), 2757–2762.

Verma, R. K., Kumari, P., Maurya, R. K., Kumar, V., Verma, R. B., and Singh, R. K. (2018). Medicinal properties of turmeric (*Curcuma longa* L.)—A review. *International Journal of Chemical Studies, 6*(4), 1354–1357.

Wani, S. A., and Kumar, P. (2018). Fenugreek: A review on its nutraceutical properties and utilization in various food products. *Journal of the Saudi Society of Agricultural Sciences, 17*(2), 97–106.

Yadav, S. K., and Sehgal, S. (1997). Effect of home processing and storage on ascorbic acid and β-carotene content of bathua (Chenopodium album) and fenugreek (Trigonella foenum graecum) leaves. *Plant Foods for Human Nutrition, 50*, 239–247.

Yakuth, S. A., Taqui, S. N., Syed, U. T., and Syed, A. A. (2021). Nutraceutical industrial chillies stalk waste as a new adsorbent for the removal of Acid Violet 49 from water and textile industrial effluent: Adsorption isotherms and kinetic models. *Desalination and Water Treatment, 155*, 94–112.

Yalcin, S. K. (2012). Enhancing citric acid production of *Yarrowialipolytica* by mutagenesis and using natural media containing carrot juice and celery byproducts. *Food Science and Biotechnology, 21*(3), 867–874.

Yang, Y. L., Al-Mahdy, D. A., Wu, M. L., Zheng, X. T., Piao, X. H., Chen, A. L., Wang, A. M., Yang, Q., and Ge, Y. W. (2022). LC-MS-based identification and antioxidant evaluation of small molecules from the cinnamon oil extraction waste. *Food Chemistry, 366*, 130576.

Yardimci, B., and Kanmaz, N. (2023). An effective-green strategy of methylene blue adsorption: Sustainable and low-cost waste cinnamon bark biomass enhanced via MnO_2. *Journal of Environmental Chemical Engineering, 11*(3), 110254.

Yusuff, A. S. (2022). Adsorptive removal of lead and cadmium ions from aqueous solutions by aluminium oxide modified onion skin wastes: Adsorbent characterization, equilibrium modelling and kinetic studies. *Energy & Environment, 33*(1), 152–169.

Zakaria, S. B., Zahari, M. S. B., and Hisamudin, S. Z. B. (2023). Development and characterization of hybrid liquid fertilizer from celery and cucumber wastes. *Materials Today: Proceedings, 75*, 116–122.

Zendrato, H. M., Devi, Y. S., Masruchin, N., &Wistara, N. J. (2021). Soda pulping of torch ginger stem: Promising source of nonwood-based cellulose. *Journal of the Korean Wood Science and Technology, 49*(4), 287–298.

Zhivkova, V. (2021). Determination of nutritional and mineral composition of wasted peels from garlic, onion and potato. *Carpathian Journal of Food Science and Technology, 13*(3), 134–146.

9

Value Addition of Confectionery Processing Industrial Waste

M. Selvamuthukumaran

9.1 Introduction

Confectionery industries generate ample amounts of solid and liquid wastes, which contain several organic compounds (Beal and Raj, 2000; Das et al., 2013; El-Kassas et al., 2015; Lafitte-Trouque and Forster, 2000). These wastes comprise valuable compounds like sugars, proteins, oils, food colors and flavors. Currently, they are being thrown away or discarded. They can be converted into aquatic feeds, bioenergy and several other value-added products (Beal and Raj, 2000; Das et al., 2013; El-Kassas et al., 2015; Lafitte-Trouque and Forster, 2000). They can be effectively utilized by means of value addition from a circular economical point of view (Karmee et al., 2015). The wastes can be effectively valorized to produce biodiesel.

9.2 Preparation of Biodiesel from Confectionery Industrial Wastes

Biodiesel can be manufactured from oils, fats and grease. Traditionally only edible feedstocks were efficiently used for commercial biodiesel production. But researchers have reported that biodiesel can also be effectively produced from non-edible waste, like sewage sludge, organic waste and food waste, which is a cheaper source for producing such biodiesel (Karmee and Chadha, 2005; Karmee and Lin, 2014). The effluent streams of confectionery processing industries can be valorized to produce biofuel, which contains oils and carbohydrates.

Biodiesel was developed from the effluent stream of lollipops. The oil from the effluent stream was extracted by using solvents like ethyl acetate and n-hexane. Around 18% oil was obtained from lollipop effluent streams. Sodium sulphate was used to remove moisture from the oil. The oil can also be produced by solvent-free methods, such as using enzyme lipase along with chemicals like $Ca(OH)_2$, CaO and KOH. Several researchers reported that the enzymatic

224   DOI: 10.1201/9781003269199-9

method is highly advantageous for generating biodiesel from confectionery industrial effluents, i.e., waste oil (Karmee, 2015, 2016a & 2016b) and for this purpose the immobilized enzymes obtained from *Candida antartica* lipase-B were found to be most suitable. Biodiesel preparation via lipase catalysis is advantageous since it is substrate specific, moisture tolerant, recyclable and operates under mild experimental conditions (Karmee, 2015, 2016a & 2016b).

9.3 Wastes from Cocoa Processing

The wastes generated from cocoa processing include pod husk, mucilage and bean shell. The pod husk contributes around 70–80% of the waste. These pod husks contains fiber and several bioactive constituents (Oddoye et al., 2013; Campos-Vega et al., 2018; Lu et al., 2018). White mass, which is covering the cocoa bean that can produce turbid liquid during its fermentation. The researchers found that this mucilage is an ample source of minerals with sugars that can be effectively used as a growing medium for certain microorganisms even commercially (Delgado-Ospina et al., 2021; Vásquez et al., 2019). The bean shell can contribute around 20% of the whole cocoa bean weight. They are currently being discarded or utilized as animal feed and fertilizers. They contain fats, fibers, antioxidants, vitamins, carbohydrates etc. The cocoa bean shell is the major by-product of chocolate processing industries that can be separated from the cotyledons either during roasting or after the roasting process (Balenti'c et al., 2018). They can be used to obtain several value-added products and have wide application in the food, pharmaceutical and cosmetic industries (Cooper et al., 2008; Okiyama et al., 2017; Rojo-poveda et al., 2020; Picchioni et al., 2020).

9.4 Production of Poly(3-hydroxybutyrate) from Cocoa Bean Shell

Cocoa bean substrate has been utilized as a fermentatable substrate to produce poly(3-hydroxybutyrate) by using the strain *Bacillus firmus* (Figure 9.1). Sánchez et al. (2023) conducted acid thermal hydrolysis of cocoa bean shell at a temperature of 135°C for 10 min to obtain broth that comprised a higher amount of fermentable sugars. They obtained around 107 mg of poly (3-hydroxybutyrate) from non-centrifuged solid broth. Their results concluded that cocoa bean shell can be effectively used to produce poly(3-hydroxybutyrate), which can be an alternative space to synthetic polypropylene.

9.5 Ingredients from Cocoa Bean Shell

The coca bean shell contains appreciable amounts of polyphenols (Table 9.1) and dietary fiber, therefore they can be effectively used as a food additive in food production (Figure 9.1). The soluble dietary fiber obtained from cocoa bean shell can be incorporated during muffin preparation, which can

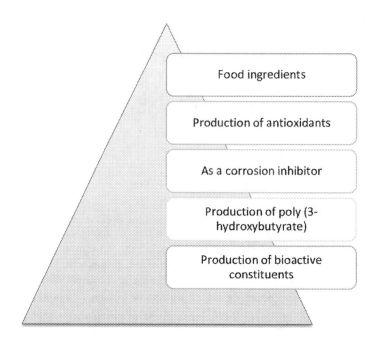

Figure 9.1 *Utilization of cocoa bean shells.*

Table 9.1 Bioactive Constituents of Cocoa Bean Shell

Name of the constituent	Content	Reference
Total tannins	2.3–25.3 mg CE/g	Nsor-Atindana et al. (2012); Barbosa-Pereira et al. (2018)
Total phenols	22–100 mg GAE/g	Nsor-Atindana et al. (2012); Barbosa-Pereira et al. (2018)
Total flavonoids	7.5–21.8 mg RU/g 1.6–43.9 mg CE/g	Nsor-Atindana et al. (2012); Barbosa-Pereira et al. (2018)
Dietary fibers	13.8–65.6 g/100 g	Rojo-Poveda et al. (2019); Agus et al. (2018)
Proteins	18.2–27.4 g/100 g	Rojo-Poveda et al. (2019); Agus et al. (2018)

successfully replace the use of 50–70% of vegetable oil during muffin production. The resultant product exhibited good texture with pleasant color and aroma (Martínez-Cervera et al., 2011; Öztürk and Ova, 2018).

9.6 Bioactive Components from Cocoa Bean Shell

The extracts obtained from cocoa bean shell found to contain dietary fiber, lipids, proteins and secondary metabolites (Sánchez et al., 2010). It contains various bioactive peptide fractions like albumin and vicilin, which exhibited strong anti-diabetic, antioxidant and anti-obesogenic activities (Domínguez-Pérez et al., 2020). The presence of phytosterols like campsterol showed anti-tumor, anti-bacterial and anti-inflammatory activities with a blood cholesterol lowering effect (Younes et al., 2022).

9.7 Utilizing Cocoa Bean Shell as a Corrosion Inhibitor

The synthetic inhibitors utilized in industries were deemed to be toxic and very expensive. The cocoa bean shell was found to contain bioactive compounds like tannins and flavonoids, which possess corrosion inhibitory properties (Singh et al., 2012; Rani and Basu, 2012). De Carvalho et al. (2021) tested various concentrations of cocoa bean shell powder and 80% (v/v) of hydroalcoholic cocoa bean shell extracts as a corrosion inhibitor for SAE 1008 carbon steel in a sodium chloride solution. Their results showed that as the inhibitor concentration of cocoa bean shell extracts increases the corrosion rate of steel was reduced, and they further concluded that corrosion inhibitory efficiency of around 96% was achieved by using the lowest inhibitory concentration of cocoa bean shell extract at 0.4 g/L, showing cocoa bean shell as an effective corrosion inhibitor for carbon steel.

9.8 Conclusions

Biodiesel can be developed from the effluent stream of lollipops. Cocoa bean shell can be used to produce antioxidants, food ingredients like dietary fiber, which can replace fat, and it can also be used as a corrosion inhibitor in steel industries, thereby efficiently replacing the synthetic ones. It is also used for producing poly(3-hydroxybutyrate), an alternative to synthetic polymer usage in packaging industries.

References

Agus, B. A. P.; Mohamad, N. N.; Hussain, N. Composition of Unfermented, Unroasted, Roasted Cocoa Beans and Cocoa Shells from Peninsular Malaysia. Journal of Food Measurement Characteristics 2018, 12, 2581–2589.

Balentić, J. P.; Ačkar, Đ.; Jokić, S.; Jozinović, A.; Babić, J.; Miličević, B.; Šubarić, D.; Pavlović, N. Cocoa Shell: A by-Product with Great Potential for Wide Application. Molecules 2018, 23, 1404.

Barbosa-Pereira, L.; Guglielmetti, A.; Zeppa, G. Pulsed Electric Field Assisted Extraction of Bioactive Compounds from Cocoa Bean Shell and Coffee Silverskin. Food Bioprocess Technology 2018, 11, 818–835.

Beal, L. J.; Raj, R. D. Sequential Two-Stage Anaerobic Treatment of Confectionery Wastewater. Journal of Agricultural Engineering Research 2000, 76, 211–217. https://doi.org/10.1006/jaer.2000.0555

Campos-Vega, R.; Nieto-Figueroa, K. H.; Oomah, B. D. Cocoa (Theobroma cacao L.) Pod Husk: Renewable Source of Bioactive Compounds. Trends in Food Science and Technology 2018, 81, 172–184.

Cooper, K. A.; Donovan, J. L.; Waterhouse, A. I.; Williamson, G. Cocoa and Health: A Decade of Research. British Journal of Nutrition 2008, 99, 1–11.

Das, B. K.; Gauri, S. S.; Bhattacharya, J. Sweetmeat Waste Fractions as Suitable Organic Carbon Source for Biological Sulfate Reduction. International Biodeterioration & Biodegradation 2013, 82, 215–223. https://doi.org/10.1016/j.ibiod.2013.03.027

de Carvalho, M. C. F.; E Silva, I. M. F. C. R.; Macedo, P. L. A.; Tokumoto, M. S.; da Cruz, R. S.; Capelossi, V. R. Assessment of the Hydroalcoholic Extract and Powder Cocoa Bean Shell as Corrosion Inhibitors for Carbon Steel in Sodium Chloride Solution. Review Materials 2021, 26, 1–17.

Delgado-Ospina, J.; Lucas-González, R.; Viuda-Martos, M.; Fernández-López, J.; Pérez-Álvarez, J. Á.; Martuscelli, M.; ChavesLópez, C. Bioactive Compounds and Techno-Functional Properties of High-Fiber Co-Products of the Cacao Agro-Industrial Chain. Heliyon 2021, 7, e06799.

Domínguez-Pérez, L. A.; Beltrán-Barrientos, L. M.; González-Córdova, A. F.; Hernández-Mendoza, A.; Vallejo-Cordoba, B. Artisanal Cocoa Bean Fermentation: From Cocoa Bean Proteins to Bioactive Peptides with Potential Health Benefits. Journal of Functional Foods 2020, 73, 104134.

El-Kassas, H. Y.; Heneash, A. M. M.; Hussein, N. R. Cultivation of Arthrospira (Spirulina) Platensis Using Confectionary Wastes for Aquaculture Feeding. Journal of Genetic Engineering and Biotechnology 2015, 13(2), 145–155. https://doi.org/10.1016/j.jgeb.2015.08.003

Karmee, S. K. Lipase Catalyzed Synthesis of Fatty Acid Methyl Esters from Crude Pongamia Oil. Energy Sources Part A 2015, 37, 536–542. https://doi.org/10.1080/15567036.2011.572131

Karmee, S. K. Liquid Biofuels from Food Waste: Current Trends, Prospect and Limitation. Renewable and Sustainable Energy Review 2016a, 53, 945–953. https://doi.org/10.1016/j.rser.2015.09.041

Karmee, S. K. Preparation of Biodiesel from Nonedible Plant Oils Using a Mixture of Used Lipases. Energy Sources Part A 2016b, 38(18), 2727–2733. https://doi.org/10.1080/15567036.2015.1098748

Karmee, S. K.; Chadha, A. Preparation of Biodiesel from Crude Oil of Pongamia Pinnata. Bioresource Technology 2005, 96, 1425–1429. https://doi.org/10.1016/j.biortech.2004.12.011

Karmee, S. K.; Lin, C. S. K. Lipids from Food Waste as Feedstock for Biodiesel Production: Case Hong Kong. Lipid Technology 2014, 26, 206–209. https://doi.org/10.1002/lite.201400044

Lafitte-Trouque, S.; Forster, C. F. Dual Anaerobic Co-Digestion of Sewage Sludge and Confectionery Waste. Bioresource Technology 2000, 71, 77–82. https://doi.org/10.1016/S0960-8524(99)00043-7

Lu, F.; Rodriguez-Garcia, J.; Van Damme, I.; Westwood, N. J.; Shaw, L.; Robinson, J. S.; Warren, G.; Chatzifragkou, A.; Mason, S. M.; Gomez, L.; et al. Valorisation Strategies for Cocoa Pod Husk and Its Fractions. Current Opinion in Green and Sustainable Chemistry 2018, 14, 80–88.

Martínez-Cervera, S.; Salvador, A.; Muguerza, B.; Moulay, L.; Fiszman, S. M. Cocoa Fibre and Its Application as a Fat Replacer in Chocolate Muffins. Lebensmittel Wissenchaft and Technology 2011, 44, 729–736.

Nsor-Atindana, J.; Zhong, F.; Mothibe, K. J.; Bangoura, M. L.; Lagnika, C. Quantification of Total Polyphenolic Content and Antimicrobial Activity of Cocoa (Thebroma cacao L.) Bean Shells. Pakistan Journal of Nutrition 2012, 11, 672–677.

Oddoye, E.; Agyente-Badu, C.; Gyedu-Akoto, E. Cocoa and Its by-Products: Identification and Utilization. Chocolate Health and Nutrition 2013, 7, 23–37.

Okiyama, D. C. G.; Navarro, S. L. B.; Rodrigues, C. E. C. Cocoa Shell and Its Compounds: Applications in the Food Industry. Trends in Food Science and Technology 2017, 63, 103–112.

Öztürk, E.; Ova, G. Evaluation of Cocoa Bean Hulls as a Fat Replacer on Functional Cake Production. Turkish Journal of Agriculture—Food Science and Technology 2018, 6, 1043.

Picchioni, F.; Warren, G. P.; Lambert, S.; Balcombe, K.; Robinson, J. S.; Srinivasan, C.; Gomez, L. D.; Faas, L.; Westwood, N. J.; Chatzifragkou, A.; et al. Valorisation of Natural Resources and the Need for Economic and Sustainability Assessment: The Case of Cocoa Pod Husk in Indonesia. Sustainability 2020, 12, 8962.

Rani, B. E. A.; Basu, B. B. J. Green Inhibitors for Corrosion Protection of Metals and Alloys: An Overview. International Journal of Corrosion 2012, 2012, 1–15.

Rojo-Poveda, O.; Barbosa-Pereira, L.; Mateus-Reguengo, L.; Bertolino, M.; Stévigny, C.; Zeppa, G. Effects of Particle Size and Extraction Methods on Cocoa Bean Shell Functional Beverage. Nutrients 2019, 11, 867.

Rojo-poveda, O.; Barbosa-pereira, L.; Zeppa, G.; St, C. Cocoa Bean Shell—A by-Product with Nutritional Properties and Biofunctional Potential. Nutrients 2020, 12, 1123.

Sánchez, D.; Moulay, L.; Muguerza, B.; Quiñones, M.; Miguel, M.; Aleixandre, A. Effect of a Soluble Cocoa Fiber-Enriched Diet in Zucker Fatty Rats. Journal of Medicinal Foods 2010, 13, 621–628.

Sánchez, M.; Laca, A.; Laca, A.; Díaz, M. Cocoa Bean Shell as Promising Feedstock for the Production of Poly (3-hydroxybutyrate) (PHB). Applied Science 2023, 13, 975. https://doi.org/10.3390/app13020975

Singh, A.; Ebenso, E. E.; Quraishi, M. A. Corrosion Inhibition of Carbon Steel in HCl Solution by Some Plant Extracts. International Journal of Corrosion 2012, 2012, 1–20.

Vásquez, Z. S.; de Carvalho Neto, D. P.; Pereira, G. V. M.; Vandenberghe, L. P. S.; de Oliveira, P. Z.; Tiburcio, P. B.; Rogez, H. L. G.; Góes Neto, A.; Soccol, C. R. Biotechnological Approaches for Cocoa Waste Management: A Review. Waste Management 2019, 90, 72–83.

Younes, A.; Li, M.; Karboune, S. Cocoa Bean Shells: A Review into the Chemical Profile, the Bioactivity and the Biotransformation to Enhance Their Potential Applications in Foods. Critical Reviews in Food Science and Nutrition 2022, 2, 1–25.

10

Value Addition of Dairy Processing Industrial Waste

Arjun Mohanakumar, Rohini Vijay Dhenge,
Amar Shankar, and Anandhu T S

10.1 Introduction

10.1.1 Overview of Dairy Industry by FAO

The global milk production for the year 2022 was 930 million tonnes, showing a slight increase of 0.6% when compared to last year 2021. This growth is primarily attributed to the expansion of dairy production in Asia, along with a small rise in Central America and the Caribbean. However, Europe is expected to experience a significant decline in milk output. Additionally, South America, Oceania, and Africa are anticipated to witness moderate decreases in production, while North America's milk production is predicted to remain stable. The dairy industry has become one of the most substantially growing industries in India over the years. As per the report of the Department of Animal Husbandry and Dairying, milk production would reach between 200 and 210 million tonnes. In India huge amount of dairy industries were existing in unorganized form except the dairy industries, which are run by state and central level organizations. The existence of cooperative sectors in milk processing is more and they implement good standardization across milk procurement, production, testing, and marketing; the organized sector can collect milk surplus from the unorganized sectors. At present, there are several training centres that have been set up by state governments for educating farmers on the best dairy and animal husbandry practices. The co-operative dairy industry under Anand Pattern follows standard practices for milk procurement, production, and marketing. These organizations are considered as the well-established dairy industry. Most of the co-operative milk producers follow a three-tier system, which includes primary societies of farmers at the village level; regional producers who convert raw milk into marketable milk and milk products at the middle level; and the main state organization, which consists of a chairman, board members from different regions, and a farmer representative at the apex level. The dairy processing industry is a critical component of the global

230

DOI: 10.1201/9781003269199-10

food and beverage sector, responsible for providing consumers with a range of high-quality dairy products.

In terms of regional distribution, the global milk production forecast for 2022 indicates variations across different parts of the world (Figure 10.1). Asia is expected to drive the overall increase in milk production, with significant volume expansions. Central America and the Caribbean are also projected to experience a slight growth in milk output. However, Europe is anticipated to face a substantial decline in milk production. Meanwhile, South America, Oceania, and Africa are expected to observe moderate decreases in milk production. On the other hand, North America's milk production is forecasted to remain steady, with no significant changes anticipated. To summarize, milk production in Asia and certain regions like Central America and the Caribbean is expected to contribute to the overall global increase, while Europe faces a decline. South America, Oceania, and Africa are also expected to witness moderate decreases, while North America's production is predicted to remain stable.

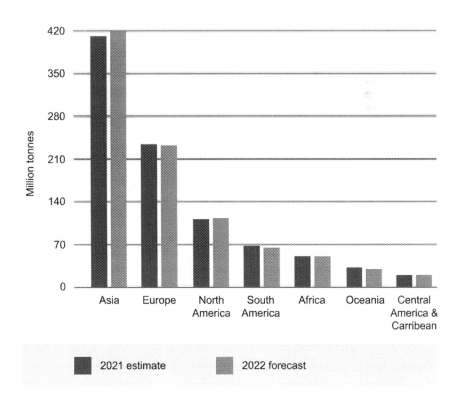

Figure 10.1 World milk production region wise (FAO, 2022).

10.1.2 Importance of Managing and Valorizing Dairy Processing Industrial Waste

Mirabella et al. (2013), reviewed the trends of waste valorization in the food manufacturing industry. They stated that food waste is a pervasive issue that encompasses the entire life cycle of food, starting from agricultural production and extending to industrial manufacturing, processing, retail, and household consumption. In developed countries, the distribution of food waste is as follows: 42% is generated by households, 39% occurs in the food manufacturing industry, 14% in the food service sector, and the remaining 5% in retail and distribution. To address this challenge, concepts rooted in industrial ecology, such as cradle to cradle and circular economy, are gaining prominence as guiding principles for eco-innovation. The objective is to achieve a "zero waste economy" where waste materials are repurposed as raw materials for new products and applications. The substantial amount of waste generated by the food industry not only represents a significant loss of valuable resources but also poses considerable management problems, both economically and environmentally. However, many of these waste materials have the potential to be effectively reused in other production systems, particularly through the utilization of biorefineries. The present study focuses specifically on the utilization of food waste originating from food manufacturing (FWm). Through an extensive review of existing literature, the authors examine the feasibility and limitations of implementing industrial symbiosis to recover waste from food processing. The emphasis is on recycling (excluding energy recovery) of solid and liquid waste generated by the food processing industry. The dairy industry is widely recognized as a significant contributor to wastewater generated during food processing activities in various countries. As a result of stricter wastewater discharge regulations, there is an escalating need for more rigorous treatment processes. Therefore, it is crucial to continually assess the environmental impact of the dairy industry and its implications on the surrounding ecosystems (Kasmi et al., 2018). Dairy effluents are characterized by several significant parameters, with the organic load being one of the key factors. This organic load primarily stems from the carbohydrates and proteins present in milk. Additionally, the content of fats, suspended solids, and nutrients also contribute to the overall level of contamination in dairy wastewater, exacerbating its environmental impact (Prazeres et al., 2012). Consequently, the discharge of dairy wastewaters into the environment can lead to various detrimental effects such as increased oxygen consumption, impairment of natural permeability, eutrophication, and potential toxicity. These impacts pose significant risks to the receiving ecosystems and their overall ecological health. Apart from the environmental concerns associated with the discharge of dairy wastewaters, the inclusion of products like milk in the waste stream is considered a loss of valuable resources. In order to mitigate these issues, dairy wastewaters are typically subjected to treatment processes that involve a combination of physico-chemical and biological methods. These treatment techniques are aimed at reducing the contaminants and pollutants present in the wastewater, promoting the removal of organic matter, and ensuring the safe and responsible management of dairy waste (Kasmi et al., 2018). Kasmi et al. (2018) stated that dairy wastewaters commonly undergo treatment using both physico-chemical and biological methods. However, due to the high costs associated with reagents and the limited removal of soluble chemical oxygen demand through physico-chemical processes, biological treatment methods are generally preferred. Among the different types of dairy waste, cheese whey has received significant attention as a potential source to produce value-added products. This review primarily examines the various sources of dairy industry releases and their environmental impact.

10.2 Valorization Techniques for Dairy Processing Industrial Waste

The dairy industry, like many other industries, has experienced rapid growth fuelled by research and technological advancements during a period of intense industrialization. In India, specifically, the dairy industry has witnessed significant progress over the past decade. However, the waste generated by the dairy industry contains a high organic load and cannot be discharged untreated. While waste treatment and management strategies are well-documented, a significant challenge lies in dealing with the sludge produced as a by-product of the treatment process. Addressing this issue requires the application of various methods for effective waste treatment, highlighting the need for technology-oriented research in this area. There has been a shift in perspectives towards sustainable waste management practices that aim to recover value-added products, including energy, to meet the rising demand for energy resources. Moreover, the sludge generated from dairy waste treatment, traditionally spread on land, can be utilized for energy generation. The volume, concentration, and composition of wastewater generated by a dairy plant can vary significantly. These variations depend on factors such as the specific type of dairy product being processed, the operational techniques employed, the plant's design, the type of wastewater treatment implemented, and the amount of water consumed. Different dairy products and processing methods result in varying amounts and characteristics of wastewater. The design of the plant, including the equipment and processes used, can also influence the wastewater characteristics. Furthermore, the type of wastewater treatment employed by the dairy plant plays a crucial role in determining the quality and composition of the wastewater discharged. Additionally, the quantity of water consumed by the dairy plant directly impacts the volume of wastewater produced. Higher water consumption generally results in larger volumes of wastewater. Taking all these factors into account, it is essential to consider the specific characteristics and quantity of wastewater generated by a dairy plant when designing appropriate treatment processes and implementing sustainable water management practices (Adesra et al., 2021). There are various physio-chemical processes for the treatment of the effluent and waste (Table 10.1).

Table 10.1 List and Advantages and Disadvantages of Various Physico-Chemical Processes **(Adesra et al., 2021)**

S. no	Physico-chemical method	Type	Advantage	Disadvantage
1	Membrane filtration	Nanofiltration (NF) Reverse osmosis Dialysis	Non-thermal Eco-friendly Easy recovery of water and other by-products	Prone to membrane fouling Expensive cleaning and regeneration scheme may be required
2	Electrochemical	Electrocoagulation Electrofloatation	Increases biodegradability Eliminates oil, grease, and metals	High capital cost and difficult to control
3	Adsorption	Charcoal Clays Clay Minerals, zeolites, and ores	Ease of use	Lack of suitable adsorbent
4	Coagulation	Ferrous sulfate Aluminium sulfate	Hastens settling of the solid. Thus, detention of water is reduced	Produces large amount of sludge

10.3 Biological Processes

Among the various methods for treating dairy waste, biological processes have emerged as the most promising. These biological treatments utilize both aerobic and anaerobic technologies, such as wetlands, trickling filters, fluidized bed reactors (FBR), up-flow sludge anaerobic blanket (USAB), completely stirred tank reactors (CSTR), moving bed biological reactors (MBBR), sequential batch reactors (SBR), and activated sludge processes (ASP) (Carvalho et al., 2013). These treatment methods have been widely reported and implemented globally and will be discussed in detail in this section. Biological processes offer effective means of treating dairy waste, providing sustainable and environmentally friendly solutions for managing and reducing the environmental impact of wastewater generated by the dairy industry.

10.4 Conversion of Whey into Essential By-Products: Recent Insights

10.4.1 Whey Processing

Whey proteins possess various properties and undergo different processes and modifications, making them suitable for diverse applications in sports and exercise

Table 10.2 Recent Research on Processing Whey Protein

Recent research area	Outcomes	References
Life cycle assessment (LCA)	—Measures the environmental impact of a product throughout its life cycle. —Identifies impact outcomes, such as eutrophication potential and ozone destruction. —Determines the Global Warming Potential (GWP) of the functional unit.	Roy et al. (2009)
Energy consumption for cheese manufacturing and storage	—Energy consumption set at 0.91 kWh/kg cheese. —Waste whey can be converted into electricity, potentially reducing overall environmental impact.	Flysjö et al. (2014)
Utilization of whey for biogas production	—Whey can be used as a substrate for biogas production. Whey can replace fossil fuels and generate electricity.	Flysjö et al. (2014)
Carbon footprint in the dairy industry	—78–85% of the carbon footprint occurs prior to the farm gate. —Methane emissions from enteric fermentation and nitrous oxide emissions from manure management and fertilizer utilization are significant contributors.	Flysjö et al. (2014)
Emissions breakdown in the dairy industry	—Packaging, transport, and fossil fuel combustion comprise 8%, 4%, and 3% of total emissions respectively from farm to consumer in a major dairy processor's value chain.	Flysjö et al. (2014)
Livestock systems and GHG emissions	—Recent assessments suggest livestock systems, including dairying, could be GHG emission net zero.	Allen et al. (2018)
Allocation of GHG emissions and environmental impact	—Allocation models consider efficiencies, raw material utilization, and product volume. —Novel approaches to reassessing milk and meat systems aim to reduce environmental impact.	Ineichen et al. (2022)
Environmental impact and whey valorization	—Valorization of waste whey can reduce environmental impact of some cheeses by up to 15%. —Lack of data availability due to fragmented industry structure and poor whey recovery recording.	González-García et al. (2013); Finnegan et al. (2018)

science, infant nutrition, medicine, and other fields (Deeth and Bansal, 2018). In recent years, there has been a growing interest in commercial whey-based food and drink products, moving away from the traditional perception of whey as a waste product or a mere ingredient in composite foods (Panghal et al., 2018). Whey and its components are increasingly being used in the production of whey-based beverages, either as plain drinks or supplemented with fruit juice, milk or milk permeate, nutraceutical compounds, and probiotics/prebiotics (Özer and Evrendilek, 2022). This has resulted in a wide variety of whey protein beverages available in the market (Table 10.2). Researchers have explored the development of naturally carbonated, whey-based probiotic drinks with antimicrobial properties against pathogenic strains (Kadyan et al., 2021). Kadyan et al. (2021) stated that the antimicrobial activity of these beverages was influenced by the fermentative strain used, primarily due to the production of bacteriocins and acids. The developed production processes can be easily integrated into existing production lines, reducing effl uent volume and treatment costs. The demand for dairy beverages, particularly healthy and functional options, is high worldwide. Fermentation of whey by-products, particularly using probiotic microorganisms, can yield beverages rich in organic acids or low-alcohol tonics, providing value-added products for the dairy industry. Examples of such beverages produced through aerobic fermentations include kombucha and kefir, which are known for their high organic acid content (Marcus et al., 2021). Marcus et al. (2021) also demonstrated the possibility of producing new value-added and sustainable organic acid or alcoholic beverages while increasing the pH of acidic by-products using reconstituted whey permeate and various yeast and mould species. In summary, whey protein–based beverages offer a range of possibilities for product development and market opportunities, catering to the increasing demand for dairy beverages that are both healthy and functional.

10.5 Biorefinery Products: Cheese Whey Valorization

Cheese Whey (CW), is an abundant and readily available substrate that is cost-effective. However, effective management practices are necessary to optimize its utilization. Currently, several biotechnological processes for valorizing CW are at medium/high Technology Readiness Levels (TRL), and integrating these processes shows promise. However, further investigation is required to fully implement the biorefinery concept in the dairy supply chain.

The characteristics of CW vary depending on the livestock species and geographical factors. The milk and resulting CW production exhibit significant seasonal variability in terms of quantity and composition, which aligns with the lactation period. To address this seasonal variation, one approach is to freeze CW during periods of peak production and thaw it as needed. Another potential solution involves assessing CW availability in a specific region and promoting consortia to ensure a consistent CW supply throughout the year (Ubando et al., 2020).

Ubando et al (2020) stated that the optimal combination of biotechnological processes for CW valorization depends on factors such as CW availability, characteristics, legislation, and market demand. Pre-treating CW, such as extracting proteins (TRL 9), can simplify downstream valorization due to the relatively high value of whey proteins (ranging from 6 to 22 € per kilogram). However, additional nutrient supplementation may be necessary for subsequent biological treatment stages. The need for post-treatments aimed at removing unwanted impurities or extracting specific compounds must be carefully evaluated, as it represents a significant cost consideration in the overall process scheme.

Fermentation of CW yields various soluble products. Acetic acid and ethanol currently have low economic values but possess larger market sizes. On the other hand, butyric and lactic acid have smaller markets but are experiencing rapid growth (15–19% compound annual growth rate, CAGR). It is important to note that obtaining individual marketable products from the mixture of carboxylic acids typically obtained from CW fermentation would require highly selective and efficient extraction systems, which are currently at TRL 2–3. Alternatively, the mixture of carboxylic acids can serve as a feedstock for biopolymer production, particularly for the production of polyhydroxyalkanoates (PHA). The technology for PHA production from biowaste is still in the developmental stage (TRL 3–5). However, the high value of PHAs (ranging from 2.8 to 3.2 € per kilogram) and the rapidly growing bioplastics market (16.5% CAGR) make biological PHA production potentially profitable soon. Tailored solutions can be explored within the dairy industry supply chain. For instance, the PHA produced from CW can be used as sustainable packaging material for dairy products. Microbial electrochemical technologies (METs) are still in the developmental phase (TRL 3–4). Due to their high cost, low power density, and limited hydrogen yield achievable through microbial fuel cells (MFCs) and microbial electrolysis cells (MECs), their use for treating undiluted CW fermentate with high carboxylic acid concentration is not economically viable. MFCs, in particular, face challenges in competing with technologies like solar and wind power for large-scale electricity production, unless multiple cells are stacked together (Gajda et al., 2018). However, both MFCs and MECs can serve as polishing stages prior to effluent disposal, thanks to their high chemical oxygen demand (COD) removal efficiencies. Among bioelectrochemical systems, microbial electrosynthesis (MES) can play a crucial role in reducing carbon emissions by recycling CO_2 from other bioprocesses and the dairy industry, converting it into carboxylic acids.

10.6 Case Study on Ghee Residue: A Recent Concern in Dairy Industries

10.6.1 M.Tech Dissertation

Ghee residue (GR) is a dairy by-product after the production of ghee that consists of high nutritive value and can be used as a human dietary supplement but is instead used for feed purposes. The GR is extracted using the direct creamy (DC) method, and the yield of the GR varies according to the method of preparation of ghee and the yield was higher. GR is the SNF portion of cream that was coagulated out during ghee preparation. The research work involves the preparation of two GR-based dairy products prepared by using GR prepared in a lab (GRP A) and the GR taken from Milma Kollam Dairy (GRP B). Three trials were conducted for the preparation of the product, and the organoleptic characteristics were compared for all three trials. The work involves a comparison of the physico-chemical properties, organoleptic properties, and antioxidant properties of the dairy product prepared using two different GRs. The values of the physico-chemical properties such as protein, fat, crude fibre, and total ash for GRP A were found to be 16.64 g/100g, 15.8 g/100g, 0.71 g/100g, and 3.2 g/100g respectively, which was compared with the values of GRP B: 13.88 g/100g, 5.33 g/100g, 0.80 g/100g, and 3.1 g/100g. GR is well known for its antixoxidant property, which is due to the presence of phospholipids and nitrogenous compounds. The value for microbial analysis for standard plate count, yeast, and mould, and coliforms were obtained for both GRP A and GRP B which was then compared. The previous parameters showed that the product from GRP A dominated the product from GRP B.

Ghee production accounts for approximately 33% of total milk production in India. The SNF part which will be left after the ghee is produced is called the ghee residue, and it accounts for one-tenth of the total ghee produced. Every year, India produces 42.20 million tonnes of ghee from its total milk output of 140 million tonnes, and in India the average amount of GR produced is 4.2 million tonnes (Varma et al., 2008). The GR obtained in large quantities is left unutilized in most of the dairy industry in India, even though they have high potential in contributing to the human dietary supplement and for the production of low-cost feed. GR can be used as an alternative source of unconventional feed ingredients (Selvamani et al., 2017). The yield of GR varies depending upon the production method of used, due to the processing conditions and raw materials used to prepare the ghee, which may vary in nonfatty serum constituents (Rafiq et al., 2019). The phospholipid content is abundant in the GR, so extracting phospholipids from the GR and using it in other products can increase the oxidative stability of the product. In one study it was found that adding 0.1% of phospholipids will enhance the oxidative stability of ghee; this can be done either by adding phospholipids extracted from the GR by solvent extraction or by heat treatment (Pruthi et al., 1980). Decades ago it was suggested that GR be used for preparing soups, candies, toffees, pastries, etc., but the idea was not commercialized due to the unknown facts on the nutritive properties of the by-product. As discussed earlier it exhibits a good antioxidant property due to the presence of phospholipids and nitrogenous compounds, and some other elements also contribute to GR's oxidative stability.

Khoa is an indigenous dairy product that is used for making many dairy-based sweet products. It has different names like khoya, palghova, kava, and mawa. The Bureau of Indian Standards defined khoa as a product obtained by the partial dehydration of milk using heat, and the final product should contain not less than 30% of the milk fat. There are different varieties of khoa available, such as Pindi, Dhap, and Danedar. In this study, khoa was used as an ingredient for the preparation of GR-based dairy products. The research deals with the study of the physico-chemical, sensory, and antioxidant properties of the two-dairy product, which was prepared using the GR obtained from the industry and the GR prepared in a lab. The relevance of the study is that it deals with the comparison of the antioxidant properties of the product which was prepared by adding GR obtained at different processing conditions; it also involves the microbial analysis of the product which is a part of the shelf-life study. The GR is now used in the preparation of confectioneries, candies, burfi like sweets, etc. but till now it has not come into the commercial product sector.

10.6.1.1 Antioxidant Properties of Ghee Residue

The antioxidant property of GR is mainly due to the presence of a high number of phospholipids and nitrogenous compounds. The other constituents exhibiting the same functions in the GR include free amino acids and some reducing substances such as free sulphydryls from denatured proteins and free sugars from lactose (Galhotra et al.,1999). The method of preparation of ghee has a direct influence on the antioxidant property of the resulting GR. The antioxidant property is high for the CB GR and is followed by DC and DB GR, and due to the presence of both lipids and non-lipids, the GR contributes to oxidative stability (Munirathnamma et al., 2017). In lipids mainly the phospholipid content is more, and it shows more antioxidant activity and other constituents consisting of α-tocopherol and vitamin A. The cephalin is the major phospholipid that exhibits the most antioxidant property (Pagote et al.,1988). The enhancement of the antioxidant property is possible by increasing the phospholipids content up to 0.1%; this can be done either using the heat treatment method or by the solvent extraction method. The maximum transfer of phospholipids from GR to ghee was observed when GR was heated with ghee in a 1:4 ratio at 130°C. These antioxidant concentrates can be applied to ghee to provide around 0.1% phospholipids, which will improve the ghee's shelf life. The non-lipids constituents present in ghee such as cysteine hydrochloride, proline, lysine, that comes under amino acids exhibit major antioxidant activity. It has been found that the incorporation of protein along with the phospholipids, glucose, and galactose to the ghee can improve the oxidative stability of the ghee.

Figure 10.2 represents the antioxidant properties of both GR products. The antioxidant activity of GR products was found using the DPPH method. The method uses the value butylated hydroxytoluene as the reference/standard value. From the graph, it is evident that the GR product lab-scale (GRP LS) shows more antioxidant property (closer to the standard value) than the GR product industrial (GRP INDS).

The khoa has no antioxidant activity, but the added GR exhibit the antioxidant property that contributes to the oxidative stability of the product. The antioxidant property of GR is mainly due to the presence of phospholipids. Some nitrogenous compounds also contribute to antioxidant activity. The phospholipids in GR vary according to the method of ghee preparation. The phospholipid content in GR varies according to different processing conditions; the content is much higher in CB GR, which is almost 17.3%, followed by 4.9% in DB GR and least in DC GR (1.6%) (Galhotra et al.,1999). The temperature maintained during the ghee preparation has a direct effect on the phospholipid content of the GR; it is evident that the industrial GR has low antioxidant property so that the product made by adding it also shows a low antioxidant property this is due to the decline in the phospholipid content. In a study, the addition of GR in ghee will improve the oxidative stability of the product, which will improve the keeping quality of ghee; if more GR is added then the oxidative stability will also increase (Ranjan et al., 2020). The CB GR exhibits more antioxidant properties than the DC method (Pagote et al.,1988).

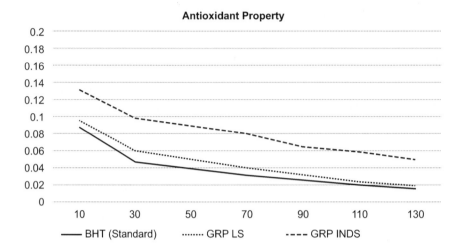

Figure 10.2 *Antioxidant activity of the GR products by DPPH method – from industrial and lab-scale GR.*

10.6.1.2 Sensory Analysis

Sensory analysis was done during the trials for the optimization of GRP A and the same formulation was used for preparing the final products of GRP A and GRP B. The acceptability of the product from GRP A was increased during successive trials. The appearance, odour, texture, taste, mouthfeel, and overall acceptance were analysed using the 9-point hedonic scale. It was found that the acceptability was high when more lab-scale GR was added. Table 10.3 represents the sensory analysis of the final products of GRP A and GRP B. GRP A was more acceptable because GRP B had a burned aftertaste because of the industrial GR added. The overall acceptance of GRP A was 8, which shows its appearance, odour, texture, taste, and mouthfeel as acceptable, but the sensory attributes were slightly liked by the people who had tasted the product.

10.6.1.3 Conclusion

The study involves the preparation of dairy products with added GR. The main importance of this study was to identify the role of GR in value addition and provide a comparative study on the effect of the addition of lab-scale GR and industrial GR on the khoa-based sweet. The observations such as protein, fat, antioxidant activity, total ash, and crude fibre show that the product GRP A dominates over GRP B. The antioxidant property was more for the lab-scale GR and hence the oxidative stability. The industry can utilize the GR because it will give value addition to the added products but due to the current processing conditions, GR obtained from the overburned industrial GR was not suitable for adding in any products.

Table 10.3 7-Scale Hedonic Scale Sensory Test for GRP A and GRP B

SI no.	Sample	Appear ance	Odour	Texture	Taste	Mouth feel	Overall accept ance
1	GRP A	8	8	8	9	9	8
2	GRP B	6	6	6	6	6	6

10.7 Conclusion

The dairy industry plays a crucial role in global milk production, and it is expected to experience growth in Asia and Central America while facing a decline in Europe. In India, the industry has witnessed substantial expansion, primarily driven by the co-operative sector. However, alongside this growth, the dairy industry also generates a significant amount of waste, which necessitates efficient management and utilization.

Managing and valorizing dairy processing industrial waste is vital due to its environmental and economic implications. Food waste, including waste from the dairy industry, represents a loss of valuable resources and presents challenges in terms of management. To address this issue and promote waste valorization, concepts from industrial ecology, such as the circular economy and cradle-to-cradle principles, are gaining prominence.

Dairy wastewater, characterized by high organic loads, poses environmental risks if not treated properly. Various physico-chemical and biological methods are commonly employed for treating dairy wastewater, with biological processes showing the most promise. These biological treatment technologies include both aerobic and anaerobic methods, which have been implemented worldwide, offering sustainable solutions for managing dairy waste.

One significant waste product in the dairy industry is whey, and recent studies have focused on its potential valorization. Whey proteins have diverse applications in various fields, and whey-based beverages have gained popularity. Furthermore, the fermentation of whey by-products using probiotic microorganisms has resulted in value-added products, such as organic acid or low-alcohol beverages.

Cheese whey has received considerable attention for its potential valorization. Biotechnological processes for cheese whey valorization have reached medium to high Technology Readiness Levels (TRL), and integrating these processes shows promise. However, further research and investigation are required to fully implement the biorefinery concept in the dairy supply chain.

The case study focused on the preparation of dairy products by incorporating ghee-residue (GR) as an added ingredient. The main objective was to explore the value addition potential of GR and conduct a comparative analysis between lab-scale GR and industrial GR on khoa-based sweets. Several parameters were evaluated to assess the impact of GR addition, including protein content, fat content, antioxidant activity, total ash, and crude fibre. The findings indicated that the product derived from lab-scale GR (GRP A) exhibited superior characteristics compared to the product derived from industrial GR (GRP B). One notable observation was the higher antioxidant activity observed in the lab-scale GR, suggesting its potential for enhancing oxidative stability in the products. This indicates that the lab-scale GR could contribute to improved product quality and shelf life. The study suggests that the dairy industry could effectively utilize GR to add value to their products. However, it should be noted that the industrial GR obtained from overburned processes was found to be unsuitable for addition to any dairy products under the current processing conditions. In summary, the scientific investigation revealed that the addition of GR to dairy products can offer value addition. The study emphasized the superiority of lab-scale GR over industrial GR, particularly in terms of antioxidant activity and oxidative stability. These findings provide insights into the potential utilization of GR by the dairy industry, with the need for appropriate processing conditions to ensure its suitability for product incorporation. In conclusion, effective management and valorization of dairy processing industrial waste, including whey, can contribute to a more sustainable and environmentally friendly dairy industry. By reducing waste and maximizing the utilization of valuable resources, the industry can move towards a more efficient and sustainable future.

References

Adesra, A., Srivastava, V. K., & Varjani, S. (2021). Valorization of dairy wastes: Integrative approaches for value added products. *Indian Journal of Microbiology, 61*(3), 270–278.

Allen, M. R., Dube, O. P., Solecki, W., Aragón-Durand, F., Cramer, W., Humphreys, S., ... & Zickfeld, K. (2018). *Framing and context. Global warming of, 1*(5).

Carvalho, F, Prazeres, A. R., & Rivas, J. (2013). Cheese whey wastewater: Characterization and treatment. *Sci Total Environ, 445–446*, 385–396. https://doi.org/10.1016/j.scitotenv. 2012.12.038.

FAO. (2022). *Dairy Market Review: Emerging trends and outlook 2022.* Rome.

Finnegan, W., Goggins, J., & Zhan, X. (2018). Assessing the environmental impact of the dairy processing industry in the Republic of Ireland. *Journal of Dairy Research, 85*(3), 396–399.

Flysjö, A., Thrane, M., & Hermansen, J. E. (2014). Method to assess the carbon footprint at product level in the dairy industry. *International Dairy Journal, 34*(1), 86–92.

Gajda, I., Greenman, J., & Ieropoulos, I. A. (2018). Recent advancements in real-world microbial fuel cell applications. *Current opinion in electrochemistry, 11*, 78–83.

Galhotra, K. K., & Wadhwa, B. K. (1993). Chemistry of ghee-residue, its significance and utilisation. *Indian Journal of Dairy Science, 46*, 142–142.

González-García, S., Hospido, A., Moreira, M. T., Feijoo, G., & Arroja, L. (2013). Environmental life cycle assessment of a Galician cheese: San Simon da Costa. *Journal of cleaner production, 52*, 253–262.

Ineichen, S., Schenker, U., Nemecek, T., & Reidy, B. (2022). Allocation of environmental burdens in dairy systems: Expanding a biophysical approach for application to larger meat-to-milk ratios. *Livestock Science, 261*, 104955.

Kadyan, S., Rashmi, H. M., Pradhan, D., Kumari, A., Chaudhari, A., & Deshwal, G. K. (2021). Effect of lactic acid bacteria and yeast fermentation on antimicrobial, antioxidative and metabolomic profile of naturally carbonated probiotic *whey drink. Lwt,* 142, 111059.

Kasmi, M. (2018). Biological processes as promoting way for both treatment and valorization of dairy industry effluents. *Waste and biomass valorization,* 9(2), 195–209.

Marcus, J. F., DeMarsh, T. A., & Alcaine, S. D. (2021). Upcycling of whey permeate through yeast-and mold-driven fermentations under anoxic and oxic conditions. *Fermentation,* 7(1), 16.

Mirabella, N., Castellani, V., & Sala, S. (2014). Current options for the valorization of food manufacturing waste: a review. *Journal of cleaner production, 65,* 28–41.

Munirathnamma, V., Gupta, V. K., & Meena, G. S. (2017). Effect of different extraction processes on the recovery of ghee residue proteins. *Indian Journal of Animal Science,* 87, 366–372.

Özer, B., & Evrendilek, G. A. (2022). Whey beverages. *In Dairy Foods* (pp. 117–137). Woodhead Publishing.

Pagote, C. N., & Bhandari, V. (1988). Antioxidant property and nutritive value of ghee residue. *Indian Dairyman, 40,* 73.

Panghal, A., Patidar, R., Jaglan, S., Chhikara, N., Khatkar, S. K., Gat, Y., & Sindhu, N. (2018). Whey valorization: current options and future scenario–a critical review. *Nutrition & Food Science, 48*(3), 520–535.

Prazeres, A. R., Carvalho, F., & Rivas, J. (2012). Cheese whey management: A review. *Journal of Environmental Management, 110,* 48–68.

Pruthi, T. D. (1980). Phospholipid content of ghee prepared at higher temperatures. *Indian Journal of Dairy Science, 33*(2), 265–267.

Rafiq, S. M., & Rafiq, S. I. (2019). Milk by-products Utilization. *In Current Issues and Challenges in the Dairy Industry. IntechOpen.*

Ranjan, R., Chauhan, A. K., Kumari, S. S., & Dubey, R. P. (2020). Nutritive value of ghee residue incorporated bakery product. *Indian Journal of Dairy Science, 73*(1).

Roy, P., Nei, D., Orikasa, T., Xu, Q., Okadome, H., Nakamura, N., & Shiina, T. (2009). A review of life cycle assessment (LCA) on some food products. *Journal of food engineering,* 90(1), 1–10.

Selvamani, J., Radhakrishnan, L., Bandeswaran, C., Gopi, H., & Valli, C. (2017). Estimation of nutritive value of ghee residue procured from western districts of Tamil Nadu, India. *Asisuan Journal of Dairy and Food Res, 36*(4), 283–287.

Ubando, A. T., Felix, C. B., & Chen, W. H. (2020). Biorefineries in circular bioeconomy: *A comprehensive review. Bioresource technology, 299,* 122585.

Varma, B. B., & Narender Raju, P. (2008). Ghee residue: Processing, properties and utilization. course compendium on "Technological advances in the utilization of dairy by-products". *Centre of Advanced Studies in Dairy Technology, NDRI, Karnal. p, 176183.*

Value Addition of Meat and Poultry Processing Industrial Waste

M. Selvamuthukumaran

11.1 Introduction

The poultry industries can generate significant amounts of wastes, especially while processing broiler chicken. The valuable components can be recovered from such waste by efficient recycling. Therefore, one has to adopt user-friendly and effective techniques to fully explore such animal resources (Voběrkova et al., 2020; Galali et al., 2020; Kannah et al., 2020). The waste and by-products generated by poultry processing industries, which includes viscera, feet, head, feathers, blood and bones.

The waste generated by poultry processing industries can be recycled by traditional processing methods like incineration and composting. Currently, various physical, microbiological and other complex methods were available for converting waste from such industries into several value-added products. The waste generated from poultry processing was converted to meat and bone meal, blood meal and fats. Hydrothermal treatment techniques were successfully implemented to treat such waste so that microorganisms will be destroyed during recycling process (Vikman et al., 2017; Volik et al., 2017; Dhillon et al., 2017; Lasekan et al., 2013).

Beyond utilizing poultry wastes for fodder production, they can be effectively employed in food as well as pharmaceutical industries as an additive (Tram et al., 2021). The global production of poultry meat is around 136 million tons per year, according to the estimate of the FAO (2021), and huge waste is generated. Therefore, one has to adopt the appropriate technology to effectively recycle poultry processing industrial wastes.

11.2 Poultry Waste as Manure

The poultry manure obtained from poultry processing waste can be used to enhance soil fertility, and it can be used along with nitrogen, phosphorus and

Table 11.1 Applications of Poultry Manure

S, no.	Applications	Reference
1	Organic fertilizer production	Feng et al. (2019); Purnomo et al. (2017)
2	Biogas and composting production	Achi et al. (2020); Hassanein et al. (2019); Ojo et al. (2018); Mehryar et al. (2017)
3	Feed supplements production	Blake & Hess (2014); Jackson et al. (2006)
4	Combustion fuel production	Zapevalov et al. (2019)
5	Biochar production	Liu et al. (2021)

potassium or alone (Samoraj et al., 2022; Fomicheva & Rabinovich, 2021). The FAO (2022) has reported that from global livestock broiler production of 18.5 billion tons, the nitrogen released into the environment with manure accounts for 6.7 million tons. There are a variety of physical, biological and chemical methods available to process poultry manure (Zapevalov et al., 2019). Recently many researchers have been indulged in finding out appropriate technologies for transforming poultry manure into several value-added products (Blake & Hess, 2014; Liu et al., 2021; Feng et al., 2019). Applications of poultry manure are provided in Table 11.1.

The prominent methods for processing poultry manure viz. fermenting raw materials in reactors of optimum capacity using accelerated fermentation process (Fomicheva & Rabinovich, 2021). The other methods also viable like using ultra-high-frequency electromagnetic field for disinfestation and drying Soboleva et al., 2017). The best recommended method is anaerobic co-digestion, gasification and fast pyrolysis (Kanani et al., 2020).

11.3 Processing of Feathers

Poultry feathers are a resource for producing materials for several industries, like food, textile, agro as well as construction (Ahmad et al., 2022). Keratin can be recovered from such poultry processing waste, and around 10 million tons of keratin-containing waste can be produced from such poultry processing wastes (Kshetri et al., 2019; Stiborova et al., 2016). Feathers contain around 85% protein, but it is quite difficult for them to convert into protein. The presence of more hydrogen bonds makes it more difficult to decompose (McKittrick et al., 2012; Bray et al., 2015). Feather decomposition can be successfully achieved by thermal and chemical methods (Stiborova et al., 2016; Stiborova et al., 2016; Brebu & Spiridon, 2011).

The textile industries consume most of the down feathers. The raw material can be used as filler in pillows, blankets and duvets. Coats can also be manufactured by using down feathers, which are lightweight and can offer very good protection. The hydrolyzed feather waste protein can be used as a

Value Addition of Meat and Poultry Processing Industrial Waste   245 modifier, which can enhance the properties of cotton fabric and also increase the susceptibility to natural dyes (Zhang et al., 2020).

11.4 Collagen

Collagen is a kind of protein that can be obtained from skin, neck, bones and poultry legs (Fisinin et al., 2017). Almeida et al. (2013) reported that gelatin of high quality can be obtained from chicken legs. Collagen can be successfully isolated from chicken legs in the solution of acetic acid with the help of enzymes like papain and pepsin (Hashim et al., 2014). High collagen can be obtained by adopting the processing stages, such as impurities removal, followed by cleaning, demineralization and degreasing (Cansu & Boran, 2015).

11.5 Waste from Livestock Meat Processing

Huge waste is generated during meat processing. Around 64% food waste is generated during meat consumption, manufacturing waste accounts for around 20%, post-harvest waste of 3.5% and distribution waste of 12%. Primary production of meat results in significant losses ascribed to rearing condition of animals, including transportation of them to the slaughterhouses. Pig processing leads to wastage of 0.2%, which occurs during slaughtering (Gustavsson et al., 2011). The improper storage condition also results in wastage of fresh meat, leads to contamination and dissatisfaction among consumers. The various waste generated during the slaughtering of animals includes bones, blood, skin, liver, kidney, gastrointestinal flushes etc. They can be recycled and converted into various value-added products, and they can be fortified for developing several food products. They can be used as an additive for several food processing industries.

11.6 Conclusions

Poultry and meat processing industrial waste can be recycled and transformed into various value-added products, which can be used in several food processing industries. The value addition of such waste can augment income and helps to maintain economic stability.

References

Achi, C. G., Hassanein, A., & Lansing, S. (2020). Enhanced biogas production of cassava wastewater using zeolite and biochar additives and manure co-digestion. Energies, 13(2), 491. http://dx.doi.org/10.3390/en13020491.

Ahmad, A., Othman, I., Tardy, B. L., Hasan, S. W., & Banat, F. (2022). Enhanced lactic acid production with indigenous microbiota from date pulp waste and keratin protein hydrolysate from chicken feather waste. Bioresource Technology Reports, 18, 101089. http://dx.doi.org/10.1016/j.biteb.2022.101089.

Almeida, P. F., Calarge, F. A., & Santana, J. C. C. (2013). Production of a product similar to gelatin from chicken feet collagen. Agricultural Engineering, 33(6), 1289–1300.

Blake, J., & Hess, J. (2014). Suitability of poultry litter ash as a feed supplement for broiler chickens. Journal of Applied Poultry Research, 23(1), 94–100. http://dx.doi.org/10.3382/japr.2013-00836.

Bray, D. J., Walsh, T. R., Noro, M. G., & Notman, R. (2015). Complete structure of an epithelial keratin dimer: Implications for intermediate filament assembly. PLoS One, 10(7), e0132706. http://dx.doi.org/10.1371/journal.pone.0132706. PMid:26181054.

Brebu, M., & Spiridon, I. (2011). Thermal degradation of keratin waste. Journal of Analytical and Applied Pyrolysis, 91(2), 288–295. http://dx.doi.org/10.1016/j.jaap.2011.03.003.

Cansu, Ü., & Boran, G. (2015). Optimization of a multi-step procedure for isolation of chicken bone collagen. Korean Journal for Food Science of Animal Resources, 35(4), 431–440. http://dx.doi.org/10.5851/kosfa.2015.35.4.431. PMid:26761863.

Dhillon, G. S., Kaur, S., Oberoi, H. S., Spier, M. R., & Brar, S. K. (2017). Agricultural-based protein by-products: Characterization and applications. In G. S. Dhillon (Ed.), Protein Byproducts: Transformation from Environmental Burden Into Value-Added Products (pp. 21–36). Edmonton: Academic Press.

Feng, G., Adeli, A., Read, J., McCarty, J., & Jenkins, J. (2019). Consequences of pelletized poultry litter applications on soil physical and hydraulic properties in reduced tillage, continuous cotton system. Soil & Tillage Research, 194, 104309. http://dx.doi.org/10.1016/j.still.2019.104309.

Fisinin, V. I., Ismailova, D. Y., Volik, V. G., Lukashenko, V. S., & Saleeva, I. P. (2017). Deep processing of collagen-rich poultry products for different use. Agricultural Biology, 52(6), 1105–1115. http://dx.doi.org/10.15389/agrobiology.2017.6.1105eng.

Fomicheva, N. V., & Rabinovich, G. Y. (2021). Technological line for processing animal waste. IOP Conference Series: Earth and Environmental Science, 677, 052004.

Food and Agriculture Organization—FAO. (2021). Gateway to Poultry Production and Products. Retrieved from www.fao.org/poultryproduction-products/production/en/

Food and Agriculture Organization—FAO. (2022). Crops and Livestock Products. Retrieved from www.fao.org/faostat/en/#data/QCL

Galali, Y., Omar, Z. A., & Sajadi, S. M. (2020). Biologically active components in by-products of food processing. Food Science & Nutrition, 8(7), 3004–3022. http://dx.doi.org/10.1002/fsn3.1665. PMid:32724565.

Gustavsson, J., Cederberg, C., Sonesson, U., & Emanuelsson, A. (2011). The Methodology of the FAO Study: Global Food Losses and Food Waste-Extent, Causes and Prevention. Rome, Italy: FAO.

Hashim, P., Ridzwan, M. S. M., & Bakar, J. (2014). Isolation and characterization of collagen from chicken feet. International Journal of Bioengineering and Life Sciences, 8(3), 250–254.

Hassanein, A., Lansing, S., & Tikekar, R. (2019). Impact of metal nanoparticles on biogas production from poultry litter. Bioresource Technology, 275, 200–206. http://dx.doi.org/10.1016/j.biortech.2018.12.048. PMid:30590206.

Jackson, D. J., Rude, B. J., Karanja, K. K., & Whitley, N. C. (2006). Utilization of poultry litter pellets in meat goat diets. Small Ruminant Research, 66(1–3), 278–281. http://dx.doi.org/10.1016/j. smallrumres.2005.09.005.

Kanani, F., Heidari, M. D., Gilroyed, B. H., & Pelletier, N. (2020). Waste valorization technology options for the egg and broiler industries: A review and recommendations. Journal of Cleaner Production, 262, 121129. http://dx.doi.org/10.1016/j.jclepro.2020.121129.

Kannah, R. Y., Merrylin, J., Devi, T. P., Kavitha, S., Sivashanmungam, P., Kumar, G., & Banu, J. R. (2020). Food waste valorization: Biofuels and value added product recovery. Bioresource Technology Reports, 11, 100524. http://dx.doi.org/10.1016/j.biteb.2020.100524.

Kshetri, P., Roy, S. S., Sharma, S. K., Singh, T. S., Ansari, M. A., Prakash, N., & Ngachan, S. (2019). Transforming chicken feather waste into feather protein hydrolysate using a newly isolated multifaceted keratinolytic bacterium Chryseobacterium sediminis RCM-SSR-7. Waste and Biomass Valorization, 10(1), 1–11. http://dx.doi.org/10.1007/s12649-017-0037-4.

Lasekan, A., Bakar, F. A., & Hashim, D. (2013). Potential of chicken byproducts as sources of useful biological resources. Waste Management, 33(3), 552–565. http://dx.doi.org/10.1016/j.wasman.2012.08.001. PMid:22985619.

Liu, C., Yin, Z., Hu, D., Mo, F., Chu, R., Zhu, L., & Hu, C. (2021). Biochar derived from chicken manure as a green adsorbent for naphthalene removal. Environmental Science and Pollution Research International, 28(27), 36585–36597. http://dx.doi.org/10.1007/s11356-021-13286-x. PMid:33704645.

McKittrick, J., Chen, P. Y., Bodde, S. G., Yang, W., Novitskaya, E. E., & Meyers, M. A. (2012). The structure, functions, and mechanical properties of keratin. JOM, 64(4), 449–468. http://dx.doi.org/10.1007/s11837-012-0302-8.

Mehryar, E., Ding, W. M., Hemmat, A., Hassan, M., Bi, J. H., Huang, H. Y., & Kafashan, J. (2017). Anaerobic co-digestion of oil refinery wastewater and chicken manure to produce biogas, and kinetic parameters determination in batch reactors. Agronomy Research, 15(5), 1983–1996.

Ojo, A. O., Taiwo, L. B., Adediran, J. A., Oyedele, A. O., Fademi, I., & Uthman, A. C. O. (2018). Physical, chemical and biological properties of an accelerated cassava based compost prepared using different ratios of cassava peels and poultry manure. Communications in Soil Science and Plant Analysis, 49(14), 1774–1786. http://dx.doi.org/10.1080/0010 3624.2018.1474914.

Purnomo, C. W., Indarti, S., Wulandari, C., Hinode, H., & Nakasaki, K. (2017). Slow release fertiliser production from poultry manure. Chemical Engineering Transactions, 56, 1531–1536.

Samoraj, M., Mironiuk, M., Izydorczyk, G., Witek-Krowiak, A., Szopa, D., Moustakas, K., & Chojnacka, K. (2022). The challenges and perspectives for anaerobic digestion of animal waste and fertilizer application of the digestate. Chemosphere, 295, 133799. http://dx.doi.org/10.1016/j.chemosphere.2022.133799. PMid:35114259.

Soboleva, O. M., Kolosova, M. M., Filipovich, L. A., & Aksenov, V. A. (2017). Electromagnetic processing as a way of increasing microbiological safety of animal waste. IOP Conference Series: Earth and Environmental Science, 66, 012025. http://dx.doi.org/10.1088/1755-1315/66/1/012025.

Stiborova, H., Branska, B., Vesela, T., Lovecka, P., Stranska, M., Hajslova, J., Jiru, M., Patakova, P., & Demnerova, K. (2016). Transformation of raw feather waste into digestible peptides and amino acids. Journal of Chemical Technology and Biotechnology, 91(6), 1629–1637. http://dx.doi.org/10.1002/jctb.4912.

Tram, N. X. T., Ishikawa, K., Minh, T. H., Benson, D., & Tsuru, K. (2021). Characterization of carbonate apatite derived from chicken bone and its in-vitro evaluation using MC3T3-E1 cells. Materials Research Express, 8(2), 025401. http://dx.doi.org/10.1088/2053-1591/abe018.

Vikman, Y. M., Siipola, V., Kanerva, H., Šližyte, R., & Wikberg, H. (2017). Poultry by-products as a potential source of nutrients. Advances in Recycling and Waste Management, 2(3), 1000142.

Voběrkova, S., Maxianová, A., Schlosserová, N., Adamcová, D., Vršanská, M., Richtera, L., Gagić, M., Zloch, J., & Vaverková, M. D. (2020). Food waste composting: Is it really so simple as stated in scientific literature? A case study. The Science of the Total Environment, 723, 138202. http://dx.doi.org/10.1016/j.scitotenv.2020.138202. PMid:32224413.

Volik, V. G., Ismailova, D. Y., Zinoviev, S. V., & Erokhina, O. N. (2017). Improving the efficiency of using secondary raw materials obtained during poultry processing. Poultry and Poultry Products, 2, 40–42.

Zapevalov, M. V., Sergeyev, N. S., Redreev, G. V., Chetyrkin, Y. B., & Zapevalov, M. S. (2019). Technology of poultry manure utilization as a renewable energy source. IOP Conference Series: Materials Science and Engineering, 582(1), 012036. http://dx.doi.org/10.1088/1757-899X/582/1/012036.

Zhang, L., Li, H., Zhu, J., & Yan, J. (2020). Structure and dyeing properties of cotton fabric modified by protein from waste feathers. IOP Conference Series. Materials Science and Engineering, 774(1), 012043. http://dx.doi.org/10.1088/1757-899X/774/1/012043.

12

Value Addition of Fish Processing Industrial Waste

Ritu Agrawal, Akansha Khati and Sabbu Sangeeta

12.1 Introduction

Fisheries are a fast-growing sector in India, providing nutrition and food security to a large population of the country, along with income and employment to more than 28 million people. At the primary level, the sector provides livelihood support to about 280 lakh people. Globally, fish represents about 16.6% of animal protein supply and 6.5% of all protein for human consumption. In India, total fish production was 162.48 lakh tons, out of which marine and inland fish production shares were 41.27 lakh tons and 121.21 lakh tons, respectively. Out of this, fisheries export was 1,369,264 tons, contributing 1.1% to total GDP (2020–21). India's contribution in world fish production is 8%, with an annual growth rate of 10.34%.

Human consumption of processed fish results in huge amount of waste in the form of skin, head, viscera, scales, bones and frames. The amount of the waste depends upon the type and size of the fish and the amount of product. In the case of industrial fish processing, there is yield of 40% edible flesh and the remaining 60% is waste, whereas the amount of annual discard from world fisheries was estimated to be 20 million tons, including waste and by-product (Figure 12.1). In India, there is a problem of fish waste, which needs to have considerable attention from fish processors. There was approximately 302,750 tons of waste generated from India's fish processing industries during 2006–07 alone. Waste generated from industrial fish processing in India, was recorded as follows: shrimp products 50%, fish fillets 70%, whole and gutted fish 10%, cuttlefish rings 50%, whole cuttlefish 30%, cuttlefish fillets 50%, whole cleaned squid 20%, squid tubes 50%, squid rings 55%. These waste products can lead to environmental pollution; thus, it is important to recycle and convert these fish wastes into useful products of higher nutritive value.

There is an intense need and scope for high-value products from fish waste; therefore, different technologies have been developed to utilize processing

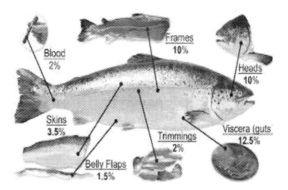

Figure 12.1 Fish waste percentage.

such waste. Fish processing waste is rich in protein, fat and minerals, whereas shellfish processing waste comprises heads and shells, which can be converted into many products. The most prominent uses of fish waste are the production of fish meal. Fish oil extraction, enzyme isolation, fish maws, isinglass, fish body and liver oil etc. are other common uses. Some other by-products are fish skin leather, amino acids, pearl essence, gelatin etc. The waste from crustaceans and shrimps can be converted into chitin, chitosan and collagen. Some other fishery by-products, like bile salts, insulin, glucosamine etc., are of great significance in the biochemical and pharmaceutical sectors.

Some of the high-value products which can be obtained from the solid waste of fish are oils, pigments, meals, minerals, enzymes and protein concentrate. Other possible uses of processing waste would be the manufacturing of valuable biomolecules such as collagen and gelatin. It is estimated that about 30% of the waste generated in fish processing consists of skin and bone, which are rich in collagen. The global collagen market is anticipated to reach USD 6.63 billion by 2025. Collagen, gelatin and their hydrolyzed products are the key product segments in this market. The market is growing due to increasing demand for collagen-based products in healthcare applications (such as wound healing, tissue engineering, bone reconstruction etc.), food and beverages and in cosmetics. Some of the valuable products developed from fish processing waste are discussed here.

12.2 Collagen

Collagen is a structural protein which is insoluble in water and fibrous in nature. The approximate molecular weight of a collagen molecule is 300KDa. Collagen collected from fish is mostly of Type I and Type III. Type I and Type III collagen are the building blocks for connective tissues, bones and skin. Collagen is not soluble in water. However, fish Type I collagen is unique in its extremely high solubility in dilute acid compared to avian and mammalian collagen.

Unique properties of marine-based collagen make it better than mammalian-based collagen. Such properties include no risk of transmitting diseases, a lack of religious constraints, a cost-effective process, low molecular weight, biocompatibility, and easy absorption by the human body. Collagen is a structural protein of the various connective tissues in the body (i.e. skin, bones, ligaments, tendons and cartilage). The applications of collagen were found in drug and gene carriers, tissue engineering, absorbable surgical suture, bone filling materials and burn/wound cover dressings. Collagen is an important component of the wound-healing process, as it acts as a natural structural scaffold or substrate for new tissue growth. Collagen has a low denaturation temperature of 25–30°C as compared to mammalian collagen (39–40°C). Collagen fibers provide structural support to all body organs and ensure the firmness, elasticity and strength needed for effective locomotion, tissue regeneration and repair. Collagen takes part in the construction of fibroblasts, which form the base for the growth of new cells. Collagen plays an active role in the protection of the skin by inhibiting the absorption and spreading of pathogenic substances, environmental toxins, micro-organisms and cancerous cells.

12.3 Gelatin

The word "gelatin" originates from the Latin word "gelatos", meaning frozen or firm. It is produced by the partial hydrolysis of collagen. Gelatin is a water soluble, fibrous, highly purified protein. Some other properties of gelatin are its gel-forming ability and water-holding capacity (Figure 12.2).

Fish solid wastes, viz. bones and scales, are high in collagen content and share approximately 75% of a fish's total body weight. This waste can be

Figure 12.2 *Fish gelatin.*

utilized for the extraction of gelatin, further used in the food and pharmaceutical industries. Fish gelatin can replace mammalian gelatin, as it contains all essential and non-amino acids. Fish gelatin replaces bovine and porcine gelatin and also increases the utilization of fish waste by reducing pollution. In different food products gelatin is used as a foaming agent, thickener, texture stabilizer and emulsifying agent. It is mainly used in dairy products, ice cream, jellies etc. Gelatin is a translucent, colorless, flavorless solid substance. Gelatin is extracted from the skin and bones of fish. It is a protein that lacks in tryptophane, so it cannot be considered a sole source of protein in animal or human nutrition. It has relatively high levels of lysine and methionine, which are deficient in cereal proteins. Gelatin is a multifunctional ingredient used in foods, pharmaceuticals, cosmetics and photographic films as a gelling agent, stabilizer, thickener, emulsifier and film former. It is a thermo-reversible hydrocolloid with a narrower gap between its melting and gelling temperatures. Gelatin is mostly produced from pig skin and cattle hides and bones. Fish skin has a significant potential for the production of high-quality gelatin, having sufficiently high gel strength and viscosity. In industry, gelatin quality is determined by its strength, viscosity, melting or gelling temperatures, water content and microbiological safety. All quality parameters depend upon the biochemical characteristics of the raw materials, manufacturing processes applied and experimental settings used for quality control tests.

Gelatin is a soluble protein produced by controlled hydrolysis of fibrous insoluble collagen. For hydrolysis of collagen into gelatin, a warm-water extraction process is generally use. Gel strength/bloom value test is used to measure the strength of the gel. The gelation process of gelatin is thermo-reversible. Gelatin gels melt by raising the temperature. The commercial value of gelatin is determined by its bloom value. Fish gelatin has a gel strength ranging from as low as 0 g to as high as 426 g. Warm water fish gelatin has higher gel strength than cold water fish gelatin. Important indices of the quality of gelatin preparation are the setting and melting points of gelatin. Being thermo-reversible, gelatin gels will start melting by increasing the temperature above a specific point, which is usually lower than the temperature of the human body. For gelatins from fish setting temperature is in the range of 8–25°C, while melting temperature is 11–28°C. The melting and gelling temperatures of gelatin can be correlated with the proportion of proline and hydroxyproline in the original collagen.

12.4 Insulin

Insulin is a hormone used for correcting the condition called diabetes mellitus in humans. Fish insulin is more stable, as it is decomposed by the protein splitting enzymes of the pancreas. Insulin can be extracted from the corpuscles of Stannius or pancreas of marine animals such as whales, tuna and bonito. These insulins have already been purified, characterized and compared with other insulins for amino acid sequences and for potency. For the treatment of diabetes, scientists have found a new source of insulin in fish that mimics human insulin more closely than any known animal source—porcine or bovine—currently in use. Biologists at the Indian Institute of Chemical Biology (IICB) found "biologically active" insulin in fat cells of catla, which is a natural alternative to the human pancreatic beta cell insulin. This insulin is highly active and has shown greater biological activity.

In the prawn industry, heads and shells leave behind a large amount of waste after processing. About 10,000 tons of processed prawn are exported annually, so there is an urgent need to find significant and economic use for this processing waste. Central Institute of Fisheries Technology (CIFT) has been working to find economic ways to utilize head and shell waste. Prawn shells are found as a good material for the preparation of chitin and chitosan.

12.5 Chitin and Chitosan

Chitin is the second most abundant biopolymer on earth, next only to cellulose. The major sources of chitin are the shell of prawns, crab, Antarctic krill, squilla, insects, clams, oysters, squid pen and fungi. The raw material for the extraction of chitin is prawn shell waste after processing (Figure 12.3). Chitin is a white, hard, inelastic, nitrogenous polysaccharide found in the outer skeleton of insects, crabs, shrimps and lobsters and in the internal structures of invertebrates (Figure 12.4). It is the second most abundant organic compound next to cellulose. It is a macro molecular linear polymer of B (1–4) N acetyl D glucosamine and is insoluble in water and many organic solvents. Chitin

Figure 12.3 *Prawn shell waste.*

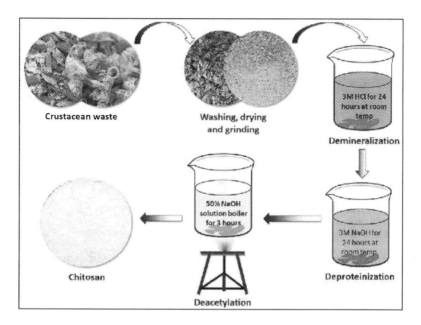

Figure 12.4 *Production of chitosan from crustacean waste.*

itself has few applications, but it acts more as a precursor of chitosan, its most familiar derivative. Chitosan is deacetylated chitin and is a liner polysaccharide of N-acetyle-2amino-2-deoxy-D-glucose and 2-amino-2-deoxy-D-glucose residues.

Chitin represents 14–27% of the dry weight of shrimp and 13–15% of crab shell waste. Recent annual availability of shrimp shell (wet basis) in India is about 100,000–125,000 tons, which is enough to produce 5,000 tons of chitin. Chitin and chitosan can be used for the clarification and purification of water and beverages, chromatography, paper and textiles, photography, food and nutrition, agriculture, pharmaceutical preparation, biodegradable membranes, biotechnology and cosmetics. Chitosan films are clear, tough, flexible and good oxygen carriers. Chitosan-based coatings can protect foods from fungal decay. Edible and biodegradable polymer films offer an alternative environmentally friendly packaging solution.

Chitosan is an important part of biotechnology, and a number of enzymes are immobilized with chitosan (Figure 12.5). It is entrapped and absorbed in the chitin. In biochemistry, it is used as a support for enzymes. The most common method of fixing enzymes to chitosan is by cross-linking them with dialdehydes like glycerol and glutraldehyde. Chitin and chitosan both have biological properties that might be beneficial to enhancing wound repair and wound healing.

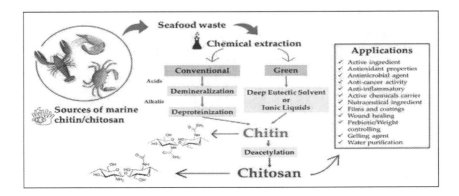

Figure 12.5 Extraction of chitin, chitosan and its application.

12.6 Fish Albumin

In physical and chemical properties, fish albumin is similar to egg albumin. The protein-rich residue of fish waste can be processed into fish albumin. Fish albumin can be produced as technical grade or food and pharmaceutical grade. It can be widely used in food and pharmaceutical products as a whipping, suspending or stabilizing agent. Food grade albumin is an additive in ice cream, soup powder, puddings, confectionery, bakery products, mayonnaise, custard powder and more.

12.7 Fish Protein Concentrate

Fish protein concentrate (FPC) is a stable protein concentrate. Protein concentration is increased by removing water, oil, bones and other materials. FPC contains 75–95% high-quality protein and has limited functional properties. Most of the processing plants produce FPC by a solvent (usually isopropyl alcohol) extraction method (Figure 12.6). The Food and Agriculture Organization (FAO) of the United Nations defines three types of FPC:

Type AA—Odorless and tasteless powder having a maximum total fat content of 0.75%. Type A—Has no specific limits of odor or flavor, but definitely has a fishy flavor and a maximum fat content of 3%. Type C—Normal fish meal produced under satisfactorily hygienic conditions.

FPC type A can be incorporated into other foods such as bread, biscuits, soups and stews at a level that does not affect their normal properties (Figure 12.7). Good results have been obtained with macaroni products, a milkshake drink, spaghetti sauce, infant foods, dietetic foods and breakfast cereals. It is stable up to 3–4 years at room temperature without any significant change in flavor. FPC is highly nutritious because of large amounts of highly digestible protein, available lysine and minerals. Though FPC

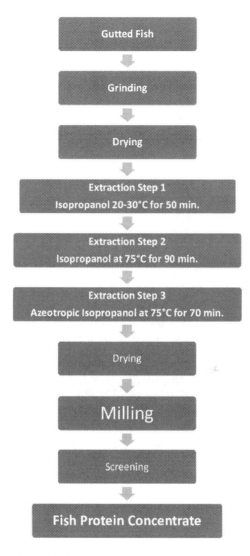

Figure 12.6 Preparation of fish protein concentrate.

is intended for human consumption, it is not relished for consumption as such. It is usually incorporated as a protein supplement in human diets at 5–10%. In bread and biscuits, the acceptable limit of FPC is 35 g per person per day.

12.8 Bioactive Peptide

The major fish waste from processing plants is fish viscera, head, tail, scales etc. Fish viscera consist of different types of food materials, having the property

Figure 12.7 Fish protein concentrate.

of enzymatic hydrolysis of proteolytic enzymes that are secreted to the gut. Enzymes and bioactive peptides can be extracted from fish waste and then can be used for fish silage, fish feed or fish sauce production. Secondary raw materials are regarded as a rich source of bioactive components, with several applications in human nutrition and well-being.

Developing peptide-based drugs is very promising for many lifestyle-mediated diseases. Aquatic organisms, including plants, fish and shellfish, are known as a rich reservoir of parent protein molecules that can offer novel sequences of amino acids in peptides, having unique bio-functional properties upon hydrolyzing with proteases from different sources. However, rather than the exploitation of fish and shellfish, secondary raw material could be a potential choice for peptide-based therapeutic development strategies. Many biologically active peptides with valuable nutritional and functional qualities are found in hydrolysates, derived from the enzymatic hydrolysis of fish proteins. Peptides with antioxidative, antihypertensive and anti-cancer characteristics have been found in secondary raw materials originating in fish. Anticoagulant and antiplatelet effects are also demonstrated by bioactive peptides derived from fish muscle. Secondary raw materials supplied from aquatic sources can be processed using chemical/enzymatic/fermentation processes to recover biochemicals.

Peptides with biological or nutritional qualities help an organism's body to function better and to increase the quality of its diet. These peptides are known as "bioactive peptides". They are formed out of short sequences of amino acids that are inactive within the parent protein's sequence. They can be released through proteolytic hydrolysis with commercially available enzymes or proteolytic microorganisms and fermentation. Enzymatic hydrolysis is a useful method for the recovery of bioactive peptides from fish by-products and improving the functional and nutritional qualities of protein without compromising its nutritional value. The basis of protein hydrolysate production is hydrolytic breakdown of high molecular weight proteins to low molecular weight proteins using

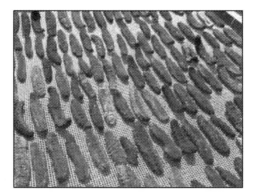

Figure 12.8 Beche-de-mer.

different enzymes, different fish as substrates and variable proteolytic factors such as pH, temperature, enzyme to substrate ratio and time.

12.9 Beche-de-mer

Beche-de-mer, also called trepang, is the boiled, dried and smoked flesh of sea cucumbers (phylum Echinodermata) used to make soups. Most beche-de-mer comes from the southwestern Pacific, where the animals (any of a dozen species of the genera (*Holothuria, Stichopus* and *Thelonota*) are obtained on coral reefs (Figure 12.8). Beche-de-mer is consumed chiefly in China. They belong to a group of marine animals, also called holothurians or sea slugs. Beche-de-mer preparation in India is a small industry that depends almost on a single species, viz. *Holothuria scabra*, found in the Gulf of Mannar, Palk Bay, Andaman and Nicobar Islands, Lakshadweep and also in the Gulf of Kutch. The animal usually burrows in sandy or muddy bottom and feeds on the nutritive material contained in it. The external sand sticking to beche-de-mer is below 0.5%, but due to faulty processing, sand content goes up to 50–60%. Such specimens are called sandy beche-de-mer (imperfect removal of chalky external coat). Some of the common quality defects are inadequate sun-drying and absorption of moisture during storage and action of fermentative micro-organisms.

12.10 Isinglass

This is obtained from an air bladder (Figure 12.9). It is a thin, almost transparent material obtained from the bladder. It consists of a protein known as collagen. Isinglass is used in the clarification of liquors and making gelatin, adhesives etc. for commercial purposes. The air bladder should be large with thick walls. For making high-quality isinglass, sturgeon has long been used.

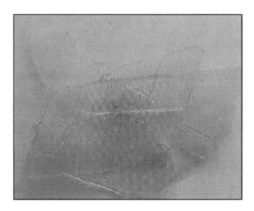

Figure 12.9 Isinglass.

Bladders of various species of eel, catfish, croaker, carp, perch and hake can be commercially used for the production of isinglass.

The fish are split open on shore, thoroughly washed and the outer membrane is removed by scraping and then air dried. The cleaned, desalted, air dried and hardened swim bladders are called fish maws. Isinglass can be used as a clarifying agent for beverages, wines, beer, and vinegar by enmeshing the suspended impurities in the fibrous structure of swollen isinglass.

12.11 Squalene

Squalene is a highly unsaturated hydrocarbon present in the liver oil of certain species of deep-sea sharks, mainly *Centrophorus* and *Squalidae* spp. The liver oil of these species contains a high percentage of squalene (90%), which can be isolated and purified and can be used as a dietary supplement. The richest source of squalene is abyssal shark liver, even though shallow sharks' livers had lower squalene content than cod livers (Figure 12.10).

Squalene has great importance in the cosmetic, food and pharmaceutical industries. It can also be found in animals, vegetables and microorganisms. Shark fishing is prohibited in some countries, so there is a need for renewable sources for squalene extraction to protect marine life. Traditional extraction methods usually involve using organic solvents such as hexane, which left residue on the extracted matrix that limited material use for human consumption. After extraction, the separation and purification stages can increase operation costs. The most interesting technology to obtain squalene is from the biological matrix by supercritical fluid extraction with CO_2 as a solvent because of its economic, safe and easy-to-remove characteristics.

Figure 12.10 Squalene.

Figure 12.11 Shark fin ray soup.

12.12 Shark Fin Rays

Sharks are a marvelous fish, as every part of the shark has some particular use. In addition to meat, several products can be obtained from it, including oil, vitamin, leather, certain cosmetic and medicinal products and fin rays. Shark fin ray is a valuable by-product. Shark fins are dried and then soaked overnight in 10% acetic acid solution. After that, the softened skin and muscles are scraped off and the rays are separated individually. The rays are then washed and dried thoroughly.

Shark fins have long been used in many Eastern countries, especially China, for making shark fin soup (Figure 12.11). Inside the fin, there are long strands with tapering ends that are light yellowish and almost transparent. These are called as shark fin rays or fin needles. The fin is covered over by the skin. Commonly used sharks for fin ray collection are hammerhead shark (*Spyrna zygaena*), grey dog shark (*Rhizoprionodon acutus*), sharp-nosed shark (*Scoliodon laticaudus*) and black-finned shark (*Carcharhinus melanopterus*).

As per records, 100 g of dried fin rays contain 83.5 g protein, 14 g water, 0.3 g fat and 2.2 g minerals. Shark fins and fin rays are in great demand in countries like China, Japan, Singapore and the Philippines. Hong Kong is the main center of the shark fin trade. Shark fin rays are an essential ingredient in some exotic soups. In many countries, serving fin ray soup during feasts has been considered a symbol of wealth and prestige. To prepare a soup, start with dried or frozen shark fin or dried shark fin rays. Soak the dried material in water for several hours to soften it. Fin rays have their own flavor. The texture of the soup and its ability to absorb flavor from other soup ingredients makes it appealing.

12.13 Conclusion

The recovery of chemical components from seafood waste materials and fish processing units, which can be used in other segments of the food industry, is a promising area of research and development for the utilization of fish waste by-products. These wastes cause a serious problem of environmental pollution and make the environmental atmosphere unhygienic and prone to various kinds of diseases. So the problem remains of what to do with these wastes. One of the best alternative ways is to convert these wastes to value-added bioactive and different products for the use of mankind and other animals. The utilization of unwanted fish wastes for production of value-added products is a need for society as a health supplement. It will lead to the control of solid waste generated from fish industries and also helps in improving fish industry economy. Hence, more research and public awareness are required for the production of value-added commodities for the betterment of human society. Developing alternative viable and cost-effective eco-friendly procedures is of utmost importance to prevent the environmental pollution caused due to dumping fish processing waste into water bodies and landfills. Biotransformation methods could be developed for effective conversion of this nutrient-rich waste into beneficial organic products.

Index

Note: Page numbers in *italics* indicate a figure and page numbers in **bold** indicate a table on the corresponding page. Page numbers followed by "n" with numbers refer to notes.

A

AD, *see* anaerobic digestion (AD)
agricultural waste, 7–8
algal SCP, 156
amygdalin, 91
amylases, 120, 152
anaerobic digestion (AD), 18, 24, 150
animal, *see also* fish
   blood, 25
   feed, 17
   hides, 25
anthocyanins, 91, 102, 152, 153, 169, 173
apple, *see also* fruit waste
   juice, 89
   peel, 90
   pulp, 93–94
   seeds, 91
apricot seeds, 92
Aspergillus niger, 28, 152, 153, 206
Aspergillus oryzae, 152, 156

B

Bacillus licheniformis, 152
Bacillus subtilis, 152
barley (*Hordeum vulgare L.*), 60, *see also*
   cereals
   in bakery products, 61–62
   bran fractions, 20
   flakes, 60–61
   in flatbread, 61
   malt, 61
   in pasta and noodles, 62
beche-de-mer/trepang, 258, *258*
b-glucan, 20
betalains, 105, 153
b-glucosidase, 152–153
bioactive compounds, 23, 88, 151

advanced extraction techniques, 151
fruit and vegetable waste, 164, 165–166,
   167, **181**
natural sources of, 151
bioactive peptides
   enzymatic hydrolysis, 257
   enzymes and, 26, 256
biodegradable waste, 5–6
biodiesel, 155–156
bioenergy, 154–156
bioethanol/ethyl alcohol, 154–155
biogas, 24
   and energy generation, 42
biohydrogen/green technology, 22–23, 155
biological conversion methods, 150,
   see also anaerobic digestion (AD);
   fermentation process
biological oxygen demand (BOD), 11
biomedical waste, 7
biomethane, 155
biosorbents, 43
blackberry, 88, 105
black gram, 79
bran, 20, 50, 63
   fortification of, 48–49, **49**
   functionality of, 43, 47–48
   and hulls, 42
   organoleptic acceptability and
      nutritional fortification, 48–49
   stabilization of, 48
   wax, 20
brewery wastewater, 11
broken grains and chaff, 42
broken rice, 66–68

C

cane trash, 30
capsaicin, 191

capsaicinoids, 191, 194
carbon dioxide (CO$_2$), 59, *59*, 100
cardamom (*Elettaria cardamomum*),
    195–198
    husk, 198
    medicinal properties, 197
    processing, 198
    types of, 196
carotenes, 105
carotenoids, 105, 109, 153, 156, 164, 169,
    171, 191
cavitation, 99
celery (*Apium graveolens* L.), 205–206
    by-products, 206
    seeds, 205
    spent, 205
cellulases, 120, 152–153
cereal germ, 21, 57
    chemical composition of, **58**
    methods of oil extraction, 57,
        59–60
cereal grains, by-products, 44–45, **45**
cereals, 20, 57, 65–66
    antioxidants and fiber extraction,
        68–69, **70**
    and grain processing waste, 11
    value additions, 42–43, *44*
    value of, 41
CGM, *see* corn gluten meal (CGM)
charcoal, 28
cheese whey valorization, 236–237
chemical conversion methods, 149–150
chemical oxygen demand (COD), 24
cherry kernels, 92
chili *(Capsicum annum)*, waste utilization
    of, 191–192
    food and pharmacological uses, 191
    oleoresins, 191
    recycling stems, 192
    seed oil, 191
    spent, 191–192
chitin and chitosan, 26, 253–254
    extraction and application, *253*,
        253–254, *254*
    sources of chitin, 253
cinnamon *(Cinnamomum zeylanicum)*,
    waste utilization of, 206–207
    bark, 207
    biowaste, 207
    wood biochar, 207
citrus, *see also* fruit waste
    essential oil, 101–102, **103–104**
    peels, 93–94
    waste, 85, 87–88

clove (*Syzygium aromaticum*), waste
    utilization of, 202–203
cocoa bean shell
    bioactive components, **226**, 226–227
    corrosion inhibitor, 227
    ingredients from, 225–226
    poly(3-hydroxybutyrate), production
        of, 225
    utilization of, *226*
coconut oil, 28
cold plasma-assisted extraction (CPAE), 101
cold pressing, 87, 101
collagen, 250–251
    properties of, 251
    Type I and Type III, 250
color extraction, fruit, 102, 105,
    106–108, 109
commercial waste, 9
composting, 17–18, 43
confectionery industrial wastes
    biodiesel preparation, 224–225
    cocoa processing, 225–226
confectionery processing, 31
cookies, 65
coriander (*Coriandrum sativum* L.), waste
    utilization of, 199
corn *(Zea mays)*, 53
    by-products, 53
    cereal germs, composition of, **58**
    cereal germs, oil extraction, *56*, 57,
        59–60
    dietary fiber, 54
    germ oil, 55–57, *56*
    germ protein, 54–55
    gluten meal extraction, *53*, 53–54
    husks, 20–21, 54
    starch, 53
    steep liquor, 55
    zein protein, 54
corn gluten meal (CGM), *53*, 53–54
corn peptides (CPs), 53–54
corn steep liquor (CSL), 55
crude fiber, 113
cultural factors, 32
cumin (*Cuminum cyminum* Linn.), 29
    waste utilization of, 201–202
customized fortification, 48

D

dairy industry
    global milk production, *231*
    overview of, 230–231
dairy processing, 10–11, 23–25, *24*

biological processes, 234
case study on ghee residue, 237–241
cheese whey valorization, 236–237
managing and valorizing, 232–233
physico-chemical processes, **234**
valorization techniques, 233
whey processing, 234–236, **235**
dairy waste, 23–24
dark-colored fruit wastes, 88
data and monitoring, 32
DBDE, *see* dielectric barrier discharge plasma extraction (DBDE)
deep eutectic solvent-assisted extraction (DESAE), 101
dielectric barrier discharge (DBD), 179
dielectric barrier discharge plasma extraction (DBDE), 179–180
dietary fiber, 54, 153–154
fruit, 113
source, 47
distillery waste, 11
DPPH method, 92, 110, 169, 206, 210, 239, *240*

E

EAE, *see* enzyme-assisted extraction (EAE)
economic impact, food waste, 14
economic opportunities, food waste, 14
edible oil processing, 26–28
edible pulses, 79
education and awareness, 32
electromagnetic waves, 99
electronic waste, 7
endoglucanase, 152–153
environmental impact, food waste, 13–14
Environment Protection Act (1986), 5
enzymatic hydrolysis, 257
enzyme-assisted extraction (EAE), 100, 180
enzymes, 152
amylases, 152
cellulases, 152–153
production of, 120–121
ethical concern, 15
exoglucanase, 152–153
extrusion, 48

F

FAO, *see* Food and Agriculture Organization (FAO)
fenugreek (*Trigonella foenum graecum*), waste utilization of, 208
fermentation process, 74, 75, 150

solid-state, 120, 150
submerged-state, 150
fertilizers and soil amendment, 43
FFA, *see* free fatty acids (FFA)
financial loss, 14
fish, *see also* insulin, fish
albumin, 254–255
collagens, 26
viscera, 256
waste, 11, 25–26, *250*
fish processing industrial waste
beche-de-mer/trepang, 258
bioactive peptide, 256–257
by-products, 250
chitin and chitosan, 253–254
collagen, 250–251
gelatin, 251–252
high-value products, 250
insulin, 252–253
isinglass, 258–259, *259*
prawn shell waste, *253*
protein concentrate, 255–256, *256*, *257*
shark fin rays, 259–261, *260*
squalene, 259, *260*
waste percentage, *250*
fish protein concentrate (FPC), 255–256, *257*
preparation of, *256*
types of, 255
flatbread, 61
flavones, 152
flavonoids, 151
flavonols, 152
flavoring and aroma ingredients, 109
FLW, *see* food loss and waste (FLW)
food, *see also* bran; cardamom (*Elettaria cardamomum*); cereals; chili (*Capsicum annum*); cinnamon (*Cinnamomum zeylanicum*); millets; rice (*Oryza sativa*); spices
insecurity, 14–15
prices, 14
production, 12
redistribution systems, 32
system inefficiency, 15
Food and Agriculture Organization (FAO), 7, 15, 42, 66, 85, 87
Food and Nutrition Program (FNP), 66
food loss and waste (FLW), 14, 121–122
food processing industry (FPI), 1, 8–9
food storage, infrastructure for, 31
food waste, 147
characterization of, 9–12, **12**
composition of, 12, **13**

Index   265

efforts to reduce, 43–44
impact of, 13–15
Food Waste Index, 78
food waste management, 15–16
challenges of, 31–32
conventional methods of, 17–18
hierarchy, *15*
practices, *16–17*
fortification, 81
four, 63
FPC, *see* fish protein concentrate (FPC)
FPI, *see* food processing industry (FPI)
free fatty acids (FFA), 28, 51
fruit and vegetable, 156–157, *see also* apple;
citrus; mango
bioactive compounds, extraction of,
164, 165–166, *167*, **181**
low concentrations of, 157
processing, 9–10, **10**, 22–23
raw materials availability, 157
sources and characteristics of, 158
fruit juice industry, 85
fruit waste, 84, 85, *see also* food waste;
vegetable waste
antioxidants and therapeutic
properties, **86**
bio-economy and techno-economy,
121–122
citrus essential oil, 101–102, **103–104**
citrus wastes, 85, 87–88
color extraction, 102, 105, **106–108**, 109
dark-colored, 88
enzymes, production of, 120–121
extraction techniques, 94, 99–101
flavoring and aroma ingredients, 109
functional ingredients, 109–110, 113,
115
green extraction techniques, 94, **95–98**,
102, 110
legislation and regulations, 122
pectin and dietary fiber extraction,
113, **114**
phenolic compounds and polyphenols
extraction, 91–92, 110, **111–112**
pomace, 89–90, 115, **116–118**, 119–120
seeds, 91–92
skin/peel, 90–91
valorization approach, 93–94

G

g-oryzanols, 69
garlic (*Allium sativa*), waste utilization of,
209–212

gasification, 149
gelatin, 25
fish, *251*, 251–252
germ oil, nutritional benefits of, 57
germ protein, 54–55
ghee residue, 237–241
antioxidant properties of, 239–240, *240*
sensory analysis, 240
gingelly seed, 65
ginger (*Zingiber officinale*), waste
utilization of, 194–195
glycemic index (GI) foods, 79
glycosylated flavones, 87
Government of India (GoI), 42
grain processing, 19–21
grape pomace, 88, 115
green chili, 191
green extraction techniques, 94, **95–98**,
102, 110, 167–168
dielectric barrier discharge plasma
extraction, 179–180
enzyme-assisted extraction, 180
guiding principles of, 167–168
hydrodynamic cavitation-assisted
extraction, 169–170
hydrostatic pressure assisted extraction,
178–179
instant controlled pressure drop, 173
microwave-assisted extraction, 170–171
pressurized hot water extraction,
173–174
pressurized liquid extraction, 174–175
pulsed electric field, 176–177
pulsed ohmic heating assisted
extraction, 177–178
supercritical fluid extraction, 171–173
ultrasound-assisted extraction, 168–169
green solvents, 101
gum, 28
gut health, 48

H

harvesting techniques, 32
hazardous waste, 7
HCAE, *see* hydrodynamic cavitation-
assisted extraction (HCAE)
heat treatment, 48
HHP-assisted extraction (HHPE), 178
application of, 178–179
high pressure processing (HPP), 178
high-voltage electrical discharge, 150, 168
Holothuria scabra, 258
hulls, 27, 75, 148

266   Index

hunger, 14–15
hydrochars, 149
hydrodynamic cavitation-assisted extraction (HCAE), 169–170
  application to extract bioactives, 170
  effectiveness of, 169
hydrostatic pressure assisted extraction, 178–179
hydrothermal carbonization, 149, *see also* hydrochars

I

incentives, 32
incineration, 149
India, 1–5, *see also* cereals; corn (*Zea mays*); millets
  dairy industry, 230, 233, 238
  fish processing, 249
  food commodities, 1, **2**
  food processing industry in, 4
  food waste management in, 31–32, 163
  fruits and vegetables production, 148
  spices production, 188, **189**
industrial citrus waste, 87
inhibition concentration (IC50), 92
instant controlled pressure drop (DIC), 173
institutional waste, 9
insulin, fish, 252–253
  biologically active, 253
  for treatment of diabetes, 253
invertase, 120
iron-rich bajra pasta, 64–65
isinglass, 258–259, *259*
  high-quality, 258
  uses, 258–259

K

kitchen waste (KW), 8

L

laccases, 121
Laminex C2K, 113, 115
landfills, 18
lemon seeds, 87
lifestyles, 32
lipid-rich food waste, 16
lipids, 57
liquid gold, 55
liquid waste, 6–7
livestock feed, 43

M

MAE, *see* microwave-assisted extraction (MAE)
mango
  by-products, 113
  peels, 91
  seed, 92
meat and poultry processing, 25, *see also* fish
  collagen, 245
  feathers processing, 244–245
  livestock meat processing, 245
  poultry manure, 243–244, **244**
meat, fish and poultry waste, 11
melon (cantaloupe), 89, 91
methane gas, 14
micronutrient fortification, 48
microwave-assisted extraction (MAE), 59–60, 99, 151, 168, 170–171
  application to extract bioactives, 171
  for bioactive compounds, 151
millets, 42, 62–63
  antioxidants and fiber extraction, 68–69, **70**
  bran, 63
  broken grains, 65–68
  flour, 63
  iron-rich bajra pasta, 64–65
  list of, **63**
  potential value additions, 42–43, **44**
  sorghum-based energy bars, 63–64
  sorghum bran peda, 64
  zinc-rich Jowar vermicelli, 64
  zinc-rich ragi cookies, 65, 66
milling, 11
Ministry of Agriculture & Farmers' Welfare, 41
molasses, 30
mushroom, 75
  cultivation, 43
mustard
  oil, 29
  waste utilization of, 204

N

nanoparticles, 43
non-biodegradable waste, 5
non-recyclable waste, 7
non-thermal plasma/cold plasma, 179
noodles, 62
Nutri Cereals, 42
nutrient content, 47

O

oil processing, 26–28
oilseed cake meal, 27
   antibiotic and antimicrobials production, 75
   antioxidants, **74**
   enzyme production, 74
   fortification, 74, *75*
   mushroom production, 75
   production of, 73
oil seed industry waste, 11–12
onion (*Allium cepa*), waste utilization of, 209–212
orange peel essential oil, 93
organic wastes, 150

P

packaging, 48
packaging materials, 43
paddy rice, 50
palm oil, 28
pasta, 62
pectin, 27
   and dietary fiber extraction, 113, **114**
pectinase enzymes, 120
pectinases, 120
PEF, *see* pulsed electric field (PEF)
pepper (*Piper nigrum* L.), waste utilization of, 193–194
personalized functional foods, 123
phenolic compounds, 110, 151–152
   antioxidant activity, 211
   extraction and recovery of, 151
   and polyphenols extraction, 91–92, 110, **111–112**
   submerged-state fermentation, 151
phenolics, 110
PHWE, *see* pressurized hot water extraction (PHWE)
pigeon pea (PPDF), 21
pigeon pea by-product flour (PPBF), 22
pigments, 153
   microbial, 153
   natural, 153
PLE, *see* pressurized liquid extraction (PLE)
POHE, *see* pulsed ohmic heating assisted extraction (POHE)
polyphenols, 151
pomace, fruit, 87, 89–90, 115, **116–118**, 119–120
pomegranate, 89
   peel, 91

prawn shell waste, *253*
pressurized hot water extraction (PHWE), 173–174
   application of, 174
   for extracting bioactives, 173–174
pressurized liquid extraction (PLE), 100, 151, 174–175
   application of, 175
   for bioactive compounds, 151
   for extracting phytochemicals, 174–175
proteases, 120–121
proteinaceous food waste, 16
protein concentrate, *256*
pulse by-products, 21–22
   bioactive constituents, 78–79
   fortification, 81
   heath benefits, 79–80, *80*
   processing, 21–22
   protein extraction, 80, **81**
pulsed electric field (PEF), 150, 176–177
pulsed electric field extraction (PEFE), 100
pulsed ohmic heating assisted extraction (POHE), 177–178
pyrolysis, 149

R

RBP, *see* rice bran protein (RBP)
ready to cook (RTC) pasta, 64
recyclable waste, 7
red berries, 88
red gram (pigeon pea), 78
regulatory challenges, 32
residential waste, 7
rice (*Oryza sativa*), 49–50
   broken, 66–68
   by-products, *50*, 50–51
   husk, 52
   starch, isolation of, *67*, 67–68
rice bran (RB), 20
   extraction of, *50*, 51–52
   nutritional composition, **51**
   stabilization of, 51
   utilization, 50–51
rice bran oil (RBO), 52
rice bran protein (RBP), 51–52
rice-husk flour, 21
rum, 30

S

Saccharomyces, 22
Saccharomyces cerevisiae, 155, 156, 211
seafood processing, 25–26

seed coat, 147–148
seeds, fruit, 87, 91–92
SFE, *see* supercritical fluid
extraction (SFE)
shark fin rays, 259–261
soup, *260*, 260–261
single-cell protein (SCP), 156
algal, 156
human consumption, 156
steps in production, 156
skin/peel, fruit, 87, 90–91
slaughterhouse waste, 25
soap stock, 28
social impact, 14–15
solid-state fermentation, 120, 150
solid waste, 6
sorghum-based energy bars, 63–64
sorghum bran peda, 64
sorghum bran stands, 69
sorting and processing, 32
sour cherry pomace, 90
soybean oil cake, 27
spices
growth rate of production, *189*
processing, 28–29
production in India, **189**
value-added products and waste
generated by, **190**
waste/by-products of, *196*
waste utilization of major, 190–191
starch, 29
granules, 66
processing, 66
steam distillation, 101
stone fruits, 89, 91–92
strawberry, 88
submerged-state fermentation, 150
sugar cane bagasse (SCB), 31
sugar cane molasses (SCM), 30
sugar processing, 30–31
supercritical fluid extraction (SFE), 59,
100–101, 151, 171–173
application to extract bioactives,
172–173
bioactive compounds, 151
supply chain, 31
synthetic pigments, 102

T

tamarind (*Tamarindus indica* L.), waste
utilization of, 199–201
food and pharmacological uses, 200
seed powder, 200–201

tannase, 121
tannins, 151–152
thermal conversion methods, 149,
see also gasification; hydrothermal
carbonization; incineration; pyrolysis
Trichoderma reesei, 153, 156
tropical fruits, 89
turmeric (*Curcuma longa* L.), 192–193
curcumin, 192
de-flavored and depigmented, 193
food and pharmacological uses, 192
processing of turmeric spent, 193
spent oleoresins, 192–193

U

ultrasonication, 99
ultrasound-assisted extraction (UAE), 60,
94, 99, 110
applications for removing
bioactives, 169
bioactive compounds, 151, 168–169,
171, 195
urbanization, 32

V

valorization approach, 93–94
vanillin, 109
vegetable by-products, 147
natural ingredient, low concentrations
of, 157
value-added ingredients, extraction
of, 158
variable sources and characteristics, 158
vegetable waste, 150–151
amount of, 151
amylases, 152
bioactive compounds, 151
biodiesel, 155–156
bioenergy, 154
bioethanol/ethyl alcohol, 154–155
biohydrogen, 155
biomethane, 155
cellulases, 152–153
dietary fiber, 153–154
enzymes, 152
phenolic compounds, 151–152
pigments, 153
single-cell protein, 156
types, 147
value-added compounds, 151
vermicelli, 64, *65*
vitamin E, 55–56, 93

Index   269

W

waste, 5, *see also* fruit waste; vegetable waste
  agricultural, 7–8
  classification of, 5–7
  definition of, 5
  generation, 5, 6
  oil seed industry, 11–12
  pulse processing industry, 80, **81**
  residential, 7
  types of, 6–7
  valorization, 19
waste conversion methods, 149–150
  biological conversion, 150
  chemical conversion, 149–150
  high-voltage electrical discharge, 150
  pulsed electric field technology, 150
  thermal conversion, 149
waste management techniques, 19
  confectionery processing, 31
  dairy processing, 23–25, *24*
  edible oil processing, 26–28
  fruit and vegetable processing, 22–23
  grain processing, 19–21
  meat and poultry processing, 25
  pulse processing, 21–22
  seafood processing, 25–26
  spices processing, 28–29
  sugar processing, 30–31
waste utilization of major spices, 190–191
  cardamom (*Elettaria cardamomum*), 195–198
  chili (*Capsicum annum*), 191–192
  ginger (*Zingiber officinale*), 194–195
  pepper (*Piper nigrum* L.), 193–194
  turmeric (*Curcuma longa* L.), 192–193

waste utilization of minor spices
  celery (*Apium graveolens* L.), 205–206
  cinnamon (*Cinnamomum zeylanicum*), 206–207
  clove (*Syzygium aromaticum*), 202–203
  coriander (*Coriandrum sativum* L.), 199
  cumin (*Cuminum cyminum* Linn.), 201–202
  fenugreek (*Trigonella foenum graecum*), 208
  garlic (*Allium sativa*), 209–212
  mustard, 204
  onion (*Allium cepa*), 209–212
  tamarind (*Tamarindus indica* L.), 199–201
wastewater sludge, 76
watermelon, 91
weight management, 48
wheat, 41
  grains, 44–45
  milling process of, 44
wheat bran, 20
  extraction of, 45, 46, 47
  nutritional composition of, **47**
whey processing, 234–236, **235**

X

xanthophylls, 105
xylanases, 121
xylo-oligosaccharides (XOs), 69

Z

zein protein, 54
zero-emission strategy, 19
zero waste economy, 232
zinc-rich Jowar vermicelli, 64
zinc-rich ragi cookies, 65, 66

Printed in the United States
by Baker & Taylor Publisher Services